Graded Syzygies

Algebra and Applications

Volume 14

Algebra and Applications aims to publish well written and carefully refereed monographs with up-to-date information about progress in all fields of algebra, its classical impact on commutative and noncommutative algebraic and differential geometry, K-theory and algebraic topology, as well as applications in related domains, such as number theory, homotopy and (co)homology theory, physics and discrete mathematics.

Particular emphasis will be put on state-of-the-art topics such as rings of differential operators, Lie algebras and super-algebras, group rings and algebras, C^*-algebras, Kac-Moody theory, arithmetic algebraic geometry, Hopf algebras and quantum groups, as well as their applications. In addition, Algebra and Applications will also publish monographs dedicated to computational aspects of these topics as well as algebraic and geometric methods in computer science.

Preface

The main goal of the book is to inspire the readers and develop their intuition about syzygies and Hilbert functions. Research on free resolutions and Hilbert functions is a core and beautiful area in Commutative Algebra.

Many examples are given in order to illustrate and develop ideas and key concepts.

The book contains open problems and conjectures. They provide a glimpse on some exciting directions in which commutative algebraists are working. We present three types of problems: Conjectures, Problems, and Open-Ended Problems. The Open-Ended problems do not describe specific problems but point to interesting directions for exploration.

I want to thank Jeff Mermin for his helpful suggestions. My work is partially supported by NSF.

September, 2010 Irena Peeva

Contents

II Hilbert Functions

III Monomial Resolutions

Notation

The following notation is used throughout the book.

k is a field
$S = k[x_1, \ldots, x_n]$ is a polynomial ring
S is standard graded by $\deg(x_i) = 1$ for $1 \leq i \leq n$
$\mathbf{m} = (x_1, \ldots, x_n)$
I is a graded ideal in S
$R = S/I$
V is a finitely generated S-module
W and U are finitely generated R-modules
$\mathbf{N} = \{0, 1, 2, 3, \ldots\}$ (since many research papers use this notation)
$\mathbf{Z}_+ = \{1, 2, 3, \ldots\}$
deg denotes degree
mdeg denotes multidegree.

Chapter I

Graded Free Resolutions

Abstract. The study of free resolutions is a core and beautiful area in Commutative Algebra. The idea to associate a free resolution to a finitely generated module was introduced in Hilbert's famous papers [Hilbert 1, Hilbert 2]. Free resolutions provide a method for describing the structure of modules. There are several challenging and exciting conjectures involving resolutions. A number of open problems on graded syzygies and Hilbert functions are listed in [Peeva-Stillman].

We are using a grading on the polynomial ring $S = k[x_1, \ldots, x_n]$ and on the objects which we are interested to study: ideals, quotient rings, modules, complexes, and free resolutions. The grading is a powerful tool. The general principle using that tool is the following: *in order to understand the properties of a graded object X, we consider X as a direct sum of vector spaces (its graded components) and we study the properties of each of these vector spaces.*

1 Graded polynomial rings

In this section we introduce some terminology and discuss a few basic applications of using a grading.

Standard Grading 1.1. We will introduce a grading on the polynomial ring $S = k[x_1, \ldots, x_n]$ over a field k. Set $\deg(x_i) = 1$ for each i. A monomial $x_1^{c_1} \ldots x_n^{c_n}$ has ***degree*** $c_1 + \ldots + c_n$. For $i \in \mathbf{N}$, we denote by S_i the k-vector space spanned by all monomials of degree i. In particular, $S_0 = k$. A polynomial $u \in S$ is called ***homogeneous***

I. Peeva, *Graded Syzygies*, Algebra and Applications 14,
DOI 10.1007/978-0-85729-177-6_1, © Springer-Verlag London Limited 2011

if $u \in S_i$ for some i, and in this case we say that u has ***degree*** i (or that u is a ***form*** of degree i) and write $\deg(u) = i$. Note that 0 is a homogeneous element with arbitrary degree. The following two properties are equivalent.

(1) $S_iS_j \subseteq S_{i+j}$ for all $i, j \in \mathbf{N}$.

(2) $\deg(uv) = \deg(u) + \deg(v)$ for every two homogeneous elements $u, v \in S$.

The above two properties hold.

Every polynomial $f \in S$ can be written uniquely as a finite sum $f = \sum_i f_i$ of non-zero elements $f_i \in S_i$, and in this case f_i is called the ***homogeneous component of*** f ***of degree*** i. Thus, we have a direct sum decomposition $S = \oplus_{i \in \mathbf{N}} S_i$ of S as a k-vector space such that $S_iS_j \subseteq S_{i+j}$ for all $i, j \in \mathbf{N}$. We say that S is ***standard graded.***

The ideal generated by all polynomials of positive degree is $\mathbf{m} = (x_1, \dots, x_n)$ and it is called the (***irrelevant***) ***maximal ideal.***

For simplicity, in the examples we usually use $x, y, z, \dots$ or $a, b, c, \dots$ instead of $x_1, x_2, \dots$.

Example 1.2. Let $A = k[x, y]$. In this case, $A_0 = k$, A_1 is the k-space of all linear forms, A_2 is the k-space of all quadrics, etc. The polynomial $x^3y^2 - 2xy^4$ is homogeneous because all of its terms have the same degree 5. The polynomial $x^3y^2 - 2xy^4 + 2y^3 - 10x^2$ is not homogeneous and has homogeneous components $x^3y^2 - 2xy^4$, $2y^3$, and $-10x^2$.

A proper ideal J in S is called ***graded*** or ***homogeneous*** if it satisfies the following equivalent conditions.

(1) If $f \in J$, then every homogeneous component of f is in J.

(2) $J = \oplus_{i \in \mathbf{N}} J_i$, where $J_i = S_i \cap J$.

(3) If $\widetilde{J}$ is the ideal generated by all homogeneous elements in J, then $J = \widetilde{J}$.

(4) J has a system of homogeneous generators.

In this case, the k-spaces J_i are called the ***homogeneous components*** of J.

Exercise 1.3. *Prove that conditions* (1),(2),(3), *and* (4) *above are equivalent.*

Let I be a graded ideal in S. Note that $S_iI_j \subseteq I_{i+j}$ for all $i, j \in \mathbf{N}$. The quotient ring $R = S/I$ inherits the grading from S by

$$R_i = S_i/I_i \quad \text{for every } i \in \mathbf{N}\,.$$

Throughout the book, R stands for S/I and is standard graded.

Proposition 1.4. *Let J be a graded ideal in R.*
(1) *The ideal* $\mathrm{rad}(J) = \{f \in R \,|\, f^r \in J$ *for some* $r\}$ *is graded.*
(2) *The ideal* $\mathrm{ann}(J) = \{f \in R \,|\, fJ = 0\}$ *is graded.*

Proof. (1) Take an element $f \in \mathrm{rad}(J)$ and let $f^r \in J$. Write $f = f_1 + \ldots + f_j$ where f_i, for $1 \leq i \leq j$, are the homogeneous components of f and $\deg(f_i) < \deg(f_{i+1})$. Therefore, f_1^r is the homogeneous component of smallest degree in f^r. Since J is graded, each homogeneous component of f^r is in J. Hence $f_1^r \in J$, and this means $f_1 \in \mathrm{rad}(J)$. Therefore

$$f - f_1 = f_2 + \ldots + f_j \in \mathrm{rad}(J)\,.$$

Applying this argument repeatedly we conclude that each homogeneous component of f is in $\mathrm{rad}(J)$.

(2) Fix a system $\mathcal{U}$ of homogeneous generators of J. Take an element $f \in \mathrm{ann}(J)$. Write $f = f_1 + \ldots + f_j$ where f_i, for $1 \leq i \leq j$, are the homogeneous components of f and $\deg(f_i) < \deg(f_{i+1})$. For each $u \in \mathcal{U}$ we have that the following holds: f_1u is the homogeneous component of smallest degree in fu, and we conclude that $f_1u = 0$ since $fu = 0$. Therefore, $f_1\mathcal{U} = 0$. Hence $f_1J = 0$, that is, $f_1 \in \mathrm{ann}(J)$. Thus,

$$f - f_1 = f_2 + \ldots + f_j \in \mathrm{ann}(J)\,.$$

Applying this argument repeatedly we conclude that each homogeneous component of f is in $\mathrm{ann}(J)$. □

Proposition 1.5. *Let J be a graded ideal in R. The following properties are equivalent.*

(1) *J is prime.*
(2) *If u and v are homogeneous elements in R and $uv \in J$, then at least one of these elements is in J.*

Proof. (1) implies (2). We have to show that (2) implies (1).

Suppose that J is not prime. Choose $f, g \in R$ with a minimal number of homogeneous components and such that $f \notin J$, $g \notin J$, $fg \in J$. Let $f = f_1 + \ldots + f_j$ and $g = g_1 + \ldots + g_s$, where f_i, for $1 \leq i \leq j$, are the homogeneous components of f, and where g_i, for $1 \leq i \leq s$, are the homogeneous components of g. We can assume that $\deg(f_i) < \deg(f_{i+1})$ and $\deg(g_i) < \deg(g_{i+1})$ for each i. Therefore, $f_1 g_1$ is the homogeneous component of smallest degree in fg. Since J is graded, each homogeneous component of fg is in J. Hence $f_1 g_1 \in J$. Therefore, at least one of the elements f_1 and g_1 is in J. Say $f_1 \in J$. Set

$$f' = f - f_1 = f_2 + \ldots + f_j \, .$$

Then $f'g \in J$ and $f' \notin J$, $g \notin J$. This contradicts to the choice of f and g since f' has fewer graded components than f. □

Definition 1.6. Since we have a grading we can measure the size of the quotient ring R by measuring the sizes of its graded components. Since $R_0 = k$, it follows that R_i is a k-vector space because $R_0 R_i \subseteq R_i$. A basis of the k-space R_i is called a ***basis in degree*** i. We have that $\dim_k(R_i) < \infty$ for all $i \in \mathbf{N}$. The generating function $i \mapsto \dim_k(R_i)$ is called the ***Hilbert function*** of R and is studied via the ***Hilbert series***

$$\mathrm{Hilb}_R(t) = \sum_{i \in \mathbf{N}} \dim_k (R_i) t^i \, .$$

The Hilbert function encodes important information about R (for example, the dimension of R); we will study Hilbert functions in detail in Chapter II.

Example 1.7. Let $A = k[x, y]$ and $J = (x^2, y^3)$. Then A/J is graded with basis $\{1\}$ in degree 0, $\{x, y\}$ in degree 1, $\{xy, y^2\}$ in degree 2, $\{xy^2\}$ in degree 3. Its Hilbert series is

$$\mathrm{Hilb}_{A/J}(t) = 1 + 2t + 2t^2 + t^3 \, .$$

Proposition 1.8. *Let M be an ideal in S generated by monomials. For each $i \in \mathbf{N}$, the k-vector space $(S/M)_i = S_i/M_i$ has the following basis*

$$\{\text{monomial } m \in S \mid m \notin M,\ \deg(m) = i\}\,.$$

Hence, $\dim_k((S/M)_i)$ *equals the number of monomials of degree i not in M.*

Example 1.9. $\quad \mathrm{Hilb}_{k[x_1]}(t) = \dfrac{1}{1-t}.$

Exercise 1.10.

$$\mathrm{Hilb}_S(t) = \frac{1}{(1-t)^n}\,,$$

$$\dim_k(S_i) = \binom{n-1+i}{i} \qquad \text{for every } i \geq 0\,.$$

2 Graded modules and homomorphisms

We will discuss modules and homomorphisms which are *graded*. The main result in this section is the foundational Theorem 2.12.

An R-module N is called ***graded***, if it has a direct sum decomposition $N = \oplus_{i\in\mathbf{Z}} N_i$ as a k-vector space and $R_i N_j \subseteq N_{i+j}$ for all $i, j \in \mathbf{Z}$. The k-spaces N_i are called the ***homogeneous components*** of N. An element $m \in N$ is called ***homogeneous*** if $m \in N_i$ for some i, and in this case we say that m has degree i and write $\deg(m) = i$. Every element $m \in N$ can be written uniquely as a finite sum $m = \sum_i m_i$, where $m_i \in N_i$, and in this case m_i is called the ***homogeneous component*** of m of ***degree*** i.

Proposition 2.1. *Let N be a graded R-module.*
(1) *There exists a system of homogeneous generators of N.*
(2) *The degrees of the elements in a system of homogeneous generators determine the grading of N.*

Proof. (1) Let $\mathcal{B}$ be a system of generators of N. The set of homogeneous components of the elements in $\mathcal{B}$ is a system of homogeneous generators of N.

(2) follows from the fact that every element in N is an R-linear combination of the generators and from the property $R_iN_j \subset N_{i+j}$ for all $i, j \in \mathbf{Z}$. □

Definition 2.2. Let N be a graded R-module. Since $R_0 = k$, it follows that N_i is a k-vector space because $R_0N_i \subseteq N_i$. A basis of the k-space N_i is called a ***basis in degree*** i. If U is a finitely generated graded R-module, then $\dim_k(U_i) < \infty$ for all $i \in \mathbf{Z}$ and $U_i = 0$ for $i \ll 0$. In this case, the generating function $i \mapsto \dim_k(U_i)$ is called the ***Hilbert function*** of U and is studied via the ***Hilbert series***

$$\mathrm{Hilb}_U(t) = \sum_{i\in\mathbf{Z}} \dim_k\,(U_i)t^i\,.$$

For $p \in \mathbf{Z}$ denote by $U(-p)$ the graded R-module such that $U(-p)_i = U_{i-p}$ for all i. We say that $U(-p)$ is the module U ***shifted*** p ***degrees***, and call p the ***shift***. Its Hilbert function is

$$\mathrm{Hilb}_{U(-p)}(t) = t^p\,\mathrm{Hilb}_U(t)\,.$$

Proposition 2.3. *The module $R(-p)$ is the free R module generated by one element in degree p.*

Proof. $R(-p)_p = R_0$. □

In this book we use the following convention: the element $1 \in R(-p)$ has degree p and is called the 1-***generator*** of $R(-p)$.

Example 2.4. Continuing Example 1.7, we have that $A/J(-7)$ has basis $\{1\}$ in degree 7, $\{x, y\}$ in degree 8, $\{xy, y^2\}$ in degree 9, $\{xy^2\}$ in degree 10. Its Hilbert series is

$$\mathrm{Hilb}_{A/J(-7)}(t) = t^7 + 2t^8 + 2t^9 + t^{10}.$$

Let N and T be graded R-modules. We say that a homomorphism $\varphi : N \to T$ has ***degree*** i if $\deg(\varphi(m)) = i + \deg(m)$ for each homogeneous element $m \in N$. Recall that 0 has arbitrary degree, thus $\deg(\varphi(m)) = i + \deg(m)$ is a condition only on the homogeneous elements outside $\operatorname{Ker}(\varphi)$. The k-space of all homomorphisms of degree i from N to T is denoted by $\operatorname{Hom}_i(N, T)$. A homomorphism $\phi : N \to T$ is called ***graded*** (or ***homogeneous***) if $\phi \in \operatorname{Hom}_i(N, T)$ for some i; we also say that ϕ is a ***homomorphism of graded modules***.

Exercise 2.5. *Let $\phi : N \to T$ be a homomorphism of graded R-modules. If $f = f_1 + \ldots + f_p \in N$ and $f_1, \ldots, f_p$ are its homogeneous components, then $\phi(f_1), \ldots, \phi(f_p)$ are the homogeneous components of $\phi(f)$.*

Example 2.6. Let $A = k[x, y]$. The homomorphism

$$A(-2) \oplus A(-5) \xrightarrow{(x^2 \;\; y^5)} A$$

is graded and has degree 0. The homomorphism $A \oplus A(-3) \xrightarrow{(x^2 \;\; y^5)} A$ is graded and has degree 2.

Let N and T be graded modules. The ***graded*** Hom from N to T is

$$\mathcal{H}(N, T) = \oplus_{i \in \mathbf{Z}} \operatorname{Hom}_i(N, T)\,.$$

In general, $\mathcal{H}(N, T)$ is a submodule of $\operatorname{Hom}(N, T)$.

Proposition 2.7. *If U is a finitely generated graded R-module and T is a graded R-module, then $\mathcal{H}(U, T) = \operatorname{Hom}(U, T)$.*

Proof. Let $\mathcal{U} = \{f_1, \ldots, f_s\}$ be a minimal system of homogeneous generators of U. Let $\varphi \in \operatorname{Hom}(U, T)$. It induces a map $\bar{\varphi} = \varphi\,|_{\mathcal{U}} : \mathcal{U} \to T$. This map can be uniquely written in the form $\bar{\varphi} = \sum_i \bar{\varphi}_i$, where $\bar{\varphi}_i$ is a map of degree i from $\mathcal{U}$ to T. The sum is finite, because $\mathcal{U}$ is a finite set. Now, for each i define a map $\varphi_i : U \to T$ by extending $\bar{\varphi}_i$ by R-linearity. We will show that φ_i is well-defined. Suppose that there exists a relation in degree t of the form $\sum_{1 \le j \le s} q_j\, f_j = 0$ for some homogeneous $q_j \in R$ and $\deg(q_j) + \deg(f_j) = t$ for $1 \le j \le s$.

We have that

$$\varphi_i\Big(\sum_{1\le j\le s} q_j f_j\Big) = \sum_{1\le j\le s} q_j\,\bar{\varphi}_i(f_j) = \sum_{1\le j\le s} q_j\,\varphi_i(f_j)_{\deg(f_j)+i}$$

$$= \Big(\sum_{1\le j\le s} q_j\,\varphi(f_j)\Big)_{i+t} = \Big(\varphi\Big(\sum_{1\le j\le s} q_j f_j\Big)\Big)_{i+t}.$$

Since $\varphi\Big(\sum_{1\le j\le s} q_j\, f_j\Big) = 0$, we obtain that $\varphi_i\Big(\sum_{1\le j\le s} q_j f_j\Big) = 0$. Thus, φ_i is well-defined. Therefore, $\varphi_i \in \operatorname{Hom}_i(U,T)$. Finally, note that $\bar{\varphi} = \sum_i \bar{\varphi}_i$ implies $\varphi = \sum_i \varphi_i$. □

A submodule T of a graded R-module N is called ***graded*** or ***homogeneous*** if it satisfies the following equivalent conditions.

(1) If $f \in T$, then every homogeneous component of f is in T.

(2) $T = \oplus_{i\in\mathbf{N}} T_i$, where $T_i = N_i \cap T$.

(3) If $\widetilde{T}$ is the submodule generated by all homogeneous elements in T, then $T = \widetilde{T}$.

(4) T has a system of homogeneous generators.

Exercise 2.8. *Prove that* (1),(2),(3), *and* (4) *above are equivalent.*

If T is a graded submodule of a graded R-module N, then N/T inherits the grading via

$$N/T = \oplus_{i\in\mathbf{Z}}\,(N/T)_i \quad \text{with } (N/T)_i = N_i/T_i\,.$$

Proposition 2.9. *If* $\alpha : N \to T$ *is a homomorphism of graded R-modules, then* $\operatorname{Ker}(\alpha)$, $\operatorname{Im}(\alpha)$, *and* $\operatorname{Coker}(\alpha)$ *are graded.*

Proof. First, we show that $\operatorname{Ker}(\alpha)$ is graded. Let $f \in \operatorname{Ker}(\alpha)$. Write $f = f_1 + \ldots + f_j$ as a sum of homogeneous components. If $\alpha(f_i) \neq 0$ for some i, then it is a homogeneous component of $\alpha(f)$. Since $\alpha(f) = 0$, we conclude that $\alpha(f_i) = 0$. Hence each homogeneous component of f is in $\operatorname{Ker}(\alpha)$. Therefore, $\operatorname{Ker}(\alpha)$ is graded.

Now, consider $\operatorname{Im}(\alpha)$. Let $g \in \operatorname{Im}(\alpha)$. Choose $f \in N$ such that $\alpha(f) = g$ and none of the homogeneous components of f is in

$\mathrm{Ker}(\alpha)$. Write $f = f_1 + \ldots + f_j$ as a sum of homogeneous components. Then $\alpha(f_1), \ldots, \alpha(f_j)$ are the homogeneous components of $\alpha(f) = g$. Therefore, each homogeneous component of g is in $\mathrm{Im}(\alpha)$. Hence $\mathrm{Im}(\alpha)$ is graded.

Finally, consider $\mathrm{Coker}(\alpha) \cong T/\mathrm{Im}(\alpha)$. This module inherits the grading of T via $\mathrm{Coker}(\alpha)_i \cong T_i/\mathrm{Im}(\alpha)_i$ for $i \in \mathbf{Z}$. □

We are ready to prove a structure theorem for graded finitely generated R-modules.

Theorem 2.10. *The following properties are equivalent.*

(1) *U is a finitely generated graded R-module.*

(2) *$U \cong W/T$, where W is a finite direct sum of shifted free R-modules, T is a graded submodule of W (called the module of* ***relations****), and the isomorphism has degree* 0.

Proof. We have to show that (1) implies (2). Let $m_1, \ldots, m_j$ be homogeneous generators of U. Let $a_1, \ldots, a_j$ be their degrees respectively. Set $W = R(-a_1) \oplus \ldots \oplus R(-a_j)$. For $1 \le i \le j$ denote by e_i the 1-generator of $R(-a_i)$ and note that it has degree a_i. Consider the homomorphism

$$\begin{aligned} \varphi : W = R(-a_1) \oplus \ldots \oplus R(-a_j) &\rightarrow U \\ e_i &\mapsto m_i \quad \text{for } 1 \le i \le j\,. \end{aligned}$$

It is graded and has degree 0. Set $T = \mathrm{Ker}(\varphi)$. By Proposition 2.9, T is graded. Clearly, $U \cong W/T$. □

Let U be a finitely generated R-module. An exact sequence of the form

$$F_1 \xrightarrow{A} F_0 \longrightarrow U \rightarrow 0\,,$$

where F_0 and F_1 are finitely generated free modules, is called a ***presentation*** of U. We have that $U \cong F_0/\mathrm{Im}A$. The matrix A is called a ***presentation matrix***. The presentation is called ***graded*** if U, F_0, F_1 are graded and the two homomorphisms have degree 0. In the proof of Theorem 2.10 we have constructed a graded presentation, which depends on the choice of a system of homogeneous generators of U.

Next, we study what choices of homogeneous systems of generators of U we have. If $\mathcal{U}$ is a set of homogeneous generators of U such that any proper subset of $\mathcal{U}$ does not generate U, then $\mathcal{U}$ is called a ***minimal system of homogeneous generators*** of U. We need the following lemma, which is very useful.

Nakayama's Lemma 2.11. *Let J be a proper graded ideal in R. Let U be a finitely generated graded R-module.*

(1) *If $U = JU$, then $U = 0$.*

(2) *If W is a graded R-submodule of U such that $U = W + JU$, then $U = W$.*

For local rings the proof of the lemma is more elaborated [Matsumura, Section 2]; for graded rings the lemma is a direct consequence of the fact that a finitely generated R-module always has a generator of minimal degree.

Proof. (1) Assume the opposite, that is $U \neq 0$. Fix a finite system $\mathcal{U}$ of homogeneous generators of U. Let $m \neq 0$ be an element in $\mathcal{U}$ of minimal degree. By the choice of m it follows that $U_j = 0$ for $j < \deg(m)$. Since J is proper and since R is positively graded, it follows that every homogeneous element in JU has degree strictly greater than $\deg(m)$. On the other hand, the equality $U = JU$ implies that $m \in JU$. This is a contradiction. We proved (1).

Applying (1) to the module U/W we get (2). □

Nakayama's Lemma 2.11 leads to the next foundational result.

Foundational Theorem 2.12. *Let U be a finitely generated graded R-module. Consider the finite dimensional graded k-space $\bar{U} = U/(x_1, \ldots, x_n)U$, and let p be its dimension.*

(1) *If we take a homogeneous basis $\{\bar{u}_1, \ldots, \bar{u}_p\}$ for $\bar{U}$ over k, and choose a homogeneous preimage $u_i \in U$ of each $\bar{u}_i$, then $\{u_1, \ldots, u_p\}$ is a minimal system of homogeneous generators of U.*

(2) *Every minimal system of homogeneous generators of U is obtained as in (1).*

(3) *Every minimal system of homogeneous generators of U has p elements. Set $q_i = \dim_k(\bar{U}_i)$ for each i (Note that only finitely*

many of the numbers q_i are non-zero). Every minimal system of homogeneous generators of the module U contains q_i elements of degree i.

(4) *Let $\{u_1, \ldots, u_p\}$ and $\{v_1, \ldots, v_p\}$ be two minimal systems of homogeneous generators of U, and let $v_s = \sum_j a_{js} u_j$ with $a_{js} \in R$ for each s. For all s, j set c_{js} to be the homogeneous component of a_{js} of degree $\deg(v_s) - \deg(u_j)$. Then the following three properties hold: $v_s = \sum_j c_{js} u_j$ for all s, $\det([c_{js}]) \in k$, and $[c_{js}]$ is an invertible matrix with homogeneous entries.*

Proof. Recall that $\mathbf{m} = (x_1, \ldots, x_n)$. (1) We have that $U = \mathbf{m}U + Ru_1 + \ldots + Ru_p$, where Ru_i is the R-module generated by the element u_i (for each i). By Nakayama's Lemma 2.11 it follows that $U = Ru_1 + \ldots + Ru_p$. Assume that the system of generators $\{u_1, \ldots, u_p\}$ is not minimal. After renumbering if necessary, we get a relation $u_1 = \sum_{2 \leq i \leq p} \alpha_i u_i$ for some coefficients $\alpha_i \in R$. Hence, $\bar{u}_1 = \sum_{2 \leq i \leq p} \bar{\alpha}_i \bar{u}_i$ (here $\bar{\alpha}_i$ is the image of α in $k = R/\mathbf{m}$), which is a contradiction because $\{\bar{u}_1, \ldots, \bar{u}_p\}$ is a basis.

(2) Suppose that $\{u_1, \ldots, u_p\}$ is a minimal system of homogeneous generators of U. Clearly, $\{\bar{u}_1, \ldots, \bar{u}_p\}$ generate $\bar{U}$. Assume that $\{\bar{u}_1, \ldots, \bar{u}_p\}$ are linearly dependent. Let $u'_{i_1}, \ldots, u'_{i_q}$ be a subset of $\{\bar{u}_1, \ldots, \bar{u}_p\}$ that is a basis of the k-vector space $\bar{U}$. By (1), it follows that the preimages of $u'_{i_1}, \ldots, u'_{i_q}$ generate U. This is a contradiction since the system of generators $\{u_1, \ldots, u_p\}$ is minimal.

(3) follows from (1) and (2).

(4) Let C be the matrix with entries c_{js}. We will show that $\det(C)$ is a non-zero element in k. The short argument that works in the local case (cf. [Matsumura, proof of Theorem 2.3(iii)]) does not work, so we have to argue using the grading.

For $1 \leq s \leq p$ the equality $v_s = \sum_j a_{js} u_j$ implies

$$v_s = (v_s)_{\deg(v_s)} = \left(\sum_j a_{js} u_j \right)_{\deg(v_s)}$$

$$= \sum_j \left(a_{js} \right)_{\deg(v_s) - \deg(u_j)} u_j = \sum_j c_{js} u_j \,.$$

We can assume that the elements in each set of generators are ordered so that their degrees increase. Denote by B_i the blocks of size $q_i \times q_i$ that are placed along the diagonal of C. Since the degrees of the generators are increasing, it follows by (3) that $\deg(u_j) \geq \deg(v_s)$ for $j > s$. Hence $c_{js} = 0$, if $j > s$ and c_{js} is outside the block $B_{\deg(v_s)}$. Therefore, there are only zeros below the blocks. If c_{js} is an entry in some block B_i, then both v_s and u_j have degree i, and therefore $c_{js} \in k$. Thus, the entries in the blocks B_i are in k. Hence, $\det(C) = \prod_i \det(B_i) \in k$.

Denote by $\bar{c}_{js}$ the image in $k = R/\mathbf{m}R$ of c_{js}, and denote by $\bar{C}$ the matrix with entries $\bar{c}_{js}$. Then, $\bar{C}$ has the blocks B_i on the diagonal and zeros everywhere else. Note that $\bar{v}_s = \sum_j \bar{c}_{js}\bar{u}_j$ for all s. Since $\bar{C}$ transforms one basis of a vector space into another, we have that $0 \neq \det(\bar{C}) = \prod_i \det(B_i)$. Hence $\det(C)$ is a non-zero element in k.

The product of C and its adjoint matrix equals the diagonal matrix with entries $\det(C)$ on the diagonal. Since $\det(C)$ is an invertible element, it follows that the matrix C is invertible. □

In this book we use the following notation. Denote by $\min(U)$ the minimal degree of an element in a minimal system of homogeneous generators of U. Denote by $\max(U)$ the maximal degree of an element in a minimal system of homogeneous generators of U.

Exercise 2.13. $\min(U)$ *and* $\max(U)$ *do not depend on the choice of a minimal system of homogeneous generators.*

3 Graded complexes

Definition 3.1. A ***complex*** $\mathbf{F}$ ***over*** R is a sequence of homomorphisms of R-modules

$$\mathbf{F} : \ldots \longrightarrow F_i \xrightarrow{d_i} F_{i-1} \longrightarrow \ldots \longrightarrow F_2 \xrightarrow{d_2} F_1 \xrightarrow{d_1} F_0 \longrightarrow \ldots ,$$

such that $d_{i-1}d_i = 0$ for $i \in \mathbf{Z}$. The collection of maps $d = \{d_i\}$ is called the ***differential*** of $\mathbf{F}$. Sometimes the complex is denoted

$(\mathbf{F}, d)$. It is called a ***left complex*** if $F_i = 0$ for all $i < 0$, that is,

$$\mathbf{F}: \quad \ldots \longrightarrow F_i \xrightarrow{d_i} F_{i-1} \longrightarrow \ldots \longrightarrow F_2 \xrightarrow{d_2} F_1 \xrightarrow{d_1} F_0 \longrightarrow 0$$

with $i \in \mathbf{N}$. Furthermore, $(\mathbf{F}, d)$ is called a ***left complex over*** W (or a complex over W) if it is a left complex and we have a homomorphism $\epsilon : F_0 \to W$, called an ***augmentation map***.

The complex is called ***graded*** if the modules F_i are graded and each d_i is a homomorphism of degree 0. In this case the module $\mathbf{F}$ is actually bigraded since

$$F_i = \oplus_{j \in \mathbf{Z}} F_{i,j} \text{ for all } \; i.$$

An element in $F_{i,j}$ is said to have ***homological degree*** i and ***internal degree*** j. We denote the homological degree by hdeg, and the internal degree by deg. Consider $\mathbf{F}$ as a module and the differential as a homomorphism $d : \mathbf{F} \to \mathbf{F}$. Then d has homological degree -1 and internal degree 0.

Construction 3.2. If each module F_i is a free finitely generated graded R-module, then we can write it as

$$F_i = \oplus_{p \in \mathbf{Z}} R(-p)^{c_{i,p}} .$$

Therefore, a graded complex of free finitely generated modules has the form

$$\mathbf{F}: \quad \ldots \longrightarrow \oplus_{p \in \mathbf{Z}} R(-p)^{c_{i,p}} \xrightarrow{d_i} \oplus_{p \in \mathbf{Z}} R(-p)^{c_{i-1,p}} \longrightarrow \ldots .$$

The numbers $c_{i,p}$ are the ***graded Betti numbers*** of the complex. We say that $c_{i,p}$ is the Betti number in homological degree i and internal degree p, or the i'th Betti number in internal degree p.

Definition 3.3. The ***homology*** of a complex $\mathbf{F}$ is defined by

$$\mathrm{H}_i(\mathbf{F}) = \mathrm{Ker}(d_i)/\mathrm{Im}(d_{i+1}) .$$

The elements in $\mathrm{Ker}(d_i)$ are called ***cycles*** and the elements in $\mathrm{Im}(d_i)$ are called ***boundaries***. The complex is ***exact*** at F_i (or at ***step*** i)

if $\mathrm{H}_i(\mathbf{F}) = 0$. The complex is ***exact*** if $\mathrm{H}_i(\mathbf{F}) = 0$ for all i. A left complex is ***acyclic*** if $\mathrm{H}_i(\mathbf{F}) = 0$ for all $i > 0$; it is ***acyclic over*** W if it is acyclic and $\mathrm{H}_0(\mathbf{F}) = W$. In the graded case, since the differential is graded, it follows that the homology is bigraded by

$$\mathrm{H}_i(\mathbf{F}) = \oplus_{j\in\mathbf{Z}}\,\mathrm{H}_i(\mathbf{F})_j \qquad \text{for all } i\,.$$

For $p \in \mathbf{Z}$ denote by $\mathbf{F}[-p]$ the homologically graded complex such that $\mathbf{F}[-p]_i = \mathbf{F}_{i-p}$ for all i. We say that $\mathbf{F}[-p]$ is the complex $\mathbf{F}$ ***homologically shifted*** p ***degrees*** (or ***twisted***), and call p the ***shift***.

The ***truncated complex*** $\mathbf{F}_{\geq p}$ is defined as

$$\mathbf{F}_{\geq p}: \quad \ldots \longrightarrow F_i \xrightarrow{d_i} F_{i-1} \longrightarrow \quad \ldots \longrightarrow F_{p+1} \xrightarrow{d_{p+1}} F_p\,.$$

Similarly, we define $\mathbf{F}_{\leq p}$.

If $(\mathbf{F}, d)$ and $(\mathbf{G}, \partial)$ are complexes of R-modules, then a ***homomorphism of complexes*** $\varphi: \mathbf{F} \to \mathbf{G}$ is a collection of homomorphisms $\varphi_i: F_i \to G_i$ for all i, such that

$$\varphi d = \partial\varphi$$

that is, the following diagram is commutative

$$\begin{array}{ccccccc} \mathbf{F}: & \ldots \longrightarrow & F_i & \xrightarrow{d_i} & F_{i-1} & \longrightarrow & \ldots \\ & & \varphi_i \downarrow & & \downarrow \varphi_{i-1} & & \\ \mathbf{G}: & \ldots \longrightarrow & G_i & \xrightarrow{\partial_i} & G_{i-1} & \longrightarrow & \ldots \quad , \end{array}$$

that is,

$$\varphi_{i-1} d_i = \partial_i \varphi_i \quad \text{for all } i\,.$$

Sometimes we say a ***map of complexes*** instead of a homomorphism of complexes. Suppose the complexes are graded; we call φ ***graded*** or a ***homomorphism of graded complexes*** if $\varphi_i: F_i \to G_i$ is a homomorphism of a fixed degree q for all i.

Lemma 3.4. *If φ is a homomorphism of complexes, then we have the inclusions $\varphi(\mathrm{Ker}(d)) \subseteq \mathrm{Ker}(\partial)$ and $\varphi(\mathrm{Im}(d)) \subseteq \mathrm{Im}(\partial)$.*

Proof. Let $f \in \varphi(\mathrm{Ker}(d_i))$ for some i. There exists a $g \in \mathrm{Ker}(d_i)$ such that $f = \varphi(g)$. Then

$$\partial(f) = \partial(\varphi(g)) = \varphi(d(g)) = 0 .$$

Hence $f \in \mathrm{Ker}(\partial)$. Therefore, $\varphi(\mathrm{Ker}(d)) \subseteq \mathrm{Ker}(\partial)$.

Let $f \in \varphi(\mathrm{Im}(d_i))$ for some i. There exists a $g \in \mathrm{Im}(d_i)$ such that $f = \varphi(g)$. Furthermore, there exists an $h \in F_i$ such that $g = d_i(h)$. Then $f = \varphi(d(h)) = \partial(\varphi(h)) \in \mathrm{Im}(\partial)$. Hence $\varphi(\mathrm{Im}(d)) \subseteq \mathrm{Im}(\partial)$. □

The above lemma implies that φ induces the following map on homology (which we also denote φ)

$$\varphi : \mathrm{H}(\mathbf{F}) = \oplus_{i\in\mathbf{Z}}\mathrm{Ker}(d_i)/\mathrm{Im}(d_{i+1}) \to \oplus_{i\in\mathbf{Z}}\mathrm{Ker}(\partial_i)/\mathrm{Im}(\partial_{i+1}) = \mathrm{H}(\mathbf{G}).$$

Thus, φ is a collection of maps $\varphi_i : \mathrm{H}_i(\mathbf{F}) \to \mathrm{H}_i(\mathbf{G})$ for all i. If φ is graded of degree q, then we have a collection of maps

$$\varphi_{i,j} : \mathrm{H}_i(\mathbf{F})_j \to \mathrm{H}_i(\mathbf{G})_{j+q}$$

for all i, j.

Definition 3.5. Let $(\mathbf{F}, d)$ and $(\mathbf{G}, \partial)$ be two complexes of free R-modules. We say that $\mathbf{G}$ is a ***subcomplex*** of $\mathbf{F}$ if $\mathbf{G} \subseteq \mathbf{F}$ and ∂ is the restriction of d on $\mathbf{G}$. In this book, we call $\mathbf{G}$ an ***essential subcomplex*** if for every i we have that $F_i = G_i \oplus T_i$ (as modules) for some free module T_i.

The next construction makes use of the grading.

Construction 3.6. Let $\mathbf{F}$ be a graded complex. Since each F_i is graded we write $F_i = \oplus_j F_{i,j}$. The differential has degree 0, therefore $d(F_{i,j}) \subseteq F_{i-1,j}$ for each i, j. Thus, the complex can be written as

$$\begin{array}{lcccccccc}
 & & \vdots & & \vdots & & \vdots & & \\
 & & \oplus & & \oplus & & \oplus & & \\
\text{j'th row:} & \ldots \to & F_{i+1,j} & \to & F_{i,j} & \to & F_{i-1,j} & \to & \ldots \\
 & & \oplus & & \oplus & & \oplus & & \\
\text{(j-1)'st row:} & \ldots \to & F_{i+1,j-1} & \to & F_{i,j-1} & \to & F_{i-1,j-1} & \to & \ldots \\
 & & \oplus & & \oplus & & \oplus & & \\
 & & \vdots & & \vdots & & \vdots & &
\end{array}$$

The j'th row is called the j'th (***graded***) ***component*** of $\mathbf{F}$. It is the sequence of k-vector spaces

$$\ldots \to F_{i+1,j} \to F_{i,j} \to F_{i-1,j} \to \ldots .$$

The complex is the direct sum of its components. Often, it is very useful to study a complex by studying its graded components.

Exercise 3.7. *A graded complex* $\mathbf{F}$ *is exact if and only if each of its graded components is an exact sequence of* k*-vector spaces.*

Example 3.8. Take $A = k[x,y]$ and $T = A/(x^5, xy)$. We have the exact graded complex

$$0 \to A(-6) \xrightarrow{\begin{pmatrix} -y \\ x^4 \end{pmatrix}} A(-5) \oplus A(-2) \xrightarrow{(x^5 \quad xy)} A \to T \to 0$$

Taking the degree 7 component, we obtain the exact sequence of k-vector spaces

$$0 \to A(-6)_7 \to A(-5)_7 \oplus A(-2)_7 \to A_7 \to T_7 \to 0 \,,$$

that is,

$$0 \to A_1 \to A_2 \oplus A_5 \to A_7 \to T_7 \to 0 \,.$$

Note that A_1 has basis (as a vector space) x, y; A_2 has basis x^2, xy, y^2; A_5 is 6-dimensional; A_7 is 8-dimensional; and T_7 has basis y^7.

4 Free resolutions

Definition 4.1. A ***free resolution*** of a finitely generated R-module U is a sequence of homomorphisms of R-modules

$$\mathbf{F}: \quad \ldots \longrightarrow F_i \xrightarrow{d_i} F_{i-1} \longrightarrow \ldots \longrightarrow F_1 \xrightarrow{d_1} F_0 \,,$$

such that

(1) $\mathbf{F}$ is a complex of finitely generated free R-modules F_i

(2) $\mathbf{F}$ is exact

(3) $U \cong F_0/\mathrm{Im}(d_1)$.

Sometimes, for convenience, we write

$$\mathbf{F}: \quad \ldots \longrightarrow F_i \xrightarrow{d_i} F_{i-1} \longrightarrow \ldots \longrightarrow F_1 \xrightarrow{d_1} F_0 \xrightarrow{d_0} U \longrightarrow 0\,.$$

In the literature the map d_0 is usually denoted ϵ and called an augmentation map. Every resolution is an acyclic left complex over U.

A resolution is ***graded*** if U is graded, $\mathbf{F}$ is a graded complex, and the isomorphism $F_0/\mathrm{Im}(d_1) \cong U$ has degree 0. In this case the differential has homological degree -1 and internal degree 0. Fix a homogeneous basis of each free module F_i. Then the differential d_i is given by a matrix D_i, whose entries are homogeneous elements in R. These matrices are called ***differential matrices*** (note that they depend on the chosen basis).

Construction 4.2. Given a graded finitely generated R-module U we will construct a graded free resolution of U by induction on homological degree.

Step 0: Set $U_0 = U$. Choose homogeneous generators $m_1, \ldots, m_r$ of U_0. Let $a_1, \ldots, a_r$ be their degrees, respectively. Set $F_0 = R(-a_1) \oplus \ldots \oplus R(-a_r)$. For $1 \le j \le r$ denote by f_j the 1-generator of $R(-a_j)$. Thus, $\deg(f_j) = a_j$. Define

$$\begin{aligned} d_0 :\ & F_0 \rightarrow U \\ & f_j \mapsto m_j \quad \text{for } 1 \le j \le r\,. \end{aligned}$$

This is a homomorphism of degree 0.

Assume by induction, that F_i and d_i are defined.

Step $i+1$: Set $U_{i+1} = \mathrm{Ker}(d_i)$. Choose homogeneous generators $l_1, \ldots, l_s$ of U_{i+1}. Let $c_1, \ldots, c_s$ be their degrees, respectively. Set $F_{i+1} = R(-c_1) \oplus \ldots \oplus R(-c_s)$. For $1 \le j \le s$ denote by g_j the 1-generator of $R(-c_j)$. Thus, $\deg(g_j) = c_j$. Define

$$\begin{aligned} d_{i+1} :\ & F_{i+1} \rightarrow U_{i+1} \subset F_i \\ & g_j \mapsto l_j \quad \text{for } 1 \le j \le r\,. \end{aligned}$$

This is a surjective homomorphism of degree 0.

The constructed complex is exact since $\mathrm{Ker}(d_i) = \mathrm{Im}(d_{i+1})$ by construction.

Example 4.3. Let $A = k[x,y]$ and $B = (x^3,\ xy,\ y^5)$. We will construct a graded free resolution of A/B over A.

Step 0: Set $F_0 = A$ and let $d_0 : A \to A/B$.

Step 1: The elements $x^3,\ xy,\ y^5$ are homogeneous generators of $\mathrm{Ker}(d_0)$. Their degrees are $3, 2, 5$ respectively. Set $F_1 = A(-3) \oplus A(-2) \oplus A(-5)$. Denote by f_1, f_2, f_3 the 1-generators of $A(-3)$, $A(-2)$, $A(-5)$. Hence $\deg(f_1) = 3$, $\deg(f_2) = 2$, $\deg(f_3) = 5$. Defining d_1 by $f_1 \mapsto x^3$, $f_2 \mapsto xy$, $f_3 \mapsto y^5$ we obtain the beginning of the resolution:

$$A(-3) \oplus A(-2) \oplus A(-5) \xrightarrow{\begin{pmatrix} x^3 & xy & y^5 \end{pmatrix}} A \to A/B \to 0\,.$$

Step 2: First, we need to find homogeneous generators of $\mathrm{Ker}(d_1)$. This requires some computation. Let $\alpha f_1 + \beta f_2 + \gamma f_3 \in \mathrm{Ker}(d_1)$, with $\alpha, \beta, \gamma \in A$. We want to solve the equation

$$\alpha x^3 + \beta xy + \gamma y^5 = 0,$$

where $\alpha, \beta, \gamma \in A$ are the unknowns. The equality $\alpha x^3 = -y(\beta x + \gamma y^4)$ implies that y divides α. The equality $\gamma y^5 = -x(\alpha x^2 + \beta y)$ implies that x divides γ. Let $\alpha = y\widetilde{\alpha}$ and $\gamma = x\widetilde{\gamma}$. It follows that $\widetilde{\alpha}x^2 + \beta + \widetilde{\gamma}y^4 = 0$. Therefore each term of β is divisible either by x^2 or by y^4. Write

$$\begin{aligned}
\widetilde{\alpha} &= \alpha' y^4 + \alpha'' && \text{where no term of } \alpha'' \text{ is divisible by } y^4\\
\beta &= \beta' y^4 + \beta'' x^2 + \bar{\beta} x^2 y^4 && \text{where no term of } \beta'' \text{ is divisible by } y^4\\
& && \text{and no term of } \beta' \text{ is divisible by } x^2\\
\widetilde{\gamma} &= \gamma' + \gamma'' x^2 && \text{where no term of } \gamma' \text{ is divisible by } x^2\,.
\end{aligned}$$

We get the equality

$$(\alpha' + \bar{\beta} + \gamma'')x^2y^4 + (\alpha'' + \beta'')x^2 + (\beta' + \gamma')y^4 = 0\,.$$

It follows that $\alpha'' + \beta'' = 0$ and $\beta' + \gamma' = 0$. Therefore, $\alpha' + \bar{\beta} + \gamma'' = 0$. It follows that all solutions $(\alpha', \alpha'', \beta', \beta'', \bar{\beta}, \gamma', \gamma'')$ are generated by $(0,1,0,-1,0,0,0)$, $(0,0,-1,0,0,1,0)$, and $(1,0,0,0,-1,0,0)$, $(0,0,0,0,-1,0,1)$. Therefore, all solutions (α, β, γ) are generated by

$$\sigma_1 = (y, -x^2, 0), \ \sigma_2 = (0, -y^4, x)$$
$$\sigma_3 = (y^5, -x^2y^4, 0), \ \sigma_4 = (0, -x^2y^4, x^3).$$

Note that $\sigma_3 = y^4\sigma_1$ and $\sigma_4 = x^2\sigma_2$. Hence all solutions (α, β, γ) of the equation $d((\alpha, \beta, \gamma)) = 0$ are minimally generated by σ_1 and σ_2.

Thus, $yf_1 - x^2f_2$ and $-y^4f_2 + xf_3$ are homogeneous generators of $\mathrm{Ker}(d_1)$. Their degrees are $4 = \deg(y) + \deg(f_1)$ and $6 = \deg(y^4) + \deg(f_2)$. Set $F_2 = A(-4) \oplus A(-6)$. Denote by g_1, g_2 the 1-generators of $A(-4)$ and $A(-6)$. Hence $\deg(g_1) = 4$ and $\deg(g_2) = 6$. Defining d_2 by $g_1 \mapsto yf_1 - x^2f_2$, $g_2 \mapsto -y^4f_2 + xf_3$ we obtain the next step in the resolution:

$$A(-4) \oplus A(-6) \xrightarrow{\begin{pmatrix} y & 0 \\ -x^2 & -y^4 \\ 0 & x \end{pmatrix}} A(-3) \oplus A(-2) \oplus A(-5)$$
$$\xrightarrow{(x^3 \ \ xy \ \ y^5)} A.$$

Step 3: First, we need to find homogeneous generators of $\mathrm{Ker}(d_2)$. Let $\mu g_1 + \nu g_2 \in \mathrm{Ker}(d_2)$ with $\mu, \nu \in A$. Hence

$$\mu y f_1 + (-\mu x^2 - \nu y^4) f_2 + \nu x f_3 = 0,$$

and therefore μ, ν satisfy the equations

$$\mu y = 0, \quad -\mu x^2 - \nu y^4 = 0, \quad \nu x = 0.$$

We conclude that $\mu = \nu = 0$. Thus, $F_3 = 0$. We obtain the graded free resolution

$$0 \to A(-4) \oplus A(-6) \xrightarrow{\begin{pmatrix} y & 0 \\ -x^2 & -y^4 \\ 0 & x \end{pmatrix}} A(-3) \oplus A(-2) \oplus A(-5)$$
$$\xrightarrow{(x^3 \ \ xy \ \ y^5)} A.$$

Example 4.4. Let $A = k[x, y]$ and $B = (x^3, xy, y^5)$. Suppose we are given (say by computer) the non-graded free resolution

$$0 \longrightarrow A^2 \xrightarrow{\begin{pmatrix} y & 0 \\ -x^2 & -y^4 \\ 0 & x \end{pmatrix}} A^3 \xrightarrow{\begin{pmatrix} x^3 & xy & y^5 \end{pmatrix}} A$$

of the module A/B over A. We will determine the grading.

Denote by f_1, f_2, f_3 the basis of A^3 with respect to which the matrix of d_1 is given. Since

$$\begin{aligned} f_1 &\mapsto x^3 \quad \text{and} \quad \deg(x^3) = 3 \\ f_2 &\mapsto xy \quad \text{and} \quad \deg(xy) = 2 \\ f_3 &\mapsto y^5 \quad \text{and} \quad \deg(y^5) = 5 \end{aligned}$$

and since we want d_1 to be homogeneous of degree 0, we set

$$\deg(f_1) = 3, \quad \deg(f_2) = 2, \quad \deg(f_3) = 5.$$

Therefore, the free A-module generated by f_1 is $A(-3)$, the free A-module generated by f_2 is $A(-2)$, and the free A-module generated by f_3 is $A(-5)$. Thus, A^3 is identified with $A(-3) \oplus A(-2) \oplus A(-5)$.

Furthermore, denote by g_1, g_2 the basis of A^2 with respect to which the matrix of d_2 is given. Since

$$\begin{aligned} g_1 \mapsto & y f_1 - x^2 f_2 \\ & \deg(y f_1 - x^2 f_2) = \deg(y f_1) = \deg(y) + \deg(f_1) = 4 \\ g_2 \mapsto & -y^4 f_2 + x f_3 \\ & \deg(-y^4 f_2 + x f_3) = \deg(y^4 f_2) = \deg(y^4) + \deg(f_2) = 6 \end{aligned}$$

and since we want d_2 to be homogeneous of degree 0, we set

$$\deg(g_1) = 4 \text{ and } \deg(g_2) = 6.$$

Hence the free A-module generated by g_1 is $A(-4)$ and the free A-module generated by g_2 is $A(-6)$. Thus, A^2 is identified with $A(-4) \oplus A(-6)$. Therefore, we obtain the graded free resolution

$$0 \longrightarrow A(-4) \oplus A(-6) \xrightarrow{\begin{pmatrix} y & 0 \\ -x^2 & -y^4 \\ 0 & x \end{pmatrix}} A(-3) \oplus A(-2) \oplus A(-5) \xrightarrow{\begin{pmatrix} x^3 & xy & y^5 \end{pmatrix}} A \,.$$

5 Resolving, that is, Repeatedly Solving

We will discuss the following interpretation of free resolutions: building a resolution consists of *repeatedly solving systems of polynomial equations.* This approach makes it possible to compute resolutions.

Consider a homomorphism $R^p \xrightarrow{B} R^q$, where B is the matrix of the map with respect to fixed bases. Let X be the column vector with entries the indeterminates $X_1, \ldots, X_p$. The following problems are equivalent.

Describe the module $\mathrm{Ker}(B)$.

$\Updownarrow$

(*) Solve the system of R-linear equations $BX = 0$ over R, (where $X_1, \ldots, X_p$ take values in R).

Example 5.1. In Example 4.3 we see that finding generators of the module $\mathrm{Ker}(d_1)$ is equivalent to solving the system of one equation $\alpha x^3 + \beta xy + \gamma y^5 = 0$ (where α, β, γ are the unknowns and take values in A). Also, we see that finding generators of the module $\mathrm{Ker}(d_2)$ is equivalent to solving the system of three equations

$$\mu y = 0, \quad -\mu x^2 - \nu y^4 = 0, \quad \nu x = 0,$$

where μ, ν are the unknowns and take values in A.

The following questions arise.

- What is the minimal number of elements that generate the kernel?
- How can we check whether a given set of solutions generates all solutions?
- Is there a formula for a matrix C, such that $R^t \xrightarrow{C} R^p \xrightarrow{B} R^q$ is exact?

In case $R = k$ is a field, these questions are answered in Linear Algebra. When R is not a field, the above questions are usually difficult. In order to have an algorithm which makes it possible to compute examples, it is important to solve the following problem.

Problem. Find a matrix C such that $R^t \xrightarrow{C} R^p \xrightarrow{B} R^q$ is exact.

In case $R = k$, we can construct the matrix C in terms of minors of B. When R is not a field, Gröbner basis theory provides an algorithm for constructing C, see Section 23.

In case $R = k$, there exists a matrix C' such that

$$0 \longrightarrow R^j \xrightarrow{C'} R^p \xrightarrow{B} R^q$$

is exact, and we can find C'. When R is not a field, it might be *impossible* to get an exact sequence starting with 0 as above, because (in every choice) the generators of the module $\operatorname{Ker}(B)$ might be dependent over R. So we only have the exact sequence $R^t \xrightarrow{C} R^p \xrightarrow{B} R^q$. Hence, if we have constructed C and want to solve problem $(*)$ completely, then we have to solve the following new problem of this type:

$$\begin{gathered}\text{Describe the module } \operatorname{Ker}(C) \\ \Updownarrow \\ \text{Solve the system of } R\text{-linear} \\ \text{equations } CY = 0 \text{ over } R,\end{gathered}$$

where Y is a column vector of indeterminates.

Set $D_0 = B$ and $D_1 = C$. Suppose we find a matrix D_2 such that $R^s \xrightarrow{D_2} R^t \xrightarrow{D_1} R^p \xrightarrow{D_0} R^q$ is exact. In order to describe $\mathrm{Ker}(D_2)$ we might need again to solve a system of equations:

$$\text{Describe the module } \mathrm{Ker}(D_2)$$
$$\Updownarrow$$
$$\text{Solve the system of } R\text{-linear}$$
$$\text{equations } D_2 Z = 0 \text{ over } R\,,$$

where Z is a column vector of indeterminates. We continue in this way ...

Thus, solving the original problem $(*)$ leads to *repeatedly solving systems of R-linear equations*

$$D_0 X = 0,\ D_1 Y = 0,\ D_2 Z = 0,\ \dots .$$

We see that this process of repeatedly solving systems of equations is the same as constructing a free resolution

$$\dots \longrightarrow R^s \xrightarrow{D_2} R^t \xrightarrow{D_1} R^p \xrightarrow{D_0} R^q\,.$$

This was illustrated in Example 4.3.

The process described above may never terminate; in this case we have an infinite resolution. In Section 15 we will prove Hilbert's Syzygy Theorem 15.2 which shows that every graded finitely generated S-module has a finite (that is, $F_i = 0$ for $i \gg 0$) graded free resolution.

6 Homotopy

Throughout the section we use the notation introduced in 6.1.

Definition 6.1. Let φ and ψ be two homomorphisms of complexes $\varphi, \psi : \mathbf{F} \to \mathbf{G}$ of finitely generated R-modules. We say that φ is ***homotopic*** to ψ (or that φ and ψ are ***homotopy equivalent***, or

homotopic) if there exists a homomorphism of homologically graded R-modules $h : \mathbf{F} \to \mathbf{G}$ of homological degree 1 such that

$$\varphi - \psi = \partial h + hd\,,$$

that is, h is a collection of homomorphisms $h_i : F_i \to G_{i+1}$, such that

$$\varphi_i - \psi_i = \partial_{i+1} h_i + h_{i-1} d_i\,.$$

It is helpful to look at the following diagram

$$\begin{array}{ccccccccc} \cdots \to & F_{i+1} & \longrightarrow & F_i & & \xrightarrow{d_i} & F_{i-1} & \to \cdots \\ & & \swarrow h_i & \downarrow \varphi_i - \psi_i & & \swarrow h_{i-1} & & \\ \cdots \to & G_{i+1} & \xrightarrow[\partial_{i+1}]{} & G_i & & \longrightarrow & G_{i-1} & \to \cdots \end{array}$$

We also consider a similar notion which we call k-homotopy. If h is not a homomorphism of R-modules, but a homomorphism of k-spaces then we call it a ***k-homotopy***. Every homotopy is a k-homotopy.

Proposition 6.2. *φ is homotopic to ψ if and only if $\varphi - \psi$ is homotopic to* 0.

Suppose that **F** and **G** are graded. If φ and ψ are graded and both have degree c, then it follows that the homotopy has internal degree c. The homotopy always has homological degree 1.

Proposition 6.3. *If φ and ψ are k-homotopy equivalent, then they induce the same map in homology.*

Proof. Replacing φ by $\varphi - \psi$ and ψ by 0, we have to show that if φ is homotopic to 0, then it induces the zero map in homology. There exists a homotopy h such that $\varphi_i = \partial_{i+1} h_i + h_{i-1} d_i$. If $x \in F_i$ is a cycle, then

$$\varphi_i(x) = \partial_{i+1} h_i(x) + h_{i-1} d_i(x) = \partial_{i+1} h_i(x) \in \mathrm{Im}(\partial_{i+1})\,.$$

Hence $\varphi_i(x)$ is a boundary. Thus, $\varphi_i(\mathrm{Ker}(d_i)) \subseteq \mathrm{Im}(\partial_{i+1})$. We conclude that the map

$$\varphi_i :\ \mathrm{H}_i(\mathbf{F}) = \mathrm{Ker}(d_i)/\mathrm{Im}(d_{i+1}) \to \mathrm{Ker}(\partial_i)/\mathrm{Im}(\partial_{i+1}) = \mathrm{H}_i(\mathbf{G})$$

is the zero map. □

Sometimes it is possible to guess how a resolution looks like, that is, guess the ranks of the free modules and a formula for the differential. Usually, straightforward computation verifies that $d_i d_{i+1} = 0$ and shows that the conjectured resolution is a complex. The difficult point is to prove that the complex is exact. There are various tools which can be used to establish exactness. The nest theorem is one of them.

Theorem 6.4. *Let*

$$\mathbf{F}: \quad \ldots \longrightarrow F_i \xrightarrow{d_i} F_{i-1} \longrightarrow \ldots \longrightarrow F_1 \xrightarrow{d_1} F_0 \,,$$

be a graded complex. If the identity and and the zero endomorphisms of $\mathbf{F}$ *(mapping* $\mathbf{F} \to \mathbf{F}$*) are* k*-homotopy equivalent, then* $\mathbf{F}$ *is exact.*

Proof. By Proposition 6.3, id and 0 induce the same map in homology, hence $\mathrm{H}_i(\mathbf{F}) = 0$ for $i > 0$. □

If we can guess a formula for the homotopy h in Theorem 6.4, then in order to show that $\mathbf{F}$ is exact we have to verify that $\mathrm{id}_i = \partial_{i+1} h_i + h_{i-1} d_i$ for all i. This method is used in order to prove that the Bar resolution 32.2 is exact.

Example 6.5. Let $A = k[x]/(x^2)$. Consider the complex

$$\mathbf{F}: \quad \ldots \to F_{i+1} = A \xrightarrow{x} F_i = A \xrightarrow{x} \ldots \xrightarrow{x} F_1 = A \xrightarrow{x} F_0 = A \,.$$

Consider $h : \mathbf{F} \to \mathbf{F}$ such that for every $i \geq 0$ we have that $h_i : F_i \to F_{i+1}$ is defined by $h_i(1) = 1$ and $h_i(x) = (1 - x)$ and extended to $F_i = A$ as a map of k-vector spaces. Then

$$(xh_{i+1} + h_i x)(1) = xh_{i+1}(1) + h_i x(1) = x + (1 - x) = 1$$
$$(xh_{i+1} + h_i x)(x) = xh_{i+1}(x) + h_i x(x) = x(1 - x) + h_i(0) = x \,.$$

Hence, h is a k-homotopy between the identity and the zero endomorphisms of $\mathbf{F}$. Therefore, the complex is exact by Theorem 6.4.

Definition 6.6. Let U and W be finitely generated R-modules. Let $\mathbf{F}$ be a complex over W, and $\mathbf{G}$ be a complex over U, and $\sigma : W \to U$

be a homomorphism. We say that a homomorphism $\varphi : \mathbf{F} \to \mathbf{G}$ is ***over*** $\sigma : W \to U$, or that φ ***induces*** σ, or that φ is a ***lifting*** of σ, if the following diagram is commutative

$$\begin{array}{cccccccc} \dots \to & F_i & \xrightarrow{d_i} & F_{i-1} & \to \dots & \to F_0 & \to W \to 0 \\ & \downarrow \varphi_i & & \downarrow \varphi_{i-1} & & \downarrow \varphi_0 & \downarrow \sigma \\ \dots \to & G_i & \xrightarrow{\partial_i} & G_{i-1} & \to \dots & \to G_0 & \to U \to 0\,. \end{array}$$

Lifting Lemma 6.7. *Let W and U be finitely generated R-modules, and $\sigma : W \to U$ be a homomorphism. Let*

$$\mathbf{F}: \quad \dots \longrightarrow F_i \xrightarrow{d_i} F_{i-1} \longrightarrow \dots \longrightarrow F_1 \xrightarrow{d_1} F_0 \xrightarrow{d_0} W \longrightarrow 0$$

be a complex with surjective d_0 and F_i finitely generated free for $i \geq 0$, and let

$$\mathbf{G}: \quad \dots \longrightarrow G_i \xrightarrow{\partial_i} G_{i-1} \longrightarrow \dots \longrightarrow G_1 \xrightarrow{\partial_1} G_0 \xrightarrow{\partial_0} U \longrightarrow 0$$

be a free resolution of U.

(1) *There exists a lifting $\varphi : \mathbf{F} \to \mathbf{G}$ which induces σ.*

(2) *If μ and ψ are two liftings of σ, then there exists a homotopy h between μ and ψ.*

If in addition $W, U, \mathbf{F}, \mathbf{G}$ are graded and σ, μ, ψ have degree p, then φ and h can be chosen to have degree p as well.

Proof. The proof is by induction on homological degree.

(1) Since F_0 is free, $G_0 \xrightarrow{\partial_0} U$ is surjective, and we have the diagram

$$\begin{array}{ccc} & & F_0 \\ & & \downarrow \sigma d_0 \\ G_0 & \xrightarrow{\partial_0} & U\,, \end{array}$$

it follows that there exists a homomorphism $\varphi_0 : F_0 \to G_0$ such that $\partial_0 \varphi_0 = \sigma d_0$. Furthermore, note that

$$\partial_0 \varphi_0(\mathrm{Im}(d_1)) \subseteq \partial_0 \varphi_0(\mathrm{Ker}(d_0)) = \sigma d_0(\mathrm{Ker}(d_0)) = 0\,.$$

Hence, $\varphi_0(\mathrm{Im}(d_1)) \subseteq \mathrm{Ker}(\partial_0) = \mathrm{Im}(\partial_1)$.

Suppose that by induction hypothesis we have that there exists a homomorphism $\varphi_{i-1} : F_{i-1} \to G_{i-1}$ such that $\varphi_{i-1}(\mathrm{Im}(d_i)) \subsetneq \mathrm{Im}(\partial_i)$. Set $\bar{W} = \mathrm{Im}(d_i)$ and $\bar{U} = \mathrm{Im}(\partial_i)$, and repeat the argument above. Since F_i is free, $G_i \xrightarrow{\partial_i} \bar{U}$ is surjective, and we have the diagram

$$\begin{array}{ccc} & & F_i \\ & & \downarrow \varphi_{i-1} d_i \\ G_i & \xrightarrow{\partial_i} & \bar{U}\,, \end{array}$$

it follows that there exists a homomorphism $\varphi_i : F_i \to G_i$ such that $\partial_i \varphi_i = \varphi_{i-1} d_i$. Furthermore, note that

$$\partial_i \varphi_i(\mathrm{Im}(d_{i+1})) \subsetneq \partial_i \varphi_i(\mathrm{Ker}(d_i)) = \varphi_{i-1} d_i(\mathrm{Ker}(d_i)) = 0\,.$$

Hence, $\varphi_i(\mathrm{Im}(d_{i+1})) \subsetneq \mathrm{Ker}(\partial_i) = \mathrm{Im}(\partial_{i+1})$.

(2) If μ and ψ are two liftings of σ, then $\mu - \psi$ and 0 are two liftings of 0. Therefore, it suffices to show that if α is a lifting of the zero map, then α is homotopic to zero.

By induction on homological degree we will construct a homotopy $h : \mathbf{F} \to \mathbf{G}$ such that $\alpha_i = h_{i-1} d_i + \partial_{i+1} h_i$. For $i = 0$, we get

$$\begin{array}{ccccccc} F_1 & \xrightarrow{d_1} & F_0 & \to W \to 0 \\ \downarrow \alpha_1 & & \downarrow \alpha_0 & \downarrow 0 \\ G_1 & \xrightarrow{\partial_1} & G_0 & \to U \to 0\,. \end{array}$$

As $\partial_0 \alpha_0 = 0$, we conclude that $\alpha_0(F_0) \subsetneq \mathrm{Im}(\partial_1)$. Hence, we have the diagram

$$\begin{array}{ccc} & & F_0 \\ & & \downarrow \alpha_0 \\ G_1 & \xrightarrow{\partial_1} & \mathrm{Im}(\partial_1)\,, \end{array}$$

so there exists a homomorphism $h_0 : F_0 \to G_1$ such that $\alpha_0 = \partial_1 h_0$. Note that

$$\partial_1(-h_0 d_1 + \alpha_1) = -\partial_1 h_0 d_1 + \alpha_0 d_1 = (-\partial_1 h_0 + \alpha_0) d_1 = 0\,,$$

so $\mathrm{Im}(-h_0 d_1 + \alpha_1) \subsetneq \mathrm{Ker}(\partial_1) = \mathrm{Im}(\partial_2)$.

Now, suppose that by induction hypothesis there exists a homomorphism h_i such that $\alpha_i = h_{i-1}d_i + \partial_{i+1}h_i$ and $\operatorname{Im}(-h_i d_{i+1} + \alpha_{i+1}) \subseteq \operatorname{Im}(d_{i+2})$. Then, we have the diagram

$$\begin{array}{ccc} & & F_{i+1} \\ & & \downarrow {\scriptstyle -h_i d_{i+1} + \alpha_{i+1}} \\ G_{i+2} & \xrightarrow{\partial_{i+2}} & \operatorname{Im}(\partial_{i+2}), \end{array}$$

so there exists an h_{i+1} such that $\partial_{i+2}h_{i+1} = -h_i d_{i+1} + \alpha_{i+1}$. Hence $\alpha_{i+1} = h_i d_{i+1} + \partial_{i+2}h_{i+1}$. Furthermore,

$$\begin{aligned} \partial_{i+2}(-h_{i+1}d_{i+2} + \alpha_{i+2}) &= -\partial_{i+2}h_{i+1}d_{i+2} + \alpha_{i+1}d_{i+2} \\ &= (-\partial_{i+2}h_{i+1} + \alpha_{i+1})d_{i+2} \\ &= (h_i d_{i+1})d_{i+2} = 0, \end{aligned}$$

so $\operatorname{Im}(-h_{i+1}d_{i+2} + \alpha_{i+2}) \subseteq \operatorname{Ker}(\partial_{i+2}) = \operatorname{Im}(\partial_{i+3})$. □

The next theorem shows that any two free resolutions of a finitely generated R-module U are homotopy equivalent.

Theorem 6.8. *If* $\mathbf{F}$ *and* $\mathbf{G}$ *are two free resolutions of a finitely generated R-nodule U, then there exist homomorphisms* $\varphi : \mathbf{F} \to \mathbf{G}$ *and* $\psi : \mathbf{G} \to \mathbf{F}$*, such that* $\varphi\psi$ *is homotopic to* $\mathrm{id} : \mathbf{G} \to \mathbf{G}$ *and* $\psi\varphi$ *is homotopic to* $\mathrm{id} : \mathbf{F} \to \mathbf{F}$*. If* U, $\mathbf{F}$, $\mathbf{G}$ *are graded, then* φ, ψ *and the homotopies can be chosen to have internal degree* 0.

Proof. By Lemma 6.7(1), the identity map $\mathrm{id} : U \to U$ lifts to to the homomorphisms of R-complexes $\varphi : \mathbf{F} \to \mathbf{G}$ and $\psi : \mathbf{G} \to \mathbf{F}$. Then $\psi\varphi : \mathbf{F} \to \mathbf{F}$ is a lifting of the identity map $\mathrm{id} : U \to U$. Another lifting of the same map is $\mathrm{id} : \mathbf{F} \to \mathbf{F}$. By Lemma 6.7(2) it follows that $\psi\varphi$ is homotopic to the identity. Similarly, we see that $\varphi\psi$ is homotopic to $\mathrm{id} : \mathbf{G} \to \mathbf{G}$. □

7 Minimal free resolutions

Most of this book is devoted to describing the properties of minimal graded free resolutions and relating them to the structure of the resolved modules. In this section we define when a graded free resolution is minimal. We will also present Theorem 7.5, which shows that the minimal graded free resolution is the smallest graded free resolution in the sense that the ranks of its free modules are less than or equal to the ranks of the corresponding free modules in an arbitrary graded free resolution of the resolved module.

Definition 7.1. A graded free resolution of a graded finitely generated R-module U is ***minimal*** if

$$d_{i+1}(F_{i+1}) \subseteq (x_1, \ldots, x_n)F_i \quad \text{for all } i \geq 0.$$

This means, that no invertible elements (non-zero constants) appear in the differential matrices.

Example 7.2. The resolution in Example 4.3 is minimal.

Recall that $\mathbf{m}$ stands for the maximal ideal $(x_1, \ldots, x_n)$.

Theorem 7.3. *The graded free resolution constructed in Construction* 4.2 *is minimal if and only if at each step we choose a minimal homogeneous system of generators of the kernel of the differential.*

Proof. We use the notation introduced in Construction 4.2 and set $\mathrm{Ker}(d_{-1}) = U$.

First, suppose that the constructed resolution is minimal. Assume now, that for some $i \geq -1$ we have chosen a non-minimal homogeneous system $l_1, \ldots, l_s$ of generators of $\mathrm{Ker}(d_i)$. After renumbering the elements $l_1, \ldots, l_s$ if necessary, we get a relation $l_1 = \sum_{2 \leq j \leq s} r_j l_j$ for some $r_j \in R$. That is, $d_{i+1}(g_1) = \sum_{2 \leq j \leq s} r_j d_{i+1}(g_j)$. Therefore,

$$g_1 - \sum_{2 \leq j \leq s} r_j g_j \in \mathrm{Ker}(d_{i+1}) = \mathrm{Im}(d_{i+2})\,.$$

Since the resolution is minimal, we have that $\mathrm{Im}(d_{i+2}) \subseteq \mathbf{m}F_{i+1}$. Hence, $g_1 - \sum_{2 \leq j \leq s} r_j g_j \in \mathbf{m}F_{i+1}$, which is a contradiction.

Now, suppose that at each step we choose a minimal homogeneous system of generators of the kernel of the differential. We want to show that the obtained resolution is minimal. Assume the contrary.

There exists an $i \geq -1$ such that $\operatorname{Im}(d_{i+2}) \not\subseteq \mathbf{m}F_{i+1}$. Therefore, $\operatorname{Ker}(d_{i+1}) = \operatorname{Im}(d_{i+2})$ contains a homogeneous element that is not in $\mathbf{m}F_{i+1}$. After renumbering the elements $g_1, \ldots, g_s$ if necessary, we can assume that $g_1 - \sum_{2\leq j\leq s} r_j g_j \in \operatorname{Ker}(d_{i+1})$ for some $r_j \in R$. Hence

$$d_{i+1}(g_1) = \sum_{2\leq j\leq s} r_j d_{i+1}(g_j)\,.$$

Hence, $l_1 = \sum_{2\leq j\leq s} r_j l_j$. This contradicts to the fact that we have chosen $l_1, \ldots, l_s$ to be a minimal homogeneous system of generators of $\operatorname{Ker}(d_i)$. □

A complex of the form

$$0 \longrightarrow R(-p) \xrightarrow{1} R(-p) \longrightarrow 0$$

is called a ***short trivial complex***. If $(\mathbf{F}, d)$ and $(\mathbf{G}, \partial)$ are complexes, then their ***direct sum*** is the complex $\mathbf{F}\oplus\mathbf{G}$ with modules $(\mathbf{F}\oplus\mathbf{G})_i = F_i \oplus G_i$ and differential $d \oplus \partial$. A direct sum of short trivial complexes (possibly placed in different homological degrees) is called a ***trivial complex***.

Example 7.4. The trivial complex

$$\begin{aligned} 0 \to R(-p) \oplus R(-q) \to R(-p) \oplus R(-q) \oplus R(-t) \\ \to R(-t) \to 0 \to R(-s) \to R(-s) \to 0 \end{aligned}$$

is the direct sum of four short trivial complexes.

Theorem 7.5. *Let U be a graded finitely generated R-module.*

(1) *There exists a minimal graded free resolution of U.*

(2) *Let $\mathbf{F}$ be a minimal graded free resolution of U. If $\mathbf{G}$ is a graded free resolution of U, then $\mathbf{G} \cong \mathbf{F} \oplus \mathbf{T}$ for some trivial complex $\mathbf{T}$, and the direct sum is a direct sum of complexes.*

(3) *Up to an isomorphism, there exists a unique minimal graded free resolution of U.*

In view of this theorem, sometimes we say "*the* minimal graded free resolution of U".

Theorem 7.5(1) holds by Theorem 7.3. Theorem 7.5(3) follows from (2). Theorem 7.5(2) is proved in Section 9. The key tool in the proof is Nakayama's Lemma 2.11.

Example 7.6. Let $A = k[x,y]$. Consider the graded free resolution

$$0 \to A(-5) \xrightarrow{\begin{pmatrix} y^2 \\ x \\ 1 \end{pmatrix}} A(-3) \oplus A(-4) \oplus A(-5) \xrightarrow{\begin{pmatrix} -y & 0 & y^3 \\ x & -y^2 & 0 \\ 0 & x & -x^2 \end{pmatrix}}$$

$$A(-2) \oplus A(-2) \oplus A(-3) \xrightarrow{(x^2 \;\; xy \;\; y^3)} A\,.$$

It is not minimal, because the last differential matrix contains the entry 1.

Let f_1, f_2, f_3 be the 1-generators of $A(-3), A(-4), A(-5)$ respectively. We change the basis in $A(-3) \oplus A(-4) \oplus A(-5)$ by

$$g_1 = f_1,\ g_2 = f_2,\ g_3 = y^2 f_1 + x f_2 + f_3\,.$$

In the new basis, the resolution is

$$0 \to A(-5) \xrightarrow{\begin{pmatrix} 0 \\ 0 \\ 1 \end{pmatrix}} A(-3) \oplus A(-4) \oplus A(-5) \xrightarrow{\begin{pmatrix} -y & 0 & 0 \\ x & -y^2 & 0 \\ 0 & x & 0 \end{pmatrix}}$$

$$A(-2) \oplus A(-2) \oplus A(-3) \xrightarrow{(x^2 \;\; xy \;\; y^3)} A\,.$$

Thus, the resolution is the direct sum of the short trivial complex

$$0 \to A(-5) \to A(-5) \to 0$$

placed in homological degrees 2 and 3, and the minimal graded free resolution

$$0 \to A(-3) \oplus A(-4) \xrightarrow{\begin{pmatrix} -y & 0 \\ x & -y^2 \\ 0 & x \end{pmatrix}} A(-2) \oplus A(-2) \oplus A(-3)$$

$$\xrightarrow{(x^2 \;\; xy \;\; y^3)} A\,.$$

Removing the short trivial complex from the original resolution is called a ***consecutive cancellation***. We say that ***the two copies of*** $A(-5)$ ***cancel***.

The theorem guarantees that in every non-minimal graded free resolution we can change basis so that after a number of consecutive cancellations we will obtain a minimal free resolution.

Free resolutions exist over an arbitrary commutative noetherian ring (they exist even over non-commutative rings). However, the concept of a minimal free resolution does not make sense over all such rings. In order to have a unique up to an isomorphism minimal free resolution (as in Theorem 7.5), one needs in particular, that each minimal system of generators of the module has the same number of elements. This follows from Nakayama's Lemma 2.11; see Theorem 2.12. Two major classes of commutative noetherian rings over which Nakayama's Lemma holds are the local noetherian rings and the positively graded finitely generated algebras over a field. Consequently, these are the two major classes of rings over which the theory of minimal free resolutions is developed.

Open-Ended Problem 7.7. (folklore) *Construct explicit minimal graded free resolutions of classes of graded modules.*

By "explicit" we mean that a resolution is described by formulas (not algorithms). Constructions requiring a choice of generators of the homology (of a certain poset, simplicial complex, or algebraic complex) are not explicit. A beautiful example of an explicit resolution is the Koszul resolution in Section 14.

8 Encoding the structure of a module

Key Point 8.1. The minimal graded free resolution $\mathbf{F}$ of a graded finitely generated R-module U is a *description of the structure of* U since it has the form

$$\ldots \to F_2 \xrightarrow{\begin{pmatrix}\text{a minimal}\\ \text{system of}\\ \text{homogeneous}\\ \text{relations}\\ \text{on the}\\ \text{relations in } d_1\end{pmatrix}}$$

$$F_1 \xrightarrow{\begin{pmatrix}\text{a minimal}\\ \text{system of}\\ \text{homogeneous}\\ \text{relations}\\ \text{on the minimal}\\ \text{generators of } U\end{pmatrix}} F_0 \xrightarrow{\begin{pmatrix}\text{a minimal}\\ \text{system of}\\ \text{homogeneous}\\ \text{generators}\\ \text{of } U\end{pmatrix}} U \to 0\,.$$

The minimality of the relations encoded in d_i follows from Theorem 7.3. Not surprisingly, *many properties of U can be read off the structure of* $\mathbf{F}$.

9 Proof of Theorem 7.5(2)

Let T and N be finitely generated R-modules. A sequence

$$T \xrightarrow{\alpha} N \xrightarrow{\beta} T$$

of homomorphisms is a ***splitting*** if $\beta\alpha = \mathrm{id}$.

Lemma 9.1. *Let T and N be finitely generated R-modules.*

(1) *If $T \xrightarrow{\alpha} N \xrightarrow{\beta} T$ is a splitting, then $N = \mathrm{Ker}(\beta) \oplus \mathrm{Im}(\alpha)$. Furthermore, if in addition T and N are graded, and α and β have degree 0, then $N = \mathrm{Ker}(\beta) \oplus \mathrm{Im}(\alpha)$ as graded modules.*

(2) *If $N \xrightarrow{\beta} T$ is surjective and T is free, then $N \cong \mathrm{Ker}(\beta) \oplus T$. Furthermore, if in addition T and N are graded, and β has degree 0, then $N \cong \mathrm{Ker}(\beta) \oplus T$ as graded modules.*

Proof. (1) Clearly, $\mathrm{Ker}(\beta) \cap \mathrm{Im}(\alpha) = 0$. Let $f \in N$. As $\beta(\alpha(\beta(f))) = \beta(f)$, we have that $f - \alpha(\beta(f)) \in \mathrm{Ker}(\beta)$. Thus, $f \in \mathrm{Ker}(\beta) \oplus \mathrm{Im}(\alpha)$. In the graded case we choose f to be homogeneous.

(2) Choose a basis $l_1, \dots, l_s$ of T. Choose preimages $f_1, \dots, f_s$ in N so that $\beta(f_i) = l_i$ for all i. Define a map $\alpha : T \to N$ by $\alpha(l_i) = f_i$. Thus, we have a splitting $\beta\alpha = \text{id}$. Hence $N = \text{Ker}(\beta) \oplus \text{Im}(\alpha)$ by (1). The splitting implies that α is injective. Therefore, $\text{Im}(\alpha) \cong T$. In the graded case we choose $l_1, \dots, l_s$ and $f_1, \dots, f_s$ to be homogeneous. □

Lemma 9.2. *If T is a graded R-module which is a direct summand of a finitely generated graded free R-module, then T is free.*

Proof. Choose a minimal system of homogeneous generators $l_1, \dots, l_s$ of T. Let $a_1, \dots, a_s$ be the degrees of these elements respectively. Let $r_1, \dots, r_s$ be the basis of $R(-a_1) \oplus \dots \oplus R(-a_s)$ consisting of the 1-generators of the shifted free modules. Consider the surjective homomorphism

$$\alpha : \; R(-a_1) \oplus \dots \oplus R(-a_s) \to T$$
$$r_i \mapsto l_i \quad \text{for } 1 \le i \le s\,.$$

We will show that α is an isomorphism. Denote by L the kernel of α and set $N = R(-a_1) \oplus \dots \oplus R(-a_s)$. Note that the map α is homogeneous and has degree 0. It follows that the module L is graded. Theorem 2.12 implies that if $\sum_i c_i l_i = 0$ with $c_i \in R$, then $c_i \in \mathbf{m}$. Therefore, $L \subseteq \mathbf{m}N$.

Suppose that $G = F \oplus T$ is a graded finitely generated free R-module. Choose a homogeneous basis $g_1, \dots, g_p$ of G. Let $\pi : \; G \to T$ be the projection map. It is a homogeneous map of degree 0. Since α is surjective, for each i there exists a homogeneous $f_i \in N$ such that $\alpha(f_i) = \pi(g_i)$. Define a graded homomorphism $\gamma : \; G \to N$ by $\gamma(g_i) = f_i$. Denote by β the restriction of γ on T. Then $\pi = \alpha\beta : \; T \to T$ is the identity map. Thus, we have a splitting. By Lemma 9.1(1), we get that

$$N = \text{Im}(\beta) \oplus \text{Ker}(\alpha) = \beta(T) \oplus L\,.$$

Therefore, the inclusion $L \subseteq \mathbf{m}N$ (established above) implies $N = \text{Im}(\beta) + \mathbf{m}N$. Apply Nakayama's Lemma 2.11 and conclude that

$N = \mathrm{Im}(\beta)$. Therefore, $L = 0$. Hence, α is an isomorphism. □

Lemma 9.3. *If*

$$\mathbf{T}: \quad \ldots \longrightarrow T_i \stackrel{d_i}{\longrightarrow} T_{i-1} \longrightarrow \ldots \longrightarrow T_1 \stackrel{d_1}{\longrightarrow} T_0 \longrightarrow 0$$

is an exact graded complex of finitely generated graded free R-modules, then it is a trivial complex.

Proof. Set $K_0 = T_0$ and $K_1 = \mathrm{Ker}(d_1)$. Since T_0 is free, by Lemma 9.1 we get $T_1 \cong K_1 \oplus K_0$. Therefore, $\mathbf{T} \cong \mathbf{T_1} \oplus \{0 \to K_0 \to K_0 \to 0\}$, where the latter complex is trivial and

$$\mathbf{T_1}: \quad \ldots \longrightarrow T_i \stackrel{d_i}{\longrightarrow} T_{i-1} \longrightarrow \ldots \longrightarrow T_2 \stackrel{d_2}{\longrightarrow} K_1 \longrightarrow 0\,.$$

Since K_1 is a graded R-module, by Lemma 9.2 it follows that K_1 is free. The equality $0 = \mathrm{H}(\mathbf{T}) = \mathrm{H}(\mathbf{T_1}) \oplus \mathrm{H}(\{0 \to K_0 \to K_0 \to 0\}) = \mathrm{H}(\mathbf{T_1})$ shows that $\mathbf{T_1}$ is exact. Therefore, we can apply the above argument to $\mathbf{T_1}$ to obtain

$$\mathbf{T_1} \cong \mathbf{T_2} \oplus \{0 \to K_1 \to K_1 \to 0\}\,,$$

where

$$\mathbf{T_2}: \ldots \to T_i \to \ldots \to T_3 \to K_2 \to 0\,.$$

Set $K_i = \mathrm{Ker}(d_i)$ for $i \geq 0$. Proceeding in the above way, we obtain that

$$\mathbf{T} \cong \oplus_{i\geq 0} \{0 \to K_i \to K_i \to 0\}$$

is a trivial complex. □

Lemma 9.4. *Let $\alpha : F \to F'$ be a homomorphism of degree 0 of graded finitely generated free modules. Choose a homogeneous basis $f_1, \ldots, f_s$ of F and a homogeneous basis $g_1, \ldots, g_p$ of F'. Let $C = [c_{ij}]$ be a matrix with entries in R such that it represents α in the given bases, that is $\alpha(f_j) = \sum_i c_{ij} g_i$ for all i, j. Set $a_{ij} = (c_{ij})_{\deg(f_j)-\deg(g_i)}$. The matrix $A = [a_{ij}]$ has homogeneous entries such that for all i, j the following two properties hold: $\alpha(f_j) = \sum_i a_{ij} g_i$, and $a_{ij} = 0$ if $\deg(a_{ij}) \neq \deg(f_j) - \deg(g_i)$.*

If in addition $F = F'$ and the degrees of the elements in each of the bases $f_1, \dots, f_s$ and $g_1, \dots, g_s$ are increasing, then A has square blocks with entries in k along the diagonal and zeros below these blocks.

Proof. Since α has degree 0, it follows that

$$\alpha(f_j) = \Big(\alpha(f_j)\Big)_{\deg(f_j)} = \Big(\sum_i c_{ij} g_i\Big)_{\deg(f_j)}$$
$$= \sum_i \Big(c_{ij}\Big)_{\deg(f_j)-\deg(g_i)} g_i = \sum_i a_{ij} g_i\,.$$

Let $F = F'$. By Theorem 2.12(3), it follows that we have the inequalities $\deg(f_j) \leq \deg(g_i)$ for $j < i$. This implies that $\deg(a_{ij}) \leq 0$ for $j < i$, and that A has square blocks with entries in k along the diagonal and zeros below these blocks. $\square$

We remark that the proof of [Eisenbud, Theorem 20.2] works in the local case, but does not work in the graded case. We present a modification of that proof which covers the graded case.

Proof of Theorem 7.5(2). By Lemma 6.8, the identity map id: $U \to U$ induces graded maps of degree 0 of complexes

$$\varphi : \mathbf{F} \longrightarrow \mathbf{G}$$
$$\psi : \mathbf{G} \longrightarrow \mathbf{F}\,,$$

and furthermore, there exists a graded homotopy h of internal degree 0 such that

$$\mathrm{id}_i - \psi_i\varphi_i = d_{i+1}h_i + h_{i-1}d_i :\; F_i \to F_i$$

for each i. Since $\mathbf{F}$ is minimal, it follows that

$$\begin{aligned}\mathrm{Im}(\mathrm{id}_i - \psi_i\varphi_i) &= \mathrm{Im}(d_{i+1}h_i + h_{i-1}d_i) \subseteq \mathrm{Im}(d_{i+1}) + h_{i-1}\mathrm{Im}(d_i)\\ &\subseteq \mathbf{m}F_i + h_{i-1}(\mathbf{m}F_{i-1}) \subseteq \mathbf{m}F_i + \mathbf{m}h_{i-1}(F_{i-1})\\ &\subseteq \mathbf{m}F_i\end{aligned}$$

for each i.

Choose a homogeneous basis $f_1, \dots, f_t$ of F_i consisting of elements whose degrees increase. Let $A = [a_{rj}]$ be the matrix constructed in Lemma 9.4 that represents the map $\psi_i\varphi_i$ in this basis. Lemma 9.4 shows that A has square blocks with entries in k along the diagonal and zeros below these blocks. The matrix of $id_i - \psi_i\varphi_i$ is $E - A$, where E is the identity matrix with 1 on the diagonal and zeros elsewhere. As $\mathrm{Im}(\mathrm{id}_i - \psi_i\varphi_i) \subset \mathbf{m}F_i$, we have that $E - A$ has entries in $\mathbf{m}$. Therefore, for all j we have that $1 - a_{jj} = 0$, so $a_{jj} = 1$. Furthermore, if $a_{rj} \in k$ and $r \neq j$, then $a_{rj} = 0$. Hence A is an upper triangular matrix. Therefore, $\det(A) = \prod_{1 \le j \le t} a_{jj} = 1$. It follows that the matrix A is invertible; its inverse matrix is the adjoint matrix.

We conclude that $\psi\varphi : \mathbf{F} \longrightarrow \mathbf{F}$ is an isomorphism. Let $\xi : \mathbf{F} \longrightarrow \mathbf{F}$ be its inverse. Then

$$\mathbf{F} \xrightarrow{\varphi} \mathbf{G} \xrightarrow{\xi\psi} \mathbf{F}$$

and $(\xi\psi)\varphi = \mathrm{id}$. This provides a splitting. Set $\mathbf{T} = \mathrm{Ker}(\xi\psi)$. By Lemma 9.1 we obtain $\mathbf{G} = \varphi(\mathbf{F}) \oplus \mathbf{T}$ as graded modules. It remains to prove that $\mathbf{G} = \varphi(\mathbf{F}) \oplus \mathbf{T}$ as complexes, that is, we have to consider how the differentials act.

Denote by d the differential of $\mathbf{F}$ and by ∂ the differential of $\mathbf{G}$. Since φ is a map of complexes, we have $\partial\varphi = \varphi d$, so $\partial(\varphi(\mathbf{F})) \subseteq \varphi(\mathbf{F})$.

We will show that $\partial(\mathbf{T}) \subseteq \mathbf{T}$, that is, $\partial(\mathbf{T}) \subseteq \mathrm{Ker}(\xi\psi)$. Let $u \in T_{i+1}$. There exist $f \in F_i$ and $v \in T_i$ such that $\partial_{i+1}(u) = \varphi(f) + v$. Then

$$\begin{array}{l} (\xi\psi)\partial(u) = \xi\psi(\varphi(f) + v) = \xi\psi(\varphi(f)) = \mathrm{id}\,(f) = f \\ \qquad\quad \| \\ \xi(\psi\partial)(u) = \xi d\psi(u) = d(\xi\psi)(u) = d(0) = 0\,. \end{array}$$

Hence, $\partial_{i+1}(u) = v \in T_i$.

Therefore, $\mathbf{G} = \varphi(\mathbf{F}) \oplus \mathbf{T} \cong \mathbf{F} \oplus \mathbf{T}$ as complexes.

We will show that $\mathbf{T}$ is a trivial complex. Clearly, $\mathrm{H}(\mathbf{G}) = \mathrm{H}(\mathbf{F}) \oplus \mathrm{H}(\mathbf{T})$. Since $\mathbf{G}$ and $\mathbf{F}$ are two graded free resolutions of U, it follows that they have the same homology $\mathrm{H}_i(\mathbf{G}) = \mathrm{H}_i(\mathbf{F}) = 0$ for $i > 0$ and $\mathrm{H}_0(\mathbf{G}) \cong U \cong \mathrm{H}_0(\mathbf{F})$. We conclude that $\mathbf{T}$ is exact. As T_i is a direct summand of G_i for all i, it follows that T_i is free by Lemma 9.2. Applying Lemma 9.3 we obtain that $\mathbf{T}$ is a trivial complex. □

10 Syzygies

Definition 10.1. Let $\mathbf{F}$ be a minimal graded free resolution of a graded finitely generated R-module U. For $i \geq 1$ the submodule

$$\mathrm{Im}(d_i) = \mathrm{Ker}(d_{i-1}) \cong \mathrm{Coker}(d_{i+1})$$

of F_{i-1} is called the i'th ***syzygy module*** of U and denoted $\mathrm{Syz}_i^R(U)$. Its elements are called i'th ***syzygies***. Often, the first syzygies are called just ***syzygies***. Set $\mathrm{Syz}_0^R(U) = U$.

Clearly,

$$\ldots \longrightarrow F_{i+1} \xrightarrow{d_{i+1}} F_i \xrightarrow{d_i} \mathrm{Syz}_i^R(U) \longrightarrow 0$$

is a minimal graded free resolution of $\mathrm{Syz}_i^R(U)$ for each i. Hence

$$\mathrm{Syz}_i^R(\mathrm{Syz}_j^R(U)) = \mathrm{Syz}_{i+j}^R(U)\,.$$

For each $i \geq 0$ we have the short exact sequence

$$\begin{array}{ccccccccc} 0 & \longrightarrow & \mathrm{Syz}_{i+1}^R(U) & \longrightarrow & F_i & \longrightarrow & \mathrm{Syz}_i^R(U) & \longrightarrow & 0 \\ & & \| & & & & \| & & \\ & & \mathrm{Ker}(d_i) & & & & \mathrm{Im}(d_i) & & \end{array}$$

and the exact complex

$$0 \to \mathrm{Syz}_i^R(U) \longrightarrow F_{i-1} \xrightarrow{d_{i-1}} F_{i-2} \longrightarrow \ldots \longrightarrow F_0 \longrightarrow U \longrightarrow 0\,.$$

Theorem 10.2. *Suppose that* $\mathbf{F}$ *is a minimal graded free resolution of* U. *Fix an* $i \geq 0$. *If* $f_1, \ldots, f_p$ *is a basis of* F_i *then the elements*

$d_i(f_1), \ldots, d_i(f_p)$ *form a minimal system of homogeneous generators of* $\mathrm{Syz}_i^R(U)$.

Proof. The homomorphism $F_i \xrightarrow{d_i} \mathrm{Syz}_i^R(U) = \mathrm{Im}(d_i)$ is surjective. Therefore, $d_i(f_1), \ldots, d_i(f_p)$ is a system of homogeneous generators of $\mathrm{Syz}_i^R(U)$. Apply Theorem 7.3 □

11 Betti numbers

Often it is very difficult to obtain a description of the differential in a graded free resolution. In such cases, we try to obtain some information about the numerical invariants of the resolution – the Betti numbers, the projective dimension, and the Poincaré series.

Throughout this section

$$\mathbf{F}: \quad \ldots \longrightarrow F_i \xrightarrow{d_i} F_{i-1} \longrightarrow \ldots \longrightarrow F_1 \xrightarrow{d_1} F_0$$

stands for a minimal graded free resolution of a graded finitely generated R-module U over R.

Definition 11.1. The i'th ***Betti number of*** U ***over*** R is

$$b_i^R(U) = \mathrm{rank}(F_i)\,.$$

By Theorem 7.5, the Betti numbers do not depend on the choice of the minimal graded free resolution of U.

Theorem 11.2.

$$\begin{aligned} b_i^R(U) &= \text{number of minimal generators of } \mathrm{Syz}_i^R(U) \\ &= \dim_k \left(\mathrm{Tor}_i^R(U,k)\right) \\ &= \dim_k \left(\mathrm{Ext}_R^i(U,k)\right) \end{aligned}$$

Proof. The first equality follows from Proposition 10.2.

Consider the complex $\mathbf{F} \otimes_R k$. We have that

$$F_i \otimes_R k = R^{b_i^R(U)} \otimes_R k = k^{b_i^R(U)}\,.$$

Since $\mathrm{Im}(d) \subset \mathbf{mF}$, it follows that $d \otimes_R k = 0$. Therefore,

$$\mathbf{F} \otimes_R k : \quad \ldots \longrightarrow k^{b_i^R(U)} \xrightarrow{0} k^{b_{i-1}^R(U)} \longrightarrow \ldots \longrightarrow k^{b_1^R(U)} \xrightarrow{0} k^{b_0^R(U)} .$$

We see that

$$\mathrm{Tor}_i^R(U,k) \cong \mathrm{H}_i(\mathbf{F} \otimes_R k) = k^{b_i^R(U)} .$$

The argument for Ext is similar. □

Definition 11.3. The ***length*** of a graded free resolution $\mathbf{G}$ is $\max\{i \mid G_i \neq 0\}$. We say that $\mathbf{G}$ is a ***finite resolution*** if its length is finite, otherwise we say that $\mathbf{G}$ is an ***infinite resolution*** . The ***projective dimension*** of U is

$$\mathrm{pd}_R(U) = \max\{i \mid b_i^R(U) \neq 0\} .$$

Thus, $\mathrm{pd}_R(U)$ is the length of the minimal free resolution $\mathbf{F}$ of U. Note that by Theorem 7.5 we have that $\mathrm{pd}_R(U)$ is the length of the shortest graded free resolution of U. The ***Poincaré series*** of U over R is

$$\mathrm{P}_U^R(t) = \sum_{i \geq 0} b_i^R(U) t^i .$$

The properties of the Poincarè series are usually of interest for *infinite* free resolutions.

Example 11.4. Let $A = k[x,y,z]$ and $B = (x^2, xy, xz, y^2)$. Computer computation shows that the minimal graded free resolution of A/B is

$$0 \to A \xrightarrow{\begin{pmatrix} z \\ x \\ -y \\ 0 \end{pmatrix}} A^4 \xrightarrow{\begin{pmatrix} y & 0 & z & 0 \\ -x & z & 0 & y \\ 0 & -y & -x & 0 \\ 0 & 0 & 0 & -x \end{pmatrix}} A^4 \xrightarrow{(x^2 \;\; xy \;\; xz \;\; y^2)} A .$$

Therefore,

$$\mathrm{pd}(A/B) = 3$$

$$\mathrm{P}_{A/B}^A(t) = 1 + 4t + 4t^2 + t^3 .$$

12 Graded Betti numbers

In this section we define the graded Betti numbers, which are a refined version of the numerical invariants Betti numbers. We use the notation introduced in the previous section.

Definition 12.1. Since $\mathbf{F}$ is graded, each free module F_i is a direct sum of modules of the form $R(-p)$. We define the ***graded Betti numbers*** of U by

$$b^R_{i,p}(U) = \text{ number of summands in } F_i \text{ of the form } R(-p)\,.$$

As in Theorem 11.2, we have that

$$b^R_{i,p}(U) = \dim_k\left(\mathrm{Tor}^R_i(U,k)_p\right) = \dim_k\left(\mathrm{Ext}^i_R(U,k)_p\right).$$

Often, for simplicity we write b_i and $b_{i,p}$ instead of $b^R_i(U)$ and $b^R_{i,p}(U)$ if it is clear what module and ring we consider. Sometimes we say "***total Betti numbers***" if we want to emphasize that we are dicussing Betti numbers, not graded Betti numbers.

The ***graded Poincaré series*** of U over R is

$$\mathrm{P}^R_U(t,z) = \sum_{i\geq 0, p\in\mathbf{Z}} b^R_{i,p}(U)t^i z^p\,.$$

Clearly,

$$\mathrm{P}^R_U(t) = \mathrm{P}^R_U(t,1)\,.$$

Similarly, for any complex $\mathbf{G}$ of free R-modules we define its ***Poincaré series*** to be $\mathrm{P}_{\mathbf{G}}(t) = \sum_i \mathrm{rank}(G_i)\,t^i$. We can also define a ***graded Poincaré series*** $\mathrm{P}_{\mathbf{G}}(t,z)$ if $\mathbf{G}$ is graded.

Example 12.2. In Example 4.3, we have that $F_0 = A$, $F_1 = A(-3)\oplus A(-2)\oplus A(-5)$, and $F_2 = A(-4)\oplus A(-6)$. Therefore, $\mathrm{pd}_A(A/B) = 2$ and

$$b_0 = 1, \quad b_1 = 3, \quad b_2 = 2$$

$$\mathrm{P}^A_{A/B}(t) = 1 + 3t + 2t^2$$

$$b_{0,0} = 1,\ b_{1,2} = 1,\ b_{1,3} = 1,\ b_{1,5} = 1,\ b_{2,4} = 1,\ b_{2,6} = 1$$

$$\mathrm{P}^A_{A/B}(t,z) = 1 + tz^2 + tz^3 + tz^5 + t^2z^4 + t^2z^6\,.$$

The Betti numbers can be given in a table in the following two ways.

(1) The labels of the rows and the columns increase upwards and to the right, respectively. The bottom row is the 0'th row and the beginning column is the 0'th column. The entry in the i'th column and the p'th row is $b_{i,p}$. Thus, the i'th column contains the data at the i'th step of the minimal graded free resolution. The vanishing Betti numbers are denoted by empty spaces or by $-$. In the previous example this table is

–	–	1	← 6'th row
–	1	–	
–	–	1	
–	1	–	
–	1	–	
–	–	–	
1	–	–	← zero'th row

(2) The labels of the columns and the rows increase to the right and downwards, respectively. The entry in the i'th column and the p'th row is $b_{i,i+p}$. The i'th column contains the data at the i'th step of the minimal graded free resolution. There is an additional row at the top; there the i'th Betti number b_i is given at the top of each column; this top row with the Betti numbers is underlined. There is an additional column to the left; it contains the labels of the rows and it is separated by a vertical line from the other columns. A vanishing Betti number is denoted by $\cdot$ or by $-$. This is the table that computer programs (such as MACAULAY and Macaulay 2) provide. We call it a ***Betti diagram*** (or ***computer table***). A Betti diagram has the form

	b_0	b_1	b_2	$\cdots$	← Betti numbers
0	$b_{0,0}$	$b_{1,1}$	$b_{2,2}$	$\cdots$	
1	$b_{0,1}$	$b_{1,2}$	$b_{2,3}$	$\cdots$	
2	$b_{0,2}$	$b_{1,3}$	$b_{2,4}$	$\cdots$	
3	$b_{0,3}$	$b_{1,4}$	$b_{2,5}$	$\cdots$	
$\vdots$	$\vdots$	$\vdots$	$\vdots$	$\vdots$	

In the previous example the Betti diagram is

	1	3	2	← Betti numbers
0	1	–	–	
1	–	1	–	
2	–	1	1	
3	–	–	–	
4	–	1	1	

Note that the Betti numbers $b_{i,i}$ appear on the diagonal in the first table and on the zero'th row in the Betti diagram.

Proposition 12.3. *Let c be the minimal degree of an element in a minimal system of homogeneous generators of U. We have that $b^R_{i,p}(U) = 0$ for $p < i + c$.*

Proof. The proof is by induction on the homological degree i. For $i = 0$ note that we have that U has a system of homogeneous generators of degrees greater or equal to c. Suppose that for some i the Betti numbers $b^R_{i,p}(U)$ vanish for $p < i + c$. We want to prove that $b^R_{i+1,p}(U) = 0$ for $p < i+1+c$. For the minimal graded free resolution $\mathbf{F}$ of U we have that $\mathrm{Im}(d_{i+1}) \subseteq \mathbf{m}F_i$. Since for $p < i + c$ we have $b^R_{i,p}(U) = 0$ by induction hypothesis, it follows that F_i has a minimal system of homogeneous generators of degrees $\geq i+c$. The elements in $\mathbf{m}$ have positive degrees. Therefore, $\mathrm{Im}(d_{i+1})$ has a minimal system of homogeneous generators of degrees $\geq i+1+c$. By Theorem 7.3, we conclude that F_{i+1} has a minimal system of homogeneous generators of degrees $\geq i+1+c$. Hence $b^R_{i+1,p}(U) = 0$ for $p < i+1+c$. □

Example 12.4. This example shows that the characteristic of the ground field matters, even if we deal with an ideal generated by monomials. The ideal

$$B = (abc, abf, ace, ahe, ahf, bch, bhe, bef, chf, cef)$$

in the polynomial ring $A = k[a, b, c, e, f, h]$ has projective dimension 2 if $\mathrm{char}(k) \neq 2$ and projective dimension 3 if $\mathrm{char}(k) = 2$. Computer computation shows that the non-vanishing graded Betti numbers of A/B are

$$b_{0,0} = 1,\ b_{1,3} = 10,\ b_{2,4} = 15,\ b_{3,5} = 6$$

if $\operatorname{char}(k) \neq 2$, and

$$b_{0,0} = 1,\ b_{1,3} = 10,\ b_{2,4} = 15,\ b_{3,5} = 6,\ b_{3,6} = 1,\ b_{4,7} = 1$$

if $\operatorname{char}(k) = 2$. Thus, the graded Betti numbers, the Betti numbers, and the projective dimension depend on the characteristic of k.

13 The connecting homomorphism

A sequence of homomorphisms of complexes

$$0 \to (\mathbf{F}, d) \xrightarrow{\varphi} (\mathbf{F}', d') \xrightarrow{\psi} (\mathbf{F}'', d'') \to 0$$

is ***exact*** if

$$0 \to F_i \to F_i' \to F_i'' \to 0$$

is exact for every i.

Construction 13.1. Given an exact sequence of complexes, we will define the ***connecting homomorphism***

$$\tau = \{\, \tau_i : \mathrm{H}_i(\mathbf{F}'') \to \mathrm{H}_{i-1}(\mathbf{F}) \,\}\,.$$

The following diagram is helpful while reading this paragraph:

$$\begin{array}{ccccccc}
0 \to & F_i & \xrightarrow{\varphi} & y \in F_i' & \xrightarrow{\psi} & x \in F_i'' & \to 0 \\
 & \downarrow & & \downarrow & & \downarrow & \\
0 \to & z \in F_{i-1} & \xrightarrow{\varphi} & d'(y) \in F_{i-1}' & \xrightarrow{\psi} & F_{i-1}'' & \to 0 \\
 & \downarrow & & \downarrow & & \downarrow & \\
0 \to & d(z) \in F_{i-2} & \xrightarrow{\varphi} & F_{i-2}' & \xrightarrow{\psi} & F_{i-2}'' & \to 0\,.
\end{array}$$

Let $\alpha \in \mathrm{H}_i(\mathbf{F}'')$. Choose a representative $x \in F_i''$ of α. Note that $d''(x) = 0$. Since ψ_i is surjective, there exists an $y \in F_i'$ such that $\psi_i(y) = x$. Then, $\psi_{i-1}(d'(y)) = d''(\psi_i(y)) = d''(x) = 0$, so $d'(y)$ is in $\mathrm{Ker}(\psi_{i-1}) = \mathrm{Im}(\varphi_{i-1})$. Hence, there exists a $z \in F_{i-1}$ such that $\varphi_{i-1}(z) = d'(y)$. Now, $\varphi_{i-2}(d(z)) = d'(\varphi_{i-1}(z)) = d'(d'(y)) = 0$. Therefore, $d(z) \in \mathrm{Ker}(\varphi_{i-2}) = 0$. So z is a cycle. Let β be the class of z in $\mathrm{H}_{i-1}(\mathbf{F})$. We define $\tau(\alpha) = \beta$.

We will check that the map is well-defined. Let x and x' be two representatives of the homology class α. Let y, z and y', z' be the elements constructed using x and x' respectively. We have to show that z and z' have the same homology class in $\mathrm{H}_{i-1}(\mathbf{F}) = \mathrm{Ker}(d_{i-1})/\mathrm{Im}(d_i)$. Thus, we have to check that $z - z' \in \mathrm{Im}(d_i)$. The following diagram is helpful while reading this paragraph:

$$\begin{array}{ccccccc}
0 \to & F_{i+1} & \xrightarrow{\varphi} & \bar{y} \in F'_{i+1} & \xrightarrow{\psi} & \bar{x} \in F''_{i+1} & \to 0 \\
 & \downarrow & & \downarrow & & \downarrow & \\
0 \to & \bar{z} \in F_i & \xrightarrow{\varphi} & y - y' - d'(\bar{y}) \in F'_i & \xrightarrow{\psi} & x - x' \in F''_i & \to 0 \\
 & \downarrow & & \downarrow & & \downarrow & \\
0 \to & z - z' \in F_{i-1} & \xrightarrow{\varphi} & d'(y) - d'(y') \in F'_{i-1} & \xrightarrow{\psi} & F''_{i-1} & \to 0\,.
\end{array}$$

First, note that $x - x'$ is a boundary; let $x - x' = d''(\bar{x})$. Since ψ_{i+1} is surjective, we can choose an $\bar{y} \in F'_{i+1}$ such that $\psi_{i+1}(\bar{y}) = \bar{x}$. Then $\psi_i(d'(\bar{y})) = d''(\psi_{i+1}(\bar{y})) = x - x' = \psi_i(y - y')$. Thus, $y - y' - d'(\bar{y}) \in \mathrm{Ker}(\psi_i) = \mathrm{Im}(\varphi_i)$, so we can choose a $\bar{z} \in F_i$ such that $\varphi_i(\bar{z}) = y - y' - d'(\bar{y})$. On the one hand, $\varphi_{i-1}(d(\bar{z})) = d'(\varphi_i(\bar{z})) = d'(y - y' - d'(\bar{y})) = d'(y) - d'(y')$ and on the other hand $\varphi_{i-1}(z - z') = d'(y) - d'(y')$. Since φ_{i-1} is injective, we conclude that $z - z' = d(\bar{z})$. Hence, $z - z' \in \mathrm{Im}(d_i)$ as desired.

The following theorem is proved in [Northcott, Section 4.6].

Theorem 13.2. *A short exact sequence of complexes*

$$0 \to (\mathbf{F}, d) \xrightarrow{\varphi} (\mathbf{F}', d') \xrightarrow{\psi} (\mathbf{F}'', d'') \to 0\,.$$

yields the homology long exact sequence

$$\ldots \to \mathrm{H}_{i+1}(\mathbf{F}'') \xrightarrow{\tau} \mathrm{H}_i(\mathbf{F}) \xrightarrow{\varphi} \mathrm{H}_i(\mathbf{F}') \xrightarrow{\psi} \mathrm{H}_i(\mathbf{F}'') \xrightarrow{\tau} \mathrm{H}_{i-1}(\mathbf{F}) \to \ldots,$$

where τ is the connecting homomorphism.

Corollary 13.3. *If any two of the complexes in Theorem 13.2 are exact, then so is the third.*

Let

$$\mathbf{U}: \quad 0 \to U \to U' \to U'' \to 0$$

be a short sequence of R-modules. A sequence of complexes

$$0 \to \mathbf{F} \to \mathbf{F}' \to \mathbf{F}'' \to 0$$

is said to be ***over*** (or a ***lifting*** of) $\mathbf{U}$, if $\mathbf{F} \to \mathbf{F}'$ is over $U \to U'$ and $\mathbf{F}' \to \mathbf{F}''$ is over $U' \to U''$.

Theorem 13.4. *Let*

$$\mathbf{U}: \quad 0 \to U \to U' \to U'' \to 0$$

be a short exact sequence of R-modules. Let $\mathbf{F}$ and $\mathbf{F}''$ be graded free resolutions of U and U'' respectively. There exists a graded free resolution $\mathbf{F}'$ of U' which can be embedded in a split short exact sequence

$$0 \to \mathbf{F} \to \mathbf{F}' \to \mathbf{F}'' \to 0$$

over $\mathbf{U}$.

Proof. Set $F_i' = F_i \oplus F_i''$ for $i \geq 0$. Denote by α the given map $U \to U'$, and by β the given map $U' \to U''$. Let d and d'' be the differentials on $\mathbf{F}$ and $\mathbf{F}''$, respectively. We will construct d' of the form

$$\begin{array}{ccc} F_i & \xrightarrow{d_i} & F_{i-1} \\ \oplus & \nearrow \varphi_i & \oplus \\ F_i'' & \xrightarrow[d_i'']{} & F_{i-1}'' \end{array}$$

So we will construct a map $\varphi = \{\varphi_i \,|\, i \geq 0\}$ with

$$\varphi_0 : F_0'' \to U' \quad \text{and} \quad \varphi_i : F_i'' \to F_{i-1} \ \text{ for } i \geq 1 ,$$

and then we will set the differential d' on $\mathbf{F}'$ to be

$$d_0' = \alpha d_0 + \varphi_0 \quad \text{and} \quad d_i' = d_i + \varphi_i + d_i'' \ \text{ for } i \geq 1 .$$

First we need to find what conditions φ must satisfy in order to produce a differential. We would like the following two diagrams to be commutative:

$$\begin{array}{ccccc} F_{i+1} & \longrightarrow & F'_{i+1} & \longrightarrow & F''_{i+1} \\ \downarrow d_{i+1} & & \downarrow d'_{i+1} & & \downarrow d''_{i+1} \\ F_i & \longrightarrow & F'_i & \longrightarrow & F''_i \end{array}$$

and

$$\begin{array}{ccccc} F_0 & \longrightarrow & F'_0 & \longrightarrow & F''_0 \\ \downarrow d_0 & & \downarrow d'_0 & & \downarrow d''_0 \\ U & \longrightarrow & U' & \longrightarrow & U''\,. \end{array}$$

Straightforward computation shows that

(1) The former diagram is always commutative.

(2) The left square in the latter diagram is always commutative.

(3) The commutativity of the right square in the latter diagram is equivalent to $d''_0 = \beta\varphi_0$.

Furthermore, we would like d' to be a differential. Straightforward computation shows that

(4) The condition $d'_0 d'_1 = 0$ is equivalent to $\alpha d_0 \varphi_1 + \varphi_0 d''_1 = 0$.

(5) For $i \geq 1$ the condition $d'_i d'_{i+1} = 0$ is equivalent to $d_i \varphi_{i+1} + \varphi_i d''_{i+1} = 0$.

Therefore, it suffices to construct a map φ that satisfies the following conditions:

- $d''_0 = \beta\varphi_0$ (from (3) above)
- $\alpha d_0 \varphi_1 + \varphi_0 d''_1 = 0$ (from (4) above)
- $d_i \varphi_{i+1} + \varphi_i d''_{i+1} = 0$ for $i \geq 1$ (from (5) above).

We will define the map φ inductively. A map $\varphi_0 : F''_0 \to U'$ satisfying $d''_0 = \beta\varphi_0 : F''_0 \to U''$ exists because the module F''_0 is free. A map $\varphi_1 : F''_1 \to F_0$ satisfying

$$(\alpha d_0)\varphi_1 = -\varphi_0 d''_1 : F''_1 \to U$$

exists because the module F''_1 is free. Suppose that we have defined φ_i. A map $\varphi_{i+1} : F''_{i+1} \to F_i$ satisfying

$$d_i \varphi_{i+1} = -\varphi_i d''_{i+1} : F''_{i+1} \to F_{i-1}$$

exists because the module F''_{i+1} is free. □

14 The Koszul complex

The minimal free resolution of an ideal generated by a regular sequence is very nicely structured and is described by the Koszul complex. An important special case is the minimal free resolution of k (equivalently, of the maximal ideal $(x_1, \dots, x_n)$) over S.

First, we recall the definition of a regular sequence. An element $r \in R$, $r \notin k$ is a ***non-zero divisor*** on a finitely generated R-module U if $ru \neq 0$ for every non-zero $u \in U$; in this case we also say that r is a ***U-regular*** element. A sequence $f_1, \dots, f_q$ of elements in R is a ***U-regular sequence*** if the following two conditions are satisfied: $(f_1, \dots, f_q)U \neq U$, and for every $1 \leq i \leq q$ we have that f_i is a non-zero divisor on the module $U/(f_1, \dots, f_{i-1})U$.

Before reading the next construction, it will be helpful if the reader thinks about the minimal free resolution of the ideal (x^2, y^2, z^2) and makes a guess (without proving it) how it looks like.

Construction 14.1. Let $f_1, \dots, f_q$ be elements in R. Let E be the exterior algebra over k on basis elements $e_1, \dots, e_q$; this means that E is the following quotient of a free algebra

$$E = k\langle e_1, \dots, e_q \rangle \,/\, (\{e_i^2 \,|\, 1 \leq i \leq q\},\ \{e_i e_j + e_j e_i \,|\, 1 \leq i < j \leq q\})\,.$$

It is graded by $\deg(e_i) = 1$ for each i. It is easy to show that $e^2 = 0$ for every linear form $e \in E$. Multiplication in E is denoted by $\wedge$. Note that $e_i \wedge e_j = -e_j \wedge e_i$ for $i \neq j$. Denote by $\mathbf{f}$ the sequence $f_1, \dots, f_q$ and by $\mathbf{K}(\mathbf{f})$ the R-module $R \otimes E$ graded homologically by $\deg(e_{j_1} \wedge \cdots \wedge e_{j_i}) = i$ and equipped with the differential

$$d(e_{j_1} \wedge \cdots \wedge e_{j_i}) = \sum_{1 \leq p \leq i} (-1)^{p+1} f_{j_p}\, e_{j_1} \wedge \cdots \wedge \widehat{e}_{j_p} \wedge \cdots \wedge e_{j_i}\ ,$$

where $\widehat{e}_{j_p}$ means that e_{j_p} is omitted in the product. The following computation shows that $d^2 = 0$. We have

$$d^2(e_{j_1} \wedge \cdots \wedge e_{j_i}) = \sum_{1 \leq p < s \leq i} \gamma_{p,s}\, e_{j_1} \wedge \cdots \wedge \widehat{e}_{j_p} \wedge \cdots \wedge \widehat{e}_{j_s} \wedge \cdots \wedge e_{j_i}$$

where the coefficient $\gamma_{p,s}$ is obtained in two steps:

(1) first we remove e_{j_s} and then we remove e_{j_p} from $e_{j_1} \wedge \cdots \wedge e_{j_i}$, so we get the coefficient $(-1)^{s+1} f_{j_s} (-1)^{p+1} f_{j_p}$

(2) first we remove e_{j_p} and then we remove e_{j_s} from $e_{j_1} \wedge \cdots \wedge e_{j_i}$, so we get coefficient $(-1)^{p+1} f_{j_p} (-1)^{s} f_{j_s}$

Therefore,

$$\gamma_{p,s} = (-1)^{s+1}(-1)^{p+1} f_{j_s} f_{j_p} + (-1)^{p+1}(-1)^{s} f_{j_p} f_{j_s} = 0 .$$

The complex $\mathbf{K}(\mathbf{f})$ is called the ***Koszul complex*** on $f_1, \ldots, f_q$. If we write the complex as

$$\mathbf{K}(\mathbf{f}) : \quad 0 \to K_q \to \ldots \to K_1 \to K_0 \to 0 ,$$

then the elements

$$\{e_{j_1} \wedge \cdots \wedge e_{j_i} \,|\, 1 \leq j_1 < \ldots < j_i \leq q\}$$

form a basis of the free R-module K_i. Hence the rank of K_i is $\binom{q}{i}$.

For a finitely generated R-module W, set $\mathbf{K}(\mathbf{f}; W) = \mathbf{K}(\mathbf{f}) \otimes_R W$. The homology of this complex is called the ***Koszul homology*** of W.

If we permute the elements $f_1, \ldots, f_q$, then we obtain another Koszul complex by simply permuting the basis of the exterior algebra.

Example 14.2. The complex $\mathbf{K}(f_1)$ is $0 \to R \xrightarrow{f_1} R \to 0$.

Example 14.3. Let $A = k[x, y, z]$. Take $f_1 = x^2$ and $f_2 = y^2$. Then K_0 has basis 1; K_1 has basis e_1, e_2; and K_2 has basis $e_1 \wedge e_2$. The differential acts as

$$d(e_1) = x^2 \quad \text{and} \quad d(e_2) = y^2$$
$$d(e_1 \wedge e_2) = x^2 e_2 - y^2 e_1 .$$

Therefore, the Koszul complex is

$$\mathbf{K}(x^2, y^2) : \quad 0 \to K_2 \xrightarrow{\begin{pmatrix} -y^2 \\ x^2 \end{pmatrix}} K_1 \xrightarrow{\begin{pmatrix} x^2 & y^2 \end{pmatrix}} K_0 .$$

Now, take $f_1 = x^2,\ f_2 = y^2,\ f_3 = z^2$. Then K_0 has basis 1; K_1 has basis e_1, e_2, e_3; K_2 has basis $e_1 \wedge e_2,\ e_1 \wedge e_3,\ e_2 \wedge e_3$; and K_3 has basis $e_1 \wedge e_2 \wedge e_3$. The differential acts as

$$d(e_1) = x^2,\ d(e_2) = y^2,\ d(e_3) = z^2$$

$$d(e_1 \wedge e_2) = x^2 e_2 - y^2 e_1$$

$$d(e_1 \wedge e_3) = x^2 e_3 - z^2 e_1$$

$$d(e_2 \wedge e_3) = y^2 e_3 - z^2 e_2$$

$$d(e_1 \wedge e_2 \wedge e_3) = x^2 e_2 \wedge e_3 - y^2 e_1 \wedge e_3 + z^2 e_1 \wedge e_2\,.$$

Therefore, the Koszul complex is

$$\mathbf{K}(x^2, y^2, z^2):\quad 0 \to K_3 \xrightarrow{\begin{pmatrix} z^2 \\ -y^2 \\ x^2 \end{pmatrix}} K_2 \xrightarrow{\begin{pmatrix} -y^2 & -z^2 & 0 \\ x^2 & 0 & -z^2 \\ 0 & x^2 & y^2 \end{pmatrix}}$$

$$K_1 \xrightarrow{(x^2\ \ y^2\ \ z^2)} K_0\,.$$

Note that $\mathbf{K}(x^2, y^2)$ is a subcomplex of $\mathbf{K}(x^2, y^2, z^2)$.

The formula for the differential implies the following formula.

Exercise 14.4. *Let $e = e_{j_1} \wedge \cdots \wedge e_{j_i}$ and $p > j_q$ for $1 \le q \le i$. We have that*

$$d(e \wedge e_p) = d(e) \wedge e_p + (-1)^i f_p\, e\,.$$

Lemma 14.5. *Let $\bar{\mathbf{f}} = \{f_1, \ldots, f_{q-1}\}$ be a sequence of elements in R, and let $f_q \in R$. Denote by $\mathbf{f}$ the sequence $\bar{\mathbf{f}}, f_q$. There is an exact sequence of complexes*

$$0 \to \mathbf{K}(\bar{\mathbf{f}}) \to \mathbf{K}(\mathbf{f}) \to \mathbf{K}(\bar{\mathbf{f}})[-1] \to 0$$

(recall that $\mathbf{K}(\bar{\mathbf{f}})[-1]$ means that the complex $\mathbf{K}(\bar{\mathbf{f}})$ is homologically shifted one degree). The homology long exact sequence obtained from this is

$$\ldots \to \mathrm{H}_i(\mathbf{K}(\bar{\mathbf{f}})) \to \mathrm{H}_i(\mathbf{K}(\mathbf{f})) \to$$

$$\mathrm{H}_{i-1}(\mathbf{K}(\bar{\mathbf{f}})) \xrightarrow{(-1)^{i+1} f_q} \mathrm{H}_{i-1}(\mathbf{K}(\bar{\mathbf{f}})) \to \ldots$$

Proof. Clearly,

$$\mathbf{K}(\mathbf{f})_i = \mathbf{K}(\bar{\mathbf{f}})_i \oplus \mathbf{K}(\bar{\mathbf{f}})_{i-1} \wedge e_q$$

for $1 \leq i \leq q$. Therefore, we have the desired short exact sequence of complexes. It yields a long exact sequence in homology by Theorem 13.2. We will compute the connecting homomorphism following Construction 13.1 (using the notation in that construction). Let $\alpha \in \mathrm{H}_i(\mathbf{K}(\bar{\mathbf{f}})[-1]) = \mathrm{H}_{i-1}(\mathbf{K}(\bar{\mathbf{f}}))$. Choose a representative $x \in \mathbf{K}(\bar{\mathbf{f}})_{i-1}$. Then $y = x \wedge e_q \in \mathbf{K}(\mathbf{f})_i$. Now $d(y) = d(x) \wedge e_q + (-1)^{i-1} f_q x$. Since $(-1)^{i+1} f_q x \in \mathbf{K}(\bar{\mathbf{f}})_{i-1}$ and $d(x) \wedge e_q \in \mathbf{K}(\bar{\mathbf{f}})_{i-2} \wedge e_q$, we conclude that $z = (-1)^{i+1} f_q x$. Therefore, the connecting homomorphism is

$$\mathrm{H}_{i-1}(\mathbf{K}(\bar{\mathbf{f}})) \xrightarrow{(-1)^{i+1} f_q} \mathrm{H}_{i-1}(\mathbf{K}(\bar{\mathbf{f}}))\,.$$

□

Proposition 14.6. *Let W be a finitely generated R-module. We have*

$$(f_1, \ldots, f_q)\, \mathrm{H}_i(\mathbf{K}(\mathbf{f}; W)) = 0 \quad \textit{for all} \;\; i \geq 0\,.$$

Proof. By symmetry, it is enough to prove that $f_q \mathrm{H}_i(\mathbf{K}(\mathbf{f}; W)) = 0$ for all $i \geq 0$. We use the notation in the proof of the previous lemma. Let $c \wedge e_q + a$ be a cycle in $\mathbf{K}(\mathbf{f}; W)_i$, where $a \in \mathbf{K}(\bar{\mathbf{f}}; W)_i$ and $c \wedge e_q \in \mathbf{K}(\bar{\mathbf{f}}; W)_{i-1} \wedge e_q$. We have to show that the homology class of $f_q(c \wedge e_q + a)$ is zero, that is, we have to show that $f_q(c \wedge e_q + a)$ is a boundary. The equalities

$$d(c \wedge e_q + a) = d(c) \wedge e_q + (-1)^{i+1} f_q c + d(a) = 0$$

imply that $d(c) = 0$ and $d(a) = (-1)^i f_q c$. Therefore,

$$\begin{aligned} d((-1)^i\, a \wedge e_q) &= (-1)^i \Big(d(a) \wedge e_q + (-1)^i f_q\, a \Big) \\ &= (-1)^i \Big((-1)^i f_q c \wedge e_q + (-1)^i f_q\, a \Big) \\ &= f_q(c \wedge e_q + a), \end{aligned}$$

where the second equality follows from $d(a) = (-1)^i f_q c$ and the first equality follows from Exercise 14.4. □

Theorem 14.7. *Let $W \neq 0$ be a finitely generated graded R-module, and $\mathbf{f} = \{f_1, \ldots, f_q\}$ be a sequence of homogeneous elements in R of positive degree. The following properties are equivalent.*

(1) $\mathrm{H}_i(\mathbf{K}(\mathbf{f}; W)) = 0$ *for* $i > 0$, *and* $\mathrm{H}_0(\mathbf{K}(\mathbf{f}; W)) = W/(\mathbf{f})W$.

(2) $\mathrm{H}_1(\mathbf{K}(\mathbf{f}; W)) = 0$.

(3) $\mathbf{f}$ *is a W-regular sequence.*

Proof. First, we will prove that (3) implies (1). By construction, $\mathrm{H}_0(\mathbf{K}(\mathbf{f}; W)) = W/(\mathbf{f})W$. We will use induction on q in order to show that $\mathrm{H}_i(\mathbf{K}(\mathbf{f}; W)) = 0$ for $i > 0$. For $q = 1$ we have the Koszul complex

$$\mathbf{K}(f_1; W): \quad 0 \to W \xrightarrow{f_1} W \to 0\,,$$

hence

$$\mathrm{H}_1(\mathbf{K}(f_1; W)) = \{m \in W \mid f_1 m = 0\} = 0\,.$$

Suppose that the assertion holds for $q - 1$, that is, the assertion holds for the sequence $\bar{\mathbf{f}} = \{f_1, \ldots, f_{q-1}\}$. Denote by $\bar{\mathbf{K}} = \mathbf{K}(\bar{\mathbf{f}}; W)$. By Lemma 14.5 we have the exact sequence

$$\begin{array}{ccccccc}
\mathrm{H}_1(\bar{\mathbf{K}}) & \to & \mathrm{H}_1(\mathbf{K}(\mathbf{f}; W)) & \to & \mathrm{H}_0(\bar{\mathbf{K}}) & \xrightarrow{f_q} & \mathrm{H}_0(\bar{\mathbf{K}})\,. \\
\| & & & & \| & & \| \\
0 & & & & W/(\bar{\mathbf{f}})W & & W/(\bar{\mathbf{f}})W
\end{array}$$

As f_q is a regular element on $W/(\bar{\mathbf{f}})W$, it follows $\mathrm{H}_1(\mathbf{K}(\mathbf{f}; W)) = 0$. Let $i > 1$. By Lemma 14.5 we have the exact sequence

$$0 = \mathrm{H}_i(\bar{\mathbf{K}}) \to \mathrm{H}_i(\mathbf{K}(\mathbf{f}; W)) \to \mathrm{H}_{i-1}(\bar{\mathbf{K}}) = 0\,.$$

Hence, $\mathrm{H}_i(\mathbf{K}(\mathbf{f}; W)) = 0$.

(1) implies (2).

We will show that (2) implies (3). We will prove by induction on q that $f_1, \ldots, f_q$ is a regular sequence on W.

First, we will check that $W/(\mathbf{f})W \neq 0$. Assume the opposite, that is, $W/(\mathbf{f})W = 0$. Since W is graded and $f_1, \ldots, f_q$ have positive degrees, we can apply Nakayama's Lemma 2.11. Therefore $W = 0$, which is a contradiction. Hence, $W/(\mathbf{f})W \neq 0$.

By Lemma 14.5 we have the exact sequence

$$\mathrm{H}_1(\bar{\mathbf{K}}) \xrightarrow{f_q} \mathrm{H}_1(\bar{\mathbf{K}}) \to \mathrm{H}_1(\mathbf{K}(\mathbf{f}; W)) = 0\,.$$

Hence, $\mathrm{H}_1(\bar{\mathbf{K}}) = (f_q)\mathrm{H}_1(\bar{\mathbf{K}})$; since the module is graded and f_q has positive degree, we can apply Nakayama's Lemma 2.11. Therefore, $\mathrm{H}_1(\bar{\mathbf{K}}) = 0$. By induction hypothesis, $\bar{\mathbf{f}} = f_1, \ldots, f_{q-1}$ is a W-regular sequence. Furthermore, by Lemma 14.5 we have the exact sequence

$$\begin{array}{ccccc} 0 = \mathrm{H}_1(\mathbf{K}(\mathbf{f}; W)) & \longrightarrow & \mathrm{H}_0(\bar{\mathbf{K}}) & \xrightarrow{f_q} & \mathrm{H}_0(\bar{\mathbf{K}})\,. \\ & & \| & & \| \\ & & W/(\bar{\mathbf{f}})W & & W/(\bar{\mathbf{f}})W \end{array}$$

It shows that f_q is a regular element on $W/(\bar{\mathbf{f}})W$. We have proved that $\mathbf{f}$ is a W-regular sequence. □

Corollary 14.8. *Let $W \neq 0$ be a finitely generated graded R-module, and $\mathbf{f} = \{f_1, \ldots, f_q\}$ be a sequence of homogeneous elements in $\mathbf{m}$. Suppose that $f_1, \ldots, f_q$ is a homogeneous W-regular sequence.*

(1) *Any permutation of $f_1, \ldots, f_q$ is a W-regular sequence.*

(2) *If $u_1 f_1 + \ldots + u_q f_q = 0$ for some $u_i \in W$, then for all i we have $u_i \in (f_1, \ldots, f_q)W$.*

Proof. (1) Let $g_1, \ldots, g_q$ be a permutation of the elements $f_1, \ldots, f_q$. The Koszul complex $\mathbf{K}(g_1, \ldots, g_q; W)$ is obtained from $\mathbf{K}(f_1, \ldots, f_q; W)$ by a change of basis in the exterior algebra. Thus, the two complexes have the same homology.

(2) Since $d(u_1 e_1 + \ldots + u_q e_q) = u_1 f_1 + \ldots + u_q f_q = 0$ we have that $u_1 e_1 + \ldots + u_q e_q \in \mathrm{Ker}(d_1)$. As the Koszul complex is exact, it follows that

$$u_1 e_1 + \ldots + u_q e_q \in \mathrm{Im}(d_2) \subseteq (f_1, \ldots, f_q)T\,,$$

where T is the module $K_1 \otimes W = W \otimes e_1 + \ldots + W \otimes e_q$. Therefore,

$u_i \in (f_1, \dots, f_q)W$ for all i. □

A major application of Theorem 14.7 is that it provides the minimal free resolution of k.

Corollary-Construction 14.9. Consider $k = S/(x_1, \dots, x_n)$ as an S-module. By Theorem 14.7 we have that the Koszul complex $\mathbf{K}(x_1, \dots, x_n)$ is the minimal graded free resolution of k over S. We denote it by $\mathbf{K}$. Then, for each $0 \le i \le n$, we have that the free S-module K_i has basis

$$\{e_{j_1} \wedge \dots \wedge e_{j_i} \,|\, 1 \le j_1 < \dots < j_i \le n\}.$$

In particular, we have $b_i^S(k) = \binom{n}{i}$ for $i \ge 0$, and $\mathrm{pd}_S(k) = n$. The differential acts as

$$d(e_{j_1} \wedge \dots \wedge e_{j_i}) = \sum_{1 \le p \le i} (-1)^{p+1} x_{j_p} e_{j_1} \wedge \dots \wedge \widehat{e}_{j_p} \wedge \dots \wedge e_{j_i} .$$

The resolution is linear, that is, the entries in the differential matrices are linear forms. Since S is standard graded, we have that K_i, as a graded module, is equal to $R(-i)^{\binom{n}{i}}$.

Theorem 14.10. *Let V be a graded finitely generated S-module. We have that*

$$b_{i,p}^S(V) = \dim_k \mathrm{H}_i(\mathbf{K} \otimes V)_p \quad \textit{for all } p \in \mathbf{Z} \textit{ and } i \ge 0 .$$

Proof. By Theorem 11.2, $b_{i,p}^S(V) = \dim_k(\mathrm{Tor}_i^S(V,k)_p)$ for all $p \in \mathbf{Z}$ and $i \ge 0$. The above construction yields $\dim_k(\mathrm{Tor}_i^S(k,V)_p) = \dim_k(\mathrm{H}_i(\mathbf{K} \otimes V)_p)$ as desired. □

Proposition 14.11. $\mathrm{Tor}_S^n(S/I, k) \cong \mathrm{soc}(S/I)$ *(recall that the socle of S/I is $\{f \in S/I \,|\, \mathbf{m}f = 0\}$).*

Proof. Compute $\mathrm{Tor}_S^n(S/I, k)$ using the Koszul resolution of k, that is, use $\mathrm{Tor}_S^n(S/I, k) \cong \mathrm{H}_n(S/I \otimes \mathbf{K})$. In the notation of 14.9 we get that $\mathrm{H}_n(S/I \otimes \mathbf{K})$ is the kernel of the homomorphism

$$S/I \otimes K_n \cong S/I \to S/I \otimes K_{n-1}$$
$$f\, e_1 \wedge \cdots \wedge e_n \mapsto \sum\nolimits_{1 \le p \le n} (-1)^{p+1}\, x_p f\, e_1 \wedge \cdots \wedge \widehat{e}_p \wedge \cdots \wedge e_n\,,$$

where $f \in S/I$. It follows that $\mathrm{H}_n(S/I \otimes \mathbf{K}) = \{f \in S/I \mid \mathbf{m}f = 0\}$, which is the socle of S/I. □

Corollary 14.12. $b_n^S(S/I) = \dim_k(\mathrm{soc}(S/I))$.

The following conjecture, in [Buchsbaum-Eisenbud], [Hartshorne], [Horrocks], is challenging and wide open. See [Charalambous-Evans 2] for a survey. In Corollary 21.6, we show that the conjecture holds for S/M if M is an ideal generated by monomials.

Conjecture 14.13. [Buchsbaum-Eisenbud], [Hartshorne], [Horrocks] *If V is an artinian graded finitely generated S-module, then*

$$b_i^S(V) \ge b_i^S(k) \quad \textit{for} \quad i \ge 0\,.$$

A more general version of 14.13 extends the conjecture to quotient rings as follows.

Conjecture 14.14. (cf. [Charalambous-Evans 2]) *If U is an artinian graded finitely generated R-module, then*

$$b_i^R(U) \ge b_i^R(k) \quad \textit{for} \quad i \ge 0\,.$$

15 Finite projective dimension

The projective dimension is a core invariant of a graded finitely generated R-module. If it is finite then it is expressed by the Auslander-Buchsbaum Formula 15.3. We also present Hilbert's Syzygy Theorem 15.2 which is a key result on resolutions over polynomial rings and states that every graded finitely generated S-module has a *finite* minimal graded free resolution.

Theorem 15.1. *If U is a graded finitely generated R-module, then* $\mathrm{pd}_R(U) \leq \mathrm{pd}_R(k)$.

Proof. Let $\mathbf{F}_k$ be the minimal graded free resolution of k. By Theorem 11.2, we have that

$$b_i^R(U) = \dim_k \left(\mathrm{Tor}_i^R(U,k)\right) = \dim_k \left(\mathrm{H}_i(U \otimes \mathbf{F}_k)\right).$$

This homology vanishes for $i > \mathrm{pd}_R(k)$ since $F_i = 0$. □

Hilbert's Syzygy Theorem 15.2. *The minimal graded free resolution of a graded finitely generated S-module is finite and its length is at most n.*

Proof. This is a corollary of Theorem 15.1 and the fact that by 14.9 we have $\mathrm{pd}_R(k) = n$. Another proof, using Gröbner basis theory, is given in [Eisenbud, 15.11]. □

A more precise version of this theorem is the Auslander-Buchsbaum Formula.

The Auslander-Buchsbaum Formula 15.3. *Let V be a graded finitely generated S-module. Its projective dimension is*

$$\mathrm{pd}_S(V) = n - \mathrm{depth}(V).$$

Proof. By Theorem 14.10 we have that

$$\mathrm{pd}_S(V) = \max\{i \,|\, \mathrm{H}_i(\mathbf{K} \otimes V) \neq 0\},$$

where $\mathbf{K}$ is the Koszul resolution of k. The proof is by induction on $j = \mathrm{depth}(V)$.

Suppose $j = 0$. Then $\mathbf{m}$ is an associated prime of V. Therefore, there exists a homogeneous element $u \in V$ such that $\mathbf{m} \cdot u = 0$. Hence $e_1 \wedge \ldots \wedge e_n \otimes u \in \mathrm{H}_n(\mathbf{K} \otimes V)$. So $\mathrm{pd}_S(V) = n$.

Suppose $j > 0$. Choose a homogeneous V-regular element $r \in \mathbf{m}$. Set $W = V/(r)V$. We have the exact sequence

$$0 \to V \xrightarrow{r} V \to W \to 0.$$

Therefore, we get the homology long exact sequence (see 38.3)

$$\ldots \to \mathrm{H}_{i+1}(\mathbf{K}\otimes W) \to \mathrm{H}_i(\mathbf{K}\otimes V) \xrightarrow{r} \mathrm{H}_i(\mathbf{K}\otimes V) \to \mathrm{H}_i(\mathbf{K}\otimes W) \to \mathrm{H}_{i-1}(\mathbf{K}\otimes V) \to \ldots$$

By Proposition 14.6, it follows that $r \cdot \mathrm{H}_i(\mathbf{K}\otimes V) = 0$ for $i \geq 0$. Therefore, for each i we have the short exact sequence

$$0 \to \mathrm{H}_i(\mathbf{K}\otimes V) \to \mathrm{H}_i(\mathbf{K}\otimes W) \to \mathrm{H}_{i-1}(\mathbf{K}\otimes V) \to 0\,.$$

It follows that $\max\{i \,|\, \mathrm{H}_i(\mathbf{K}\otimes W) \neq 0\} = 1+\max\{i \,|\, \mathrm{H}_i(\mathbf{K}\otimes V) \neq 0\}$. Hence, $\mathrm{pd}_S(W) = \mathrm{pd}_S(V)+1$. By induction hypothesis, we have that

$$\mathrm{pd}_S(W) = n - \mathrm{depth}(W) = n - \mathrm{depth}(V) + 1\,.$$

Therefore, $\mathrm{pd}_S(V) = n - \mathrm{depth}(V)$.

We present another proof in Section 20. □

Corollary 15.4. *Let V be a graded finitely generated S-module. We have*

$$\mathrm{pd}_S(V) \geq \mathrm{codim}(V)\,.$$

If V is artinian then $\mathrm{pd}_S(V) = n$.

Proof. $\mathrm{pd}_S(V) = n - \mathrm{depth}(V) \geq n - \dim(V) = \mathrm{codim}(V)$. Equality holds if V is artinian. □

The following result is proved in [Bayer-Stillman].

Theorem 15.5. *Let J be a graded ideal in S. Suppose that we have generic coordinates (i.e., generic variables). The variables $x_s, \ldots, x_n$ form a regular sequence on S/J if and only if they form a regular sequence on $S/\mathrm{in}_{rlex}(J)$, where $\mathrm{in}_{rlex}(J)$ is the initial ideal of J with respect to the revlex order.*

Combining 15.5 with the Auslander-Buchsbaum Formula yields:

Theorem 15.6. *If J is a graded ideal in S, then*

$$\mathrm{pd}(S/J) = \mathrm{pd}(S/\mathrm{in}_{rlex}(J))\,,$$

where $\mathrm{in}_{rlex}(J)$ *is the initial ideal of* J *with respect to the revlex order in generic coordinates.*

The following direction of research is interesting.

Open-Ended Problem 15.7. (folklore) *Obtain bounds on the projective dimension for classes of ideals.*

See Theorem 68.2 for a sharp bound on the projective dimension of toric rings. The following problem raised by Stillman is wide open.

Problem 15.8. (Stillman) *Suppose that* $\mathrm{char}(k) = 0$. *Fix a sequence of natural numbers* $a_1, \ldots, a_s$. *Does there exist a number* p, *such that* $\mathrm{pd}_T(T/J) \leq p$ *if* T *is a polynomial ring and* J *is a graded ideal with a minimal system of homogeneous generators of degrees* $a_1, \ldots, a_s$? *Note that the number of variables in the polynomial ring* T *is not fixed.*

The crucial assumption in this problem is that the degrees $a_1, \ldots, a_r$ are fixed. If the degrees can vary, then the following result of Burch (cf, also [Bruns]) implies that there exists no upper bound on the projective dimension.

Theorem 15.9. *The projective dimension of a graded ideal generated by 3 elements can be arbitrarily large.*

It is natural to ask: When (over what R) does every graded finitely generated module have finite projective dimension? The answer is given by the following fundamental result in Commutative Algebra.

Serre's Theorem 15.10. *The following properties are equivalent.*

(1) *Every graded finitely generated* R*-module has finite projective dimension.*
(2) $\mathrm{pd}_R(k) < \infty$.
(3) R *is a polynomial ring, that is,* $R = S/I$ *for some ideal* I *generated by linear forms.*

See [Matsumura, Theorem 19.2] for a proof of Serre's Theorem.

Note that S/I is a graded regular ring if and only if I is generated by linear forms.

16 Hilbert functions

In this section we present Hilbert's method for computing Hilbert functions using resolutions.

Proposition 16.1. *The Hilbert function is additive on short exact sequences, that is, if*

$$0 \to K \to N \to W \to 0$$

is a short exact sequence of graded finitely generated R-modules and homomorphisms of degree 0, then

$$\mathrm{Hilb}_N(t) = \mathrm{Hilb}_K(t) + \mathrm{Hilb}_W(t)\,.$$

Proof. For each $q \geq 0$, we have the short exact sequence of k-vector spaces

$$0 \to K_q \to N_q \to W_q \to 0\,.$$

□

Free resolutions were introduced by Hilbert. His motivation was to compute the Hilbert function of a finitely generated graded module using a resolution. His method of such computations is presented below.

Theorem 16.2. (Hilbert) *Let* $\mathbf{F}$ *be a graded free resolution of a finitely generated graded R-module U. Write*

$$F_i = \oplus_{p\in\mathbf{Z}} R^{c_{i,p}}(-p)\,.$$

For each p suppose that $c_{i,p} = 0$ for $i \gg 0$. Then

$$\mathrm{Hilb}_U(t) = \mathrm{Hilb}_R(t) \sum_{i\geq 0} \sum_{p\in\mathbf{Z}} (-1)^i c_{i,p} t^p\,.$$

If $R = S$, then

$$\mathrm{Hilb}_U(t) = \frac{\sum_{i\geq 0} \sum_{p\in\mathbf{Z}} (-1)^i c_{i,p} t^p}{(1-t)^n}\,.$$

The following cases are the main two cases when for each p we have $c_{i,p} = 0$ for $i \gg 0$, and Theorem 16.2 can be applied.

(1) The resolution $\mathbf{F}$ is finite.

(2) The resolution $\mathbf{F}$ is minimal. In this case $c_{i,p} = b^R_{i,p}(U)$ and we apply Proposition 12.3.

Proof. First, we present a short proof that works in the case when the resolution $\mathbf{F}$ is finite. The proof is by induction on the length of $\mathbf{F}$. Let K be the image of the first differential map. We have the short exact sequence

$$0 \to K \to F_0 \to U \to 0 .$$

Since the differential map has degree 0, it follows that

$$\mathrm{Hilb}_U(t) = \mathrm{Hilb}_{F_0}(t) - \mathrm{Hilb}_K(t) .$$

Now, F_0 is the free module $\oplus_p R(-p)^{c_{0,p}}$, hence

$$\mathrm{Hilb}_{F_0}(t) = \sum_{p\in\mathbf{Z}} c_{0,p} t^p \,\mathrm{Hilb}_R(t) = \mathrm{Hilb}_R(t) \sum_{p\in\mathbf{Z}} c_{0,p} t^p .$$

By induction, we have that

$$\mathrm{Hilb}_K(t) = \mathrm{Hilb}_R(t) \sum_{i\geq 1} \sum_{p\in\mathbf{Z}} (-1)^{i-1} c_{i,p} t^p .$$

Therefore, we obtain the desired

$$\begin{aligned}
\mathrm{Hilb}_U(t) &= \mathrm{Hilb}_{F_0}(t) - \mathrm{Hilb}_K(t) \\
&= \mathrm{Hilb}_R(t) \sum_{p\in\mathbf{Z}} c_{0,p} t^p - \mathrm{Hilb}_R(t) \sum_{i\geq 1} \sum_{p\in\mathbf{Z}} (-1)^{i-1} c_{i,p} t^p \\
&= \mathrm{Hilb}_R(t) \sum_{i\geq 0} \sum_{p\in\mathbf{Z}} (-1)^{i} c_{i,p} t^p .
\end{aligned}$$

Now, we present a longer proof that works for infinite resolutions as well. Since each F_i is graded we write $F_i = \oplus_j F_{i,j}$. The condition, that for each p we have $c_{i,p} = 0$ for $i \gg 0$, implies that for each j we have $F_{i,j} = 0$ for $i \gg 0$.

Fix an internal degree j. As in Construction 3.6, take the j'th graded component of $\mathbf{F}$. We get the exact sequence of k-vector spaces

$$0 \to \ldots \to F_{i,j} \to F_{i-1,j} \to \ldots \to F_{0,j} \to U_j \to 0\,.$$

Therefore,

$$(*) \qquad \dim_k(U_j) = \sum_{i \geq 0} (-1)^i \dim_k(F_{i,j})\,.$$

Note that the sum above is finite. Thus,

$$\dim_k(U_j)t^j = \sum_{i \geq 0} (-1)^i \dim_k(F_{i,j})t^j\,.$$

Summing over all j we get

$$\begin{aligned} \mathrm{Hilb}_U(t) &= \sum_j \dim_k(U_j)t^j = \sum_j \Big(\sum_{i \geq 0} (-1)^i \dim_k(F_{i,j})t^j \Big) \\ &= \sum_{i \geq 0} (-1)^i \Big(\sum_j \dim_k(F_{i,j})t^j \Big) = \sum_{i \geq 0} (-1)^i \mathrm{Hilb}_{F_i}(t)\,. \end{aligned}$$

Furthermore, we have that $\mathrm{Hilb}_{R(-p)}(t) = t^p \,\mathrm{Hilb}_R(t)$. Recall 3.2 and obtain the equality

$$\begin{aligned} \mathrm{Hilb}_{F_i}(t) &= \sum_p c_{i,p} \,\mathrm{Hilb}_R(R(-p)) = \sum_p c_{i,p}\, t^p \,\mathrm{Hilb}_R(t) \\ &= \mathrm{Hilb}_R(t) \sum_p c_{i,p}\, t^p\,. \end{aligned}$$

Hence

$$\mathrm{Hilb}_U(t) = \mathrm{Hilb}_R(t) \sum_{i \geq 0} \sum_{p \in \mathbf{Z}} (-1)^i c_{i,p} t^p\,.$$

According to 1.10, we can substitute $\mathrm{Hilb}_S(t) = \dfrac{1}{(1-t)^n}$ and obtain the desired formula in the case when $R = S$. □

Example 16.3. We will illustrate the argument above using Example 4.3. Let $A = k[x,y]$ and $U = A/(x^3, xy, y^5)$. We have the graded free resolution

$$0 \to A(-4) \oplus A(-6) \to A(-3) \oplus A(-2) \oplus A(-5) \to A \to U \ .$$

The formula in the theorem yields

$$\mathrm{Hilb}_U(t) = \frac{1 - t^3 - t^2 - t^5 + t^4 + t^6}{(1-t)^2} = 1 + 2t + 2t^2 + t^3 + t^4 \,.$$

Of course, in this simple example, it is easier to calculate $\mathrm{Hilb}_U(t)$ by listing a monomial basis of $U = k[x,y]/(x^3, xy, y^5)$.

Consider U_4; it has basis $\{y^4\}$. We will illustrate how to get $\dim_k(U_4) = 1$ using $(*)$. Take the internal degree 4 component of the resolution in Example 4.3 and obtain the exact sequence of k-vector spaces

$$\begin{aligned} 0 \to A(-4)_4 \oplus A(-6)_4 \to A(-3)_4 &\oplus A(-2)_4 \oplus A(-5)_4 \\ &\to A_4 \to U_4 \to 0 \ , \end{aligned}$$

that is,

$$0 \to A_0 \oplus 0 \to A_1 \oplus A_2 \oplus 0 \to A_4 \to U_4 \to 0 \ .$$

Therefore,

$$\begin{aligned} \dim_k(U_4) &= \dim_k(A_4) - \big(\dim_k(A_1) + \dim_k(A_2)\big) + \dim_k(A_0) \\ &= 5 - 2 - 3 + 1 = 1 \end{aligned}$$

as desired.

Exercise 16.4. *Let $f_1, \ldots, f_q$ be a homogeneous S-regular sequence of forms of degrees $a_1, \ldots, a_q$. The Hilbert series of $S/(f_1, \ldots, f_q)$ is*

$$\frac{\prod_{1 \le i \le q} (1 - t^{a_i})}{(1-t)^n} \,.$$

Expanding $\dfrac{1}{(1-t)^n}$ in Theorem 16.2 yields the following result.

Corollary 16.5. *Let V be a graded finitely generated S-module. There exists a polynomial $g(i) \in \mathbf{Q}[i]$ of degree $< n$ such that $g(j) = \dim_k(V_j)$ for all $j \gg 0$.*

The polynomial $g(i)$ in Corollary 16.5 is called the ***Hilbert polynomial*** of V. The following theorem can be proved by induction on the dimension, cf. [Bruns-Herzog, Theorem 4.1.3].

Theorem 16.6. *Let V be a graded finitely generated S-module. Its Hilbert polynomial has degree $\dim(V) - 1$.*

Let V be a graded finitely generated S-module. Fix a finite graded free resolution $\mathbf{F}$ of V. The sum in Theorem 16.2 is finite, so

$$\mathrm{E}_{\mathbf{F}}(t) = \sum_{i \geq 0} \sum_{p \in \mathbf{Z}} (-1)^i c_{i,p} t^p$$

is a polynomial in $\mathbf{Z}[t]$. We say that $\mathrm{E}_{\mathbf{F}}(t)$ is the ***Euler polynomial*** of the resolution $\mathbf{F}$. Since every graded free resolution of V is isomorphic to the direct sum of the minimal graded free resolution and a trivial complex, it follows that the Euler polynomial does not depend on the choice of the resolution; so we call it the ***Euler polynomial*** of V. The rank of the i'th free module in the resolution is $c_i = \sum_{p \in \mathbf{Z}} c_{i,p}$. The alternating sum $\mathrm{E}_{\mathbf{F}}(1) = \sum_{i \geq 0} (-1)^i c_i$ of the ranks of the free modules in the resolution is called the ***Euler characteristic***. When we choose the resolution $\mathbf{F}$ to be minimal we get the sum

$$\mathrm{E}_V(1) = \sum_{i \geq 0} (-1)^i b_i^S(V)\,,$$

which is called the ***Euler characteristic*** of V.

By Theorem 16.2 we have

$$\mathrm{Hilb}_V(t) = \frac{\mathrm{E}_{\mathbf{F}}(t)}{(1-t)^n}\,.$$

Set $h(t) = \dfrac{\mathrm{E}_{\mathbf{F}}(t)}{(1-t)^q}$, where q is the maximal power such that $(1-t)^q$ divides $\mathrm{E}_{\mathbf{F}}(t)$. Set $s = n - q$. Thus, $h(1) \neq 0$ and

$$\mathrm{Hilb}_V(t) = \frac{h(t)}{(1-t)^s}\,.$$

The number $h(1)$ is called the ***multiplicity*** (or the ***degree***) of the module V, and is denoted $\mathrm{mult}(V)$.

Theorem 16.7. *Let V be a graded finitely generated S-module. If V is artinian, then* $\mathrm{Hilb}_V(t) = h(t)$ *and $h(1)$ is its length. If V is not artinian, then (using the preceding notation) we have*

(1) $\mathrm{Hilb}_V(t) = \dfrac{h(t)}{(1-t)^{\dim(V)}}\,.$

(2) *The leading coefficient of the Hilbert polynomial $g(i)$ is*

$$\frac{h(1)}{(\dim(V)-1)!}\,.$$

(3) *If $h(t) = h_r t^r + \ldots + h_1 t + h_0$, then*

$$g(i) = \sum_{0 \leq j \leq r} h_j \binom{\dim(V) - 1 + i - j}{\dim(V) - 1}\,.$$

(4) $\dim_k(V_i) = g(i)$ *for* $i \geq \deg(h(t))$.

We use the convention $\binom{p}{0} = 1$ *for every p.*

If we can compute the graded Betti numbers of V, then we can compute the following invariants:

- the Hilbert series of V; by Theorem 16.2
- the dimension of V; by Theorem 16.7(1)
- the Hilbert polynomial of V; by Theorem 16.7(4)
- the multiplicity of V; by Theorem 16.7(3).

Proof. We have that $\mathrm{Hilb}_V(t) = \dfrac{h(t)}{(1-t)^s}$ for some $s \leq n$. Suppose that V is not artinian; then $s \geq 1$. Let $h(t) = h_r t^r + \ldots + h_1 t + h_0$.

Hence

$$\begin{aligned}\mathrm{Hilb}_V(t) &= h(t)\sum_{i\ge 0}\binom{s-1+i}{s-1}t^i\\ &= \text{terms of degree less than } r\\ &\quad+\sum_{i\ge 0}\left(h_r\binom{s-1+i}{s-1}+h_{r-1}\binom{s-1+i+1}{s-1}+\dots\right.\\ &\qquad\left.+h_0\binom{s-1+i+r}{s-1}\right)t^{i+r}.\end{aligned}$$

It follows that the Hilbert polynomial is

$$\begin{aligned}g(i+r) = h_r\binom{s-1+i}{s-1}+h_{r-1}\binom{s-1+i+1}{s-1}\\ +\dots+h_0\binom{s-1+i+r}{s-1},\end{aligned}$$

(here i is the variable and r is a constant) and that (4) holds. Hence

$$g(i) = \sum_{0\le j\le r} h_j\binom{s-1+i-j}{s-1}$$

Its degree is $s-1$. By Theorem 16.6, we get that $s-1 = \dim(V)-1$. Hence $s = \dim(V)$ as desired. We proved (1) and obtained the formula in (3). The leading coefficient of $g(i)$ is $(h_r+\dots+h_0)\dfrac{1}{(s-1)!} = \dfrac{h(1)}{(\dim(V)-1)!}$, so (2) holds.

Now suppose that $V \neq 0$ is artinian. We have that $\mathrm{Hilb}_V(t) = \dfrac{h(t)}{(1-t)^s}$ for some $s \le n$. The argument above implies that $s = 0$. Now

$$h(1) = \mathrm{Hilb}_V(1) = \sum_{i\ge 0}\dim_k(V_i)$$

is the length of V. □

Example 16.8. Let $A = k[a,b,c,h]$ and $B = (a^2, ab, bc)$. Computer computation shows that the graded Betti numbers of A/B are

$$b_{0,0} = 1, \ b_{1,2} = 3, \ b_{2,3} = 2\,.$$

By Theorem 16.2, the Hilbert series of A/B is

$$\begin{aligned} \mathrm{Hilb}_{A/B}(t) &= \frac{1-3t^2+2t^3}{(1-t)^4} = \frac{2t+1}{(1-t)^2} = (2t+1)\sum_{i\geq 0}(1+i)t^i \\ &= 1 + \sum_{i\geq 1}(3i+1)t^i\,. \end{aligned}$$

The Hilbert polynomial is $3i+1$. The dimension of A/B is 2, and the codimension is 2. The multiplicity is $\mathrm{mult}(A/B) = 3$.

Exercise 16.9. *Let $f_1, \ldots, f_q$ be a homogeneous regular sequence of forms of degrees $a_1, \ldots, a_q$. Compute the Hilbert polynomial of $S/(f_1, \ldots, f_q)$ and show that* $\mathrm{mult}\,(S/(f_1, \ldots, f_q)) = a_1 \ldots a_q$.

Corollary 16.10. *Let V be a graded finitely generated S-module. If $\dim(V) \neq n$ then its Euler characteristic is 0.*

Proof. The Euler characteristic equals $\mathrm{E}_{\mathbf{F}}(1)$ and $\mathrm{Hilb}_V(t) = \dfrac{\mathrm{E}_{\mathbf{F}}(t)}{(1-t)^n}$. By Theorem 16.7(1), it follows that $1-t$ divides $\mathrm{E}_{\mathbf{F}}(t)$. Hence $\mathrm{E}_{\mathbf{F}}(1) = 0$. □

The coefficients of the polynomial $h(t)$ form the ***h*-vector**. There are several problems about the h-vector in Algebraic Combinatorics. We list three of them.

Conjecture 16.11. (Stanley) *If S/I is a Cohen-Macaulay integral domain, then its h-vector is unimodal.* (Recall that a sequence $c_0, \ldots, c_r$ of real numbers is ***unimodal*** if there exists an $0 \leq s \leq r$ so that $c_0 \leq \ldots \leq c_{s-1} \leq c_s \geq c_{s+1} \geq \ldots \geq c_r$.)

A monomial is ***squarefree*** if it is not divisible by the square of any of the variables.

Conjecture 16.12. (Charney-Davis-Stanley), cf. [Stanley 2, Problem 4] *Let I be an ideal generated by quadratic squarefree monomials and such that S/I is Gorenstein with h-vector $(h_0, \ldots, h_{2e})$. Is it true that*

$$(-1)^e(h_0 - h_1 + h_2 - \ldots + h_{2e}) \geq 0?$$

Problem 16.13. cf. [Stanley 2, Problem 1] *If I is an ideal generated by squarefree monomials and such that the quotient S/I is Gorenstein with an h-vector $(h_0, \ldots, h_c)$, then is it true that*

$$h_0 \leq h_1 \leq \ldots \leq h_{\frac{c}{2}}\,?$$

It might be reasonable to drop/replace the assumption that the ideal is generated by monomials.

17 Pure resolutions

In this section we consider free resolutions in which each differential matrix has entries of the same degree. Especially interesting are the linear free resolutions in which all differentials have linear entries.

Throughout the section we consider a graded free resolution $\mathbf{F}$ of a finitely generated graded R-module U, and we use the following notation. Let $c_{i,p}$ be the number of copies of $R(-p)$ in F_i. Note that if the resolution is minimal then $c_{i,p}$ coincide with the graded Betti numbers $b_{i,p}$.

The set of i'th ***shifts*** in $\mathbf{F}$ is

$$\{p\,|\,c_{i,p} \neq 0\}\,.$$

Denote by t_i the minimal i'th shift, and by T_i the maximal i'th shift, that is

$$t_i = \min\{p\,|\,c_{i,p} \neq 0\} \quad \text{and} \quad T_i = \max\{p\,|\,c_{i,p} \neq 0\}\,.$$

Example 17.1. In Example 4.3 we have that the zeroth shift is 0, the first shifts are 2, 3, 5, and the second shifts are 4, 6.

Proposition 17.2. *Let $f_1, \ldots, f_q$ be a homogeneous regular sequence. Then*

$$\frac{(\prod_{i=1}^{q} t_i)}{q!} \le \operatorname{mult}(S/(f_1, \ldots, f_q)) \le \frac{(\prod_{i=1}^{q} T_i)}{q!},$$

where t_i and T_i are the minimal and the maximal shifts, respectively, in the minimal free resolution of $S/(f_1, \ldots, f_q)$.

Proof. Let $f_1, \ldots, f_q$ be a homogeneous regular sequence of forms of degrees $a_1 \le \ldots \le a_q$. The minimal free resolution of $S/(f_1, \ldots, f_q)$ is the Koszul resolution. Therefore, the shifts are

$$t_i = a_1 + \ldots + a_i \quad \text{and} \quad T_i = a_{q-i+1} + \ldots + a_q \quad \text{for } 1 \le i \le q\,.$$

The multiplicity is $\operatorname{mult}(S/(f_1, \ldots, f_q)) = a_1 \ldots a_q$ by Exercise 16.9. Therefore, we have the inequalities

$$\begin{aligned}
\frac{\prod_{1 \le i \le q} t_i}{q!} &= \frac{\prod_{1 \le i \le q} (a_1 + \ldots + a_i)}{q!} \\
&\le \frac{\prod_{1 \le i \le q} i\, a_i}{q!} = \frac{q! a_1 \ldots a_q}{q!} \\
&= \operatorname{mult} S/(f_1, \ldots, f_q) \\
\frac{\prod_{1 \le i \le q} T_i}{q!} &= \frac{\prod_{1 \le i \le q} (a_{q-i+1} + \ldots + a_q)}{q!} \\
&\ge \frac{\prod_{1 \le i \le q} i\, a_{q-i+1}}{q!} = \frac{q! a_1 \ldots a_q}{q!} \\
&= \operatorname{mult}(S/(f_1, \ldots, f_q))\,.
\end{aligned}$$

□

Definition 17.3. We say that $\mathbf{F}$ is ***pure*** if it has the form

$$\mathbf{F}: \quad \ldots \longrightarrow R(-p_i)^{c_{i,p_i}} \xrightarrow{d_i} R(-p_{i-1})^{c_{i-1,p_{i-1}}} \longrightarrow \ldots,$$

that is, for each i the set of i'th shifts consists of one number denoted p_i, that is $t_i = T_i = p_i$. A pure resolution is ***q*-*linear*** if $p_i = q + i$ for all i, that is

$$\begin{aligned}\mathbf{F}: \quad \dots \to R(-q-i)^{c_{i,q+i}} &\to R(-q-i+1)^{c_{i-1,q+i-1}} \\ &\to \dots \to R(-q-1)^{c_{1,q}} \to R(-q)^{c_{0,q}}.\end{aligned}$$

Such a resolution is called linear because the entries in the differential matrices are linear forms. Furthermore, $\mathbf{F}$ is ***linear*** if $p_i = i$ for all i, that is

$$\mathbf{F}: \quad \dots \to R(-i)^{c_{i,i}} \to R(-i+1)^{c_{i-1,i-1}} \longrightarrow \dots \to R^{c_{0,0}}.$$

The ring R is called ***Koszul*** if the R-module k has a linear resolution; see Section 34.

Example 17.4. Let $A = k[x, y]$. The resolution

$$0 \to A(-4)^2 \xrightarrow{\begin{pmatrix} y & 0 \\ -x & y \\ 0 & -x \end{pmatrix}} A(-3)^3 \xrightarrow{(x^3\ x^2y\ xy^2)} A$$

is pure with $p_0 = 0$, $p_1 = 3$, $p_2 = 4$. Its truncation

$$0 \to A(-4)^2 \xrightarrow{\begin{pmatrix} y & 0 \\ -x & y \\ 0 & -x \end{pmatrix}} A(-3)^3$$

is a 3-linear resolution of the ideal (x^3, x^2y, xy^2).

The properties defined in Definition 17.3 can be expressed by vanishing of graded Betti numbers as follows.

Proposition 17.5. *Let $\mathbf{F}$ be a graded free resolution of a finitely generated graded R-module U.*

(1) *$\mathbf{F}$ is pure if and only if for each i there exists a number p_i such that $c_{i,r} = 0$ for $r \neq p_i$.*

(2) $\mathbf{F}$ *is q-linear if and only if* $c_{i,r} = 0$ *for* $r \neq q + i$.
(3) $\mathbf{F}$ *is linear if and only if* $c_{i,r} = 0$ *for* $r \neq i$.

Corollary 17.6. *If there exists a pure, or q-linear, or linear, graded free resolution of U, then the minimal graded free resolution of U is pure, q-linear, or linear, respectively.*

Proof. Suppose that the numbers $c_{i,r}$ satisfy some of the vanishing properties in the above proposition. By Theorem 7.5(2), we have that $b^R_{i,r}(U) \leq c_{i,r}$ for all i, r. Therefore, the graded Betti numbers $b^R_{i,r}(U)$ satisfy the same vanishing property. □

The Boij-Söderberg theory originated from the conjectures in [Boij-Söderberg]. The key idea is that pure resolutions can be used as building blocks in order to obtain any Betti diagram. First, we consider the question what are the possible Betti diagrams of pure minimal free resolutions. We say that $\mathbf{t} = (t_0, \ldots, t_s) \in \mathbf{Z}^s$ is a ***degree sequence*** if $t_{i+1} > t_i$ for each i, and $s \leq n$. The ***distinguished pure diagram*** $P_\mathbf{t}$ ***for*** $\mathbf{t}$ is the diagram with entries $b_{i,j} = 0$ for $j \neq t_i$ and $b_{i,t_i} = \prod_{q \neq i} \frac{1}{|t_i - t_q|}$ for all i. A Betti diagram of a graded finitely generated S-module has ***type*** $\mathbf{t}$ if

$$b_{i,j} \neq 0 \iff j = t_i \,.$$

It is proved by Herzog-Kühl that such a diagram is a rational multiple of $P_\mathbf{t}$. On the other hand, it is proved by Eisenbud-Fløystad-Weyman that there exists a rational multiple of $P_\mathbf{t}$ that is the Betti diagram of a pure minimal free resolution of a graded finitely generated S-module.

The following strong result is proved in [Boij-Söderberg 2, Eisenbud-Schreyer].

Theorem 17.7. *The Betti diagram of a graded finitely generated S-module is a positive* $\mathbf{Q}$*-linear combination of Betti diagrams of graded finitely generated modules with pure resolutions.*

A corollary from the Boij-Söderberg theory is that the Multiplicity Conjecture holds. The following result is proved in [Boij-Söderberg 2, Eisenbud-Schreyer]; we considered a special case in Proposition 17.2.

Theorem 17.8. *Let V be a graded finitely generated S-module generated in degree 0. Set $q = \operatorname{codim}(V)$ and $r = \operatorname{pd}(V)$. Then*

$$b_0^S(V)\frac{\prod_{i=1}^r t_i}{r!} \leq \operatorname{mult}(V) \leq b_0^S(V)\frac{\prod_{i=1}^q T_i}{q!}.$$

In the rest of this section, we discuss linear resolutions. For simplicity, sometimes we say a "linear resolution" instead of a "q-linear resolution" when q can be suppressed.

Proposition 17.9. *Let $\mathbf{F}$ be a graded free resolution of U. The resolution is q-linear if and only if $F_0 = R(-q)^{c_{0,q}}$ and the entries in the differential matrices are linear forms. In particular for $q = 0$, we have that $\mathbf{F}$ is linear if and only if $F_0 = R^{c_{0,0}}$ and the entries in the differential matrices are linear forms.*

Proof. The proof is by induction on the homological degree i. For $i = 0$ we have that $F_0 = R(-q)^{c_{0,q}}$. Suppose that $F_i = R(-q-i)^{c_{i,q+i}}$. Since the entries in the differential are linear forms and d_{i+1} has internal degree 0, it follows that F_{i+1} has a system of homogeneous generators of degree $q + i + 1$. Hence $F_{i+1} = R(-q-i-1)^{c_{i+1,q+i+1}}$. □

Note that the condition $F_0 = R(-q)^{c_{0,q}}$ implies that U is generated by homogeneous elements of degree q. Therefore:

Corollary 17.10. *Let $\mathbf{F}$ be the minimal graded free resolution of U. The resolution is q-linear if and only if U is generated by elements of degree q and the entries in the differential matrices are linear forms. In particular for $q = 0$, we have that $\mathbf{F}$ is linear if and only if U is generated by elements of degree 0 and the entries in the differential matrices are linear forms.*

By Theorem 16.2 the graded Betti numbers determine the Hilbert series $\operatorname{Hilb}_U(t)$. In general, the Hilbert series $\operatorname{Hilb}_U(t)$ does not determine the graded Betti numbers of U. However, we have the following result.

Proposition 17.11. *If a finitely generated graded R-module U has a q-linear minimal free resolution, then the graded Betti numbers of U*

are determined by its Hilbert series.

Proof. Let $\mathbf{F}$ be a q-linear minimal free resolution of U. By Theorem 16.2 it follows that

$$\mathrm{Hilb}_U(t) = \mathrm{Hilb}_R(t) \sum_{i \geq 0} (-1)^i b^R_{i,i+q}(U) t^{i+q}\,.$$

Theorem 16.2 applied to the S-module R yields that

$$\mathrm{Hilb}_R(t) = \frac{w(t)}{(1-t)^n}$$

for some polynomial w. Hence

$$(1-t)^n \mathrm{Hilb}_U(t) = w(t) \sum_{i \geq 0} (-1)^i b^R_{i,i+q}(U) t^{i+q}\,.$$

Hence, $b^R_{i,i+q}(U)$ is determined by $\mathrm{Hilb}_U(t)$, $w(t)$, and the numbers $\{b^R_{j,j+q}(U) \,|\, j < i\}$. Given $\mathrm{Hilb}_U(t)$ and $\mathrm{Hilb}_R(t)$, we can compute inductively (by induction on i) $b^R_{i,i+q}(U)$. □

Definition 17.12. Let $q = \min(U)$ be the minimal degree of an element in a minimal homogeneous system of generators of U. The subcomplex

$$L(\mathbf{F}):\ \ldots \to R(-q-i)^{b_{i,i+q}} \xrightarrow{d_i} R(-q-(i-1))^{b_{i-1,i-1+q}} \to \ldots \to R(-q)^{b_{0,q}}$$

of the graded minimal free resolution of U is called the ***linear strand*** of U. All entries in the differential matrices in the linear strand are linear forms. If the resolution is q-linear, then it coincides with its linear strand. The linear strand is an interesting invariant of the module U. The ***length of the linear strand*** is

$$\max\{i \,|\, b^R_{i,i+q}(U) \neq 0\}\,.$$

Open-Ended Problem 17.13. (folklore) *How is the length of the linear strand related to the other invariants of U?*

In order to obtain a reasonable answer, one has to impose assumptions on the properties of the module U.

If $q = \min(U)$ is the minimal degree of an element in a minimal homogeneous system of generators of U, then the elements of $\mathrm{Syz}_i^R(U)_{i-1+q}$ are called ***linear syzygies*** for $i \geq 2$. The paper [Eisenbud-Koh] is on linear syzygies.

See [Römer] for results and discussion on the following conjecture by Herzog.

Conjecture 17.14. (Herzog) *If V is an r'th syzygy of a graded finitely generated S-module and its linear strand has length p, then*

$$b^S_{i,i+\min(V)}(V) \geq \binom{p+r}{i+r} \quad \textit{for } 0 \leq i \leq p\,.$$

Fix an integer number p. Let $J \subseteq \mathbf{m}^2$ be a graded ideal in S. We say that S/J has the ***property*** N_p if

$$b^S_{i,i+1+j}(S/J) = 0 \quad \text{for all } i \leq p,\ j > 0\,.$$

Thus, S/J has N_1 if and only if J is generated by quadrics. In Algebraic Geometry we often also consider the property N_0, which says that S/J is projectively normal. In Section 66 we discuss N_p for Veronese rings. The following is an interesting direction of research.

Open-Ended Problem 17.15. (folklore) *Establish the property N_p for interesting classes of rings.*

18 Regularity

We would like to know the degrees in which the Betti numbers are located, that is, to know the sets of shifts. Often it is impossible to obtain such precise information, and then it is useful to have an upper bound for the degrees of the non-vanishing Betti numbers. Such bounds are provided by the notions regularity and rate; rate is used when the regularity is infinite. See [Chardin] and [Bayer-Mumford] for a survey on regularity.

Definition 18.1. The ***Castelnuovo-Mumford regularity*** (or simply ***regularity***) of a graded finitely generated R-module U is

$$\operatorname{reg}_R(U) = \max\{j \mid b^R_{i,i+j}(U) \neq 0 \text{ for some } i\}\,.$$

In particular:

Proposition 18.2.
(1) *If U has a q-linear resolution, then* $\operatorname{reg}(U) = q$.
(2) *R is Koszul if and only if* $\operatorname{reg}(k) = 0$.

Note that

$$\operatorname{reg}(I) = \operatorname{reg}(S/I) + 1\,.$$

Hilbert's Syzygy Theorem 15.2 implies that every graded finitely generated S-module has finite regularity. In a Betti diagram, we see the regularity as the label of the last (bottom) row in which we have a non-zero Betti number. Thus, *the projective dimension is the length of the Betti diagram, while the regularity is its width.*

Example 18.3. Consider the Betti diagram in Example 12.2. It is

	1	3	2
0	1	-	-
1	-	1	-
2	-	1	1
3	-	-	-
4	-	1	1

Thus,

$$b_{0,0} = 1,$$
$$b_{1,1+1} = 1,\ b_{1,1+2} = 1,\ b_{1,1+4} = 1,$$
$$b_{2,2+2} = 1,\ b_{2,2+4} = 1$$

are the non-zero graded Betti numbers. The regularity is 4 and the projective dimension is 2.

Theorem 18.4. *Let J be a graded artinian ideal in S. Set*

$$q = \max\{\, i \,|\, (S/J)_i \neq 0 \,\}\,.$$

Then $\operatorname{reg}(S/J) = q$.

Proof. Computing $\operatorname{Tor}_S^n(S/J, k)$ using the Koszul resolution $\mathbf{K}$ of k we see that

$$\operatorname{Tor}_{i,i+j}^S(S/J, k) \cong \operatorname{H}_{i,i+j}(S/J \otimes \mathbf{K}) = 0 \quad \text{for } j > q$$

since $\left(S/J \otimes \mathbf{K}_i\right)_{i+j} = 0$ for $j > q$. Hence $\operatorname{reg}(S/J) \leq q$. By Corollary 14.11 and its proof it follows that $\operatorname{Tor}_{n,n+q}^S(S/J, k) \neq 0$. □

Example 18.5. Consider the ideal $B = (x^2, y^2, z^3, v^3, xy, yz)$ in the ring $A = k[x, y, z, v]$. We have that $xz^2v^2 \neq 0$ in A/B, and $(A/B)_i = 0$ for $i \geq 6$. Hence, $\operatorname{reg}(A/B) = 5$.

The next proposition shows how regularity changes in a short exact sequence.

Proposition 18.6. *Suppose that*

$$0 \to U \to U' \to U'' \to 0$$

is a short exact sequence of graded finitely generated R-modules with homomorphisms of degree 0. Then
(1) *If* $\operatorname{reg}(U') > \operatorname{reg}(U'')$, *then* $\operatorname{reg}(U) = \operatorname{reg}(U')$.
(2) *If* $\operatorname{reg}(U') < \operatorname{reg}(U'')$, *then* $\operatorname{reg}(U) = \operatorname{reg}(U'') + 1$.
(3) *If* $\operatorname{reg}(U') = \operatorname{reg}(U'')$, *then* $\operatorname{reg}(U) \leq \operatorname{reg}(U'') + 1$.

Proof. Fix an internal degree j. The short exact sequence

$$0 \to U \to U' \to U'' \to 0$$

yields a long exact sequence (see 38.3)

$$\begin{aligned} \ldots \to\ & \operatorname{Tor}_{i+1}^R(U'', k)_j \\ & \to \operatorname{Tor}_i^R(U, k)_j \to \operatorname{Tor}_i^R(U', k)_j \to \operatorname{Tor}_i^R(U'', k)_j \to \ldots \end{aligned}$$

We will use this long exact sequence in order to prove the proposition.

First, we will prove (1). Fix an i such that $\operatorname{Tor}_i^R(U',k)_{i+\operatorname{reg}(U')} \neq 0$. Set $j = i + \operatorname{reg}(U')$. From the long exact sequence we get that

$$\ldots \to \operatorname{Tor}_i^R(U,k)_j \to \operatorname{Tor}_i^R(U',k)_j \to \operatorname{Tor}_i^R(U'',k)_j = 0$$

is exact. Hence $\operatorname{Tor}_i^R(U,k)_j \neq 0$. Therefore, $\operatorname{reg}(U) \geq \operatorname{reg}(U')$.

Let $\operatorname{reg}(U) > \operatorname{reg}(U')$. Fix an i such that $\operatorname{Tor}_i^R(U,k)_{i+\operatorname{reg}(U)} \neq 0$. Set $j = i + \operatorname{reg}(U)$. From the long exact sequence we get that

$$\operatorname{Tor}_{i+1}^R(U'',k)_j \to \operatorname{Tor}_i^R(U,k)_j \to \operatorname{Tor}_i^R(U',k)_j = 0$$

is exact. Since $\operatorname{reg}(U) > \operatorname{reg}(U') > \operatorname{reg}(U'')$, it follows that $\operatorname{reg}(U'') \leq \operatorname{reg}(U) - 2$, so $\operatorname{Tor}_{i+1}^R(U'',k)_j = 0$. As $\operatorname{Tor}_i^R(U,k)_j \neq 0$, we have a contradiction.

Thus, $\operatorname{reg}(U) = \operatorname{reg}(U')$. We proved (1).

Next, we will prove that if $\operatorname{reg}(U') \leq \operatorname{reg}(U'')$, then $\operatorname{reg}(U) \leq \operatorname{reg}(U'') + 1$. In particular, (3) holds.

Suppose that $\operatorname{reg}(U) > \operatorname{reg}(U'') + 1$. We fix a number i such that $\operatorname{Tor}_i^R(U,k)_{i+\operatorname{reg}(U)} \neq 0$. Set $j = i + \operatorname{reg}(U)$. From the long exact sequence we get that

$$0 = \operatorname{Tor}_{i+1}^R(U'',k)_j \to \operatorname{Tor}_i^R(U,k)_j \to \operatorname{Tor}_i^R(U',k)_j$$

is exact. Since $\operatorname{reg}(U) > \operatorname{reg}(U'')+1 \geq \operatorname{reg}(U')+1$ we have $\operatorname{Tor}_i^R(U',k)_j = 0$. Hence $\operatorname{Tor}_i^R(U,k)_j = 0$, which is a contradiction. Thus, $\operatorname{reg}(U) \leq \operatorname{reg}(U'') + 1$.

It remains to finish the proof of (2).

Fix an i such that $\operatorname{Tor}_{i+1}^R(U'',k)_{i+1+\operatorname{reg}(U'')} \neq 0$. Set $j = i + 1 + \operatorname{reg}(U'')$. From the long exact sequence we get that

$$\operatorname{Tor}_{i+1}^R(U',k)_j \to \operatorname{Tor}_{i+1}^R(U'',k)_j \to \operatorname{Tor}_i^R(U,k)_j$$

is exact. Since $\operatorname{reg}(U') < \operatorname{reg}(U'')$, it follows that $\operatorname{Tor}_{i+1}^R(U',k)_j = 0$. As $\operatorname{Tor}_{i+1}^R(U'',k)_j \neq 0$, we conclude that $\operatorname{Tor}_i^R(U,k)_j \neq 0$. Therefore, $\operatorname{reg}(U) \geq \operatorname{reg}(U'') + 1$.

We have proved the inequality $\operatorname{reg}(U) \leq \operatorname{reg}(U'') + 1$. Thus, $\operatorname{reg}(U) = \operatorname{reg}(U'') + 1$. □

Corollary 18.7. *Suppose that* $0 \to U \to U' \to U'' \to 0$ *is a short exact sequence of graded finitely generated* R*-modules with homomorphisms of degree* 0*. Then*

(1) $\operatorname{reg}(U') \leq \max\{\operatorname{reg}(U), \operatorname{reg}(U'')\}$.
(2) $\operatorname{reg}(U) \leq \max\{\operatorname{reg}(U'), \operatorname{reg}(U'') + 1\}$.
(3) $\operatorname{reg}(U'') \leq \max\{\operatorname{reg}(U'), \operatorname{reg}(U) - 1\}$.

Exercise 18.8. *Let* V *be a graded finitely generated* S*-module.*

(1) *Let* $l \in S_1$ *be a linear form, and let* $(0 : l)_V = \{m \in V \mid lm = 0\}$*. Then*

$$\operatorname{reg}(V) \leq \max\{\operatorname{reg}(V/lV), \operatorname{reg}((0 : l)_V)\},$$

and equality holds if $\dim((0 : l)_V) \leq 1$.

(2) *Let* $f \in S_p$ *be a form of degree* p*, and let* $(0 : f)_V = \{m \in V \mid lm = 0\}$*. Then*

$$\operatorname{reg}(V) \leq \max\{\operatorname{reg}(V/lV) - p + 1, \operatorname{reg}((0 : f)_V)\},$$

and equality holds if $\dim((0 : l)_V) \leq 1$.

The following exact sequences are helpful in computing regularity of ideals.

Exercise 18.9. *Let* I *and* J *be graded ideals in* S*. We have exact sequences*

$$0 \to I \cap J \to I \oplus J \to I + J \to 0$$

$$0 \to S/(I \cap J) \to S/I \oplus S/J \to S/(I + J) \to 0$$

$$0 \to \operatorname{Tor}_1^S(S/I, S/J) \to S/IJ \to S/(I \cap J) \to 0$$

and isomorphisms

$$S/(I + J) \cong S/I \otimes_S S/J \cong \operatorname{Tor}_0^S(S/I, S/J).$$

The following result, proved in [Bayer-Stillman], is helpful when we study regularity.

Theorem 18.10. *Let J be a graded ideal in S. Denote by T the initial ideal of J with respect to the revlex order in generic coordinates. We have that* $\operatorname{reg}(J) = \operatorname{reg}(T)$.

An interesting direction of research is to obtain sharp upper bounds on regularity.

Theorem 18.11. [Bayer-Mumford, Theorem 3.7 and Proposition 3.8] *Suppose that* $\operatorname{char}(k) = 0$. *Let J be a graded ideal in S, and* $r = \max(J)$ *be the maximal degree of an element in a minimal system of homogeneous generators of J. We have the upper bound*

$$\operatorname{reg}(J) \leq (2r)^{2^{n-2}}\,.$$

The next example shows that the above theorem is nearly the best possible in general.

Theorem 18.12. [Mayr-Meyer] *For every $p \in \mathbf{N}$, there exists a graded ideal J in a polynomial ring with* $10p$ *variables generated by polynomials of degree less or equal to* 4 *and with* $\operatorname{reg}(J) \geq 2^{2^p} + 1$.

The Mayr-Meyer's examples provide the only known family of ideals for which the regularity is doubly exponential in the number of variables, while the maximal degree of an element in a minimal system of homogeneous generators of the ideal is fixed (it is 4). Eisenbud has pointed out recently that it is of interest to construct and study more such examples.

Problem 18.13. (folklore) *Find families (or examples) of gra-ded ideals in S with large regularity (doubly exponential, or exponential, or polynomial) in the number of variables, while the maximal degree of an element in the minimal systems of homogeneous generators of the considered ideals is bounded by a constant.*

A much smaller upper bound on regularity is expected for prime ideals. The next conjecture from [Eisenbud-Goto] has its roots in the work of Castelnuovo. It is one of the most interesting, challenging, and important conjectures on graded free resolutions.

The Regularity Conjecture 18.14. *If $P \subset (x_1, \ldots, x_n)^2$ is a prime graded ideal in S, then*

$$\operatorname{reg}(P) \leq \operatorname{mult}(S/P) - \operatorname{codim}(P) + 1 .$$

Recall that $\operatorname{codim}(P) = n - \dim(S/P)$ is equal to the maximal length of a chain of prime ideals contained in P.

The conjecture is sharp: for example, the equality holds for the defining ideal of the twisted cubic curve. A weaker form of the conjecture, also wide open, is given below.

Conjecture 18.15. *If $P \subset (x_1, \ldots, x_n)^2$ is a prime graded ideal in S, then*

$$\max(P) \leq \operatorname{mult}(S/P) .$$

The following problem has inspired a lot of research.

Open-Ended Problem 18.16. (folklore) *Let J be a graded ideal in S. Assuming J satisfies some special conditions, find a sharp upper bound for* $\operatorname{reg}(J)$ *in terms of the maximal degree of an element in a minimal system of homogeneous generators of J.*

For example, the following result is of this type.

Theorem 18.17. *Let $J_1, \ldots, J_p$ be ideals in S generated by linear forms.*

(1) [Conca-Herzog] $\operatorname{reg}(J_1 \cdots J_p) = p$.
(2) [Derksen-Sidman] $\operatorname{reg}(J_1 \cap \ldots \cap J_p) = p$.

Another trend in studying regularity is to find some asymptotic properties in the behavior of the regularities of powers of a fixed graded ideal. For example, we have the following result.

Theorem 18.18. [Cutkosky-Herzog-Trung, Theorem 3.1], [Kodiyalam, Corollary 3] *Let J be a graded ideal in S. There exist constants*

$e \geq 0$ and $c \leq \max(J)$ such that

$$\operatorname{reg}(J^p) = cp + e \quad \text{for} \quad p \gg 0 .$$

19 Truncation

In this section we study the asymptotic behavior of the structure of a graded ideal.

Definition 19.1. If I is a graded ideal in S and $p \geq 0$, we consider the ideal $I_{\geq p} = \oplus_{i \geq p} I_i$ and call it a ***truncation*** of I.

Example 19.2. Let $T = (a^3, b^3, c^3, ab, ac)$ in the ring $A = k[a, b, c]$. We have that

$$T_{\geq 3} = (\, a^3,\, b^3,\, c^3,\, abc,\, a^2b,\, ab^2,\, a^2c,\, ac^2\,) .$$

Proposition 19.3. *Let I be a graded ideal in S. Fix a $p \geq 0$. We have that $\operatorname{mult}(S/I) \leq \operatorname{mult}(S/I_{\geq p})$, and equality holds if S/I is not artinian.*

Proof. If S/I is artinian, then so is $S/I_{\geq p}$. In this case

$$\operatorname{mult}(S/I) = \dim_k(S/I) \leq \dim_k(S/I_{\geq p}) = \operatorname{mult}(S/I_{\geq p}) .$$

If S/I is not artinian, then $\dim_k(S/I)_j = \dim_k(S/I_{\geq p})_j \neq 0$ for all $j \geq p$. Therefore, S/I and $S/I_{\geq p}$ have the same Hilbert polynomial. So, they have the same multiplicity. □

Theorem 19.4. *Let I be a graded ideal in S. For every $i \geq 0$, we have the following relations between the Betti numbers of $I_{\geq p}$ and $I_{\geq p+1}$ over S:*

$$b_{i,i+j}(I_{\geq p+1}) = 0 \quad \textit{for } j \leq p$$

$$b_{i,i+p+1}(I_{\geq p+1}) = b_{i,i+p+1}(I_{\geq p}) + \binom{n}{i+1} \dim_k(I_p) - b_{i+1,i+p+1}(I_{\geq p})$$

$$b_{i,i+j}(I_{\geq p+1}) = b_{i,i+j}(I_{\geq p}) \quad \textit{for } j \geq p+2 .$$

Proof. The short exact sequence

$$0 \to I_{\geq p+1} \to I_{\geq p} \to k(-p)^{\dim_k(I_p)} \to 0$$

yields the long exact sequence (see 38.3)
(*)

$$\ldots \to \operatorname{Tor}_{i+1}^S(k(-p)^{\dim_k(I_p)}, k)_{i+j}$$
$$\to \operatorname{Tor}_i^S(I_{\geq p+1}, k)_{i+j} \to \operatorname{Tor}_i^S(I_{\geq p}, k)_{i+j} \to \operatorname{Tor}_i^S(k(-p)^{\dim_k(I_p)}, k)_{i+j}$$
$$\to \ldots$$

for each j. The minimal graded free resolution of k is the Koszul complex. Therefore,

$$b_{i,i+p}(k(-p)^{\dim_k(I_p)}) = \binom{n}{i}\dim_k(I_p) \quad \text{for } 1 \leq i \leq n$$
$$b_{i,i+j}(k(-p)^{\dim_k(I_p)}) = 0 \quad \text{for } j \neq p,\ 1 \leq i \leq n.$$

Therefore, (*) implies that $b_{i,i+j}(I_{\geq p+1}) = b_{i,i+j}(I_{\geq p})$ for $j > p+1$. We have $b_{i,i+j}(I_{\geq p+1}) = 0$ for $j \leq p$ by Proposition 12.3. In degree $(i, i+p+1)$ we get by (*) that

$$0 \to \operatorname{Tor}_{i+1}^S(I_{\geq p}, k)_{i+p+1} \to \operatorname{Tor}_{i+1}^S(k(-p)^{\dim_k(I_p)}, k)_{i+p+1}$$
$$\to \operatorname{Tor}_i^S(I_{\geq p+1}, k)_{i+p+1} \to \operatorname{Tor}_i^S(I_{\geq p}, k)_{i+p+1} \to 0\,.$$

is exact. Hence,

$$b_{i,i+p+1}(I_{\geq p+1}) = b_{i,i+p+1}(I_{\geq p}) + \binom{n}{i+1}\dim_k(I_p) - b_{i+1,i+p+1}(I_{\geq p})\,.$$

□

Corollary 19.5. *Let I and J be graded ideals in S. Fix a $p \geq 0$. If $I_{\geq p}$ and $J_{\geq p}$ have the same graded Betti numbers, then so do $I_{\geq p+1}$ and $J_{\geq p+1}$.*

Corollary 19.6. *Let I be a graded ideal in S. Fix a $p \geq 0$. For every $i \geq 0$ we have*

$$b_{i,i+j}^S(I_{\geq p}) = 0 \quad \text{for } j \leq p-1$$
$$b_{i,i+j}^S(I_{\geq p}) = b_{i,i+j}^S(I) \quad \text{for every } j \geq p+1.$$

The next theorem is an immediate consequence of the above corollary.

Theorem 19.7. *Let I be a graded ideal in S. If $p \geq \operatorname{reg}(I)$, then $\operatorname{reg}(I_{\geq p}) = p$ (that is, $I_{\geq p}$ has a p-linear resolution).*

20 Regular elements

In this section we discuss the following helpful technique: we can study a minimal free resolution by first factoring out a maximal graded regular sequence.

First, we remark that if $\operatorname{char}(k) = 0$ then we can use a regular sequence of linear forms; this follows from the following result proved in [Bruns-Herzog, Propositions 1.5.11 and 1.5.12].

Proposition 20.1. *Let U be a finitely generated graded R-module of depth s. There exists a U-regular sequence $u_1, \dots, u_s$ of homogeneous elements. If in addition k is infinite, then there exists a U-regular sequence $u_1, \dots, u_s$ of linear forms.*

Theorem 20.2. *Let U be a finitely generated graded R-module, and let u be a homogeneous U-regular element. We have that*

$$\operatorname{reg}(U/uU) = \operatorname{reg}(U) + \deg(u) - 1$$

and

$$\operatorname{Hilb}_{U/uU}(t) = (1 - t^{\deg(u)})\operatorname{Hilb}_U(t)\,.$$

Proof. We will use the short exact sequence

$$0 \to U(-p) \xrightarrow{u} U \to U/uU \to 0$$

of homomorphisms of degree 0.

Applying Proposition 16.1 to the exact sequence above we get $\operatorname{Hilb}_{U/uU}(t) = (1 - t^{\deg(u)})\operatorname{Hilb}_U(t)\,.$

Set $p = \deg(u)$. Apply Lemma 18.6 to the exact sequence above. Since $\operatorname{reg}(U(-p)) = \operatorname{reg}(U) + p$, it follows that we have case (2) or (3) in Proposition 18.6. Hence

$$\operatorname{reg}(U(-p)) = \operatorname{reg}(U) + p \leq \operatorname{reg}(U/uU) + 1\,.$$

We conclude that $\operatorname{reg}(U/uU) \geq \operatorname{reg}(U)+p-1$. If $p \geq 2$, then we are in case (2) in Proposition 18.6, so we get the desired equality. Let $p = 1$. In case (3) of Proposition 18.6 we get $\operatorname{reg}(U) = \operatorname{reg}(U/uU)$. In case (2) we get $\operatorname{reg}(U(-1)) = \operatorname{reg}(U/uU) + 1$ so $\operatorname{reg}(U) = \operatorname{reg}(U/uU)$. □

Theorem 20.3. *Let $u \in R$ be both R-regular and U-regular. If $\mathbf{F}$ is a free resolution of U over R, then $\mathbf{F} \otimes_R R/(u)$ is a free resolution of U/uU over $R/(u)$. In addition, if u is homogeneous and $\mathbf{F}$ is graded, then $\mathbf{F} \otimes_R R/(u)$ is graded. Furthermore, if $\mathbf{F}$ is minimal and $u \in \mathbf{m}$, then $\mathbf{F} \otimes_R R/(u)$ is minimal.*

Proof. We have to show that the complex $\mathbf{F} \otimes_R R/u$ is exact, that is, $\mathrm{H}_i(\mathbf{F} \otimes_R R/(u)) = 0$ for $i > 0$. By definition $\mathrm{H}_i(\mathbf{F} \otimes_R R/(u)) = \operatorname{Tor}_i^R(U, R/(u))$. Since u is R-regular, we have that $\mathbf{G}:\ 0 \to R \xrightarrow{u} R$ is a free resolution of R/u over R. Therefore, $\operatorname{Tor}_i^R(U, R/(u)) = \mathrm{H}_i(U \otimes_R \mathbf{G})$. But $U \otimes_R \mathbf{G}:\ 0 \to U \xrightarrow{u} U$ is exact since u is U-regular. Hence, $\mathrm{H}_i(U \otimes_R \mathbf{G}) = 0$ for $i > 0$. □

Corollary 20.4. *Suppose that U and W are graded finitely generated R-modules. Let u be both R-regular and U-regular. Suppose that $uW = 0$.*

(1) $\operatorname{Tor}_i^R(U, W) \cong \operatorname{Tor}_i^{R/(u)}(U/uU, W)$ *for all* $i \geq 0$.

(2) $\operatorname{Ext}_R^i(U, W) \cong \operatorname{Ext}_{R/(u)}^i(U/uU, W)$ *for all* $i \geq 0$.

Proof. (1) Let $\mathbf{F}$ be a graded free resolution of U over R. By Theorem 20.3, $\mathbf{F} \otimes_R R/(u)$ is a graded free resolution of U/uU. Now

$$\begin{aligned}\operatorname{Tor}_i^R(U, W) &\cong \mathrm{H}_i(\mathbf{F} \otimes W) = \mathrm{H}_i(\mathbf{F} \otimes (R/(u) \otimes W)) \\ &= \mathrm{H}_i\big((\mathbf{F} \otimes R/(u)) \otimes W\big) \cong \operatorname{Tor}_i^{R/(u)}(U/uU, W)\,.\end{aligned}$$

A similar argument proves (2). □

Theorem 20.5. *Let U be a graded finitely generated R-module. Then*

$$\operatorname{depth}(\operatorname{Syz}_1^R(U)) = \begin{cases} \operatorname{depth}(U)+1 & \textit{if } \operatorname{depth}(U) < \operatorname{depth}(R) \\ \operatorname{depth}(R) & \textit{otherwise.} \end{cases}$$

Proof. By 10.1, we have the short exact sequence

$$0 \to \operatorname{Syz}_1^R(U) \to F_0 \to U \to 0\,.$$

It yields the long exact sequence (see 38.3)

$$\begin{aligned} \ldots \to \operatorname{Ext}_R^{i-1}(k,U) \to \\ \to \operatorname{Ext}_R^i(k,\operatorname{Syz}_1^R(U)) \to \operatorname{Ext}_R^i(k,F_0) \to \operatorname{Ext}_R^i(k,U) \to \ldots\,. \end{aligned}$$

By [Eisenbud, Proposition 18.4], $\operatorname{depth}(U)$ is the smallest number j such that $\operatorname{Ext}_R^j(k,U) \neq 0$. Note that F_0 is a free module, so the smallest number t, such that $\operatorname{Ext}_R^t(k,F_0) \neq 0$, is $t = \operatorname{depth}(R)$.

If $j < t$, then the long exact sequence implies that

$$0 = \operatorname{Ext}_R^j(k,F_0) \to \operatorname{Ext}_R^j(k,U) \to \operatorname{Ext}_R^{j+1}(k,\operatorname{Syz}_1^R(U))$$

is exact, so the smallest number s, such that $\operatorname{Ext}_R^s(k,\operatorname{Syz}_1^R(U)) \neq 0$, is $s = j+1$.

If $j = t$, then the long exact sequence implies that

$$0 = \operatorname{Ext}_R^{i-1}(k,U) \to \operatorname{Ext}_R^i(k,\operatorname{Syz}_1^R(U)) \to \operatorname{Ext}_R^i(k,F_0)$$

is exact, so $\operatorname{Ext}_R^i(k,\operatorname{Syz}_1^R(U)) = 0$ for $i < j = t$. □

The following result is proved in [Avramov, Proposition 3.3.5] for the minimal free resolution of k over R.

Theorem 20.6. *If u is a homogeneous R-regular element, then*

$$\begin{aligned} \mathrm{P}_k^R(t) &= (1+t)\,\mathrm{P}_k^{R/u}(t) && \textit{if } \deg(u) = 1 \\ \mathrm{P}_k^R(t) &= (1-t^2)\,\mathrm{P}_k^{R/u}(t) && \textit{if } \deg(u) > 1\,. \end{aligned}$$

At the end of this section, we present a second proof of the Auslander-Buchsbaum Formula 15.3.

Proof. The proof is by induction on $i = n - \mathrm{depth}(V)$.

Suppose that $i = 0$. There exists a sequence α of n homogeneous elements that is regular both on S and V. Denote by $\mathbf{F}_V$ the minimal graded free resolution of V over S and let $j = \mathrm{pd}(V)$. We will show that $j = 0$. Assume the opposite, namely, $j > 0$. By Theorem 20.3 $\bar{\mathbf{F}} = \mathbf{F}_V \otimes_S S/(\alpha)$ is the minimal graded free resolution of the module $V/(\alpha)V$. Since $\bar{\mathbf{F}}$ has length j, we conclude that the last differential $\bar{d}_j : \bar{F}_j \to \bar{F}_{j-1}$ is injective. On the other hand: The ring $S/(\alpha)$ has depth equal to $\mathrm{depth}(S) - |\alpha| = n - n = 0$. Hence, there exists a homogeneous element $0 \neq f \in S/(\alpha)$ such that $f\mathbf{m} = 0$. As $\bar{\mathbf{F}}$ is minimal, we have that $\bar{d}_j(\bar{F}_j) \subseteq \mathbf{m}\bar{F}_{j-1}$. Therefore, $\bar{d}_j(f\bar{F}_j) = 0$. Since $\bar{d}_j$ is injective, we conclude that $f\bar{F}_j = 0$. This contradicts to $f \neq 0$ because $\bar{F}_j$ is a free $S/(\alpha)$-module. Therefore, $j = 0$.

Suppose that $i > 0$, and that we have proved the Auslander-Buchsbaum Formula for $i - 1$. By Theorem 20.5, $\mathrm{depth}(\mathrm{Syz}_1^S(V)) = \mathrm{depth}(V) + 1$. Therefore, $n - \mathrm{depth}(\mathrm{Syz}_1^S(V)) = i - 1$. By induction hypothesis, we have that

$$\mathrm{pd}(\mathrm{Syz}_1^S(V)) = n - \mathrm{depth}(\mathrm{Syz}_1^S(V)) = n - \mathrm{depth}(V) - 1\,.$$

Now note that $\mathrm{pd}(\mathrm{Syz}_1^S(V)) = \mathrm{pd}(V) - 1$ by the construction of syzygies. □

One can also define almost regular sequences and study how they change a resolution, see [Aramova-Herzog 2].

21 Polarization

As an application of Theorem 20.3 we will discuss the construction of polarization, which is used in order to reduce the study of free resolutions of monomial ideals to the study of free resolutions of squarefree monomial ideals. The advantage of the squarefree case is that one can use a correspondence between squarefree monomial ideals and simplicial complexes (called the Stanley-Reisner correspondence); see Section 62.

An ideal is ***monomial*** if it can be generated by monomials. A monomial ideal is ***squarefree*** if it is generated by monomials not divisible by the square of any of the variables. For a monomial m in S we set $\mathrm{rad}(m) = \prod_{\substack{1\le i\le n \\ x_i/m}} x_i$; this is the largest squarefree monomial which divides m.

Throughout this section M stands for an ideal in S generated minimally by monomials $m_1, \dots, m_r$.

First, we introduce the construction of partial polarization.

Construction 21.1. Let

$$\{x_{j_1}, \dots, x_{j_s}\} = \{x_i \in S \,|\, x_i^2 \text{ divides some of } m_1, \dots, m_r\};$$

here, we assume $j_1 < \dots < j_s$ and clearly $s \le n$. Consider the ring $X = k[x_1, \dots, x_n, t_{j_1}, \dots, t_{j_s}]$. If m is a monomial we write $\bar{m}$ for the monomial $\dfrac{m}{\mathrm{rad}(m)}$ written in the variables t_i. Set

$$P = \Big(\mathrm{rad}(m_1)\cdot \bar{m}_1\,,\ \dots,\ \mathrm{rad}(m_r)\cdot \bar{m}_r\Big).$$

Then

$$X/(P + (\{t_{j_i} - x_{j_i} | 1 \le j \le s\}) \cong S/M$$
$$X/(P + (t_{j_1} - 1, \dots, t_{j_s} - 1)) \cong S/\mathrm{rad}(M)\,.$$

The ideal P is called the ***partial polarization*** of M.

Example 21.2. If $M = (x_1^3, x_1x_2^2x_3, x_2^2x_3^3)$ then

$$P = (x_1t_1^2, x_1x_2x_3t_2, x_2x_3t_2t_3^2)$$

is the partial polarization.

Theorem 21.3. *The elements* $t_{j_{i+1}} - x_{j_{i+1}}$ *and* $t_{j_{i+1}} - 1$ *(for* $0 \le i \le s-1$*) are non-zero divisors on* $X/(P + (t_{j_1} - x_{j_1}, \dots, t_{j_i} - x_{j_i}))$ *and* $X/(P + (t_{j_1} - 1, \dots, t_{j_i} - 1))$*, respectively.*

Proof. Since $X/(P + (t_{j_1} - 1, \dots, t_{j_i} - 1))$ is graded, it follows that the non-homogeneous element $t_{j_{i+1}} - 1$ is a non-zero divisor.

We will prove that $t_{j_{i+1}} - x_{i+1}$ is a non-zero divisor on the quotient $X/(P + (t_{j_1} - x_{j_1}, \ldots, t_{j_i} - x_{j_i}))$. For simplicity, we set

$$x = x_{j_{i+1}}, \quad t = t_{j_{i+1}}, \quad T = k[x_1, \ldots, x_n, t_{j_{i+1}}, \ldots, t_{j_s}].$$

Let Q be the monomial ideal in the polynomial ring T such that

$$T/Q \cong X/(P + (t_{j_1} - x_{j_1}, \ldots, t_{j_i} - x_{j_i})).$$

Construction 21.1 shows that Q is squarefree in the variable x, that is, no minimal monomial generator of Q is divisible by x^2.

Let f be a polynomial such that $(t-x)f \in Q$ and f has no terms in Q. We will show that $xf \in Q$. Assume the opposite. Choose a lexicographic order on T for which x is the smallest variable. Let m be the smallest monomial in f such that $xm \notin Q$. Therefore there is another term m' in f for which $xm = tm'$. Hence there is a monomial q so that $m = tq$ and $m' = xq$ are terms of f. As Q is squarefree in the variable x it follows that $m' = xq \notin Q$ implies $x^2 q \notin Q$. Therefore m' is a smaller term than m with $xm' \notin Q$ which is a contradiction.

Thus, $xf \in Q$, and therefore $tf \in Q$. We have to show that $f \in Q$. Since Q is a monomial ideal, it suffices to show that if a monomial f satisfies $xf \in Q$ and $tf \in Q$ then $f \in Q$. We now assume that f is a monomial. By Construction 21.1, if some minimal generator of Q is divisible by t, then it is divisible by x as well. So the condition $tf \in Q$ implies that x divides f or $f \in Q$. But since Q is squarefree in the variable x, the condition $xf \in Q$ implies that $f \in Q$. □

We will prove a lower bound on the Betti numbers of a monomial ideal. Note that the radical $\mathrm{rad}(M)$ of M is generated by the monomials $\mathrm{rad}(m_i)$.

Theorem 21.4. *The Betti numbers of S/M are greater or equal to those of $S/\mathrm{rad}(M)$. The regularity of $S/\mathrm{rad}(M)$ is less or equal than that of S/M.*

Proof. Let $\mathbf{F}$ be the minimal free resolution of X/P over the ring X. By Theorem 21.3 and Theorem 20.3 it follows that $\mathbf{F} \otimes X/(t_{j_1} -$

$x_{j_1}, \ldots, t_{j_s} - x_{j_s})$ and $\mathbf{F} \otimes X/(t_{j_1} - 1, \ldots, t_{j_s} - 1)$ are exact complexes. The former is the minimal free resolution of S/M over S, while the latter is a (possibly non-minimal) free resolution of $S/\mathrm{rad}(M)$. □

Theorem 21.3 also makes it possible to prove the following result for infinite free resolutions.

Exercise 21.5. *The Betti numbers of k over $S/\mathrm{rad}(M)$ are less or equal than the Betti numbers of k over S/M.*

Corollary 21.6. (Charalambous) *The Betti numbers for an artinian monomial ideal M satisfy*

$$b_i^S(S/M) \geq \binom{n}{i}.$$

Proof. If M is artinian, then its radical is $M' = (x_1, \ldots, x_n)$. The minimal free resolution of S/M' is the Koszul complex, so its i'th Betti number is $\binom{n}{i}$. □

The above corollary proves a special case of Conjecture 14.13.

Next, we describe the construction of polarization.

Construction 21.7. Recall that M stands for an ideal minimally generated by monomials $m_1, \ldots, m_r$. We will introduce new variables denoted $t_{j,i}$.

Let m be a monomial in S. Let $m = q_1 \cdots q_n$, where $q_j = x_j^{c_j}$ for $1 \leq j \leq n$. We say that

$$\tilde{q}_j = \begin{cases} 1 & \text{if } c_j = 0 \\ x_j \prod_{1 \leq i \leq c_j - 1} t_{j,i} & \text{if } c_j \neq 0 \end{cases}$$

is the ***polarization*** of q_j. The ***polarization*** of $m = q_1 \cdots q_n$ is $\widetilde{m} = \tilde{q}_1 \cdots \tilde{q}_n$. The ***polarization*** of a monomial ideal $M = (m_1, \ldots, m_r)$ is $M_{\mathrm{pol}} = (\widetilde{m}_1, \ldots, \widetilde{m}_r)$.

We denote by

$$S_{\mathrm{pol}} = S[t_{1,1}, \ldots, t_{1,p_1}, t_{2,1}, \ldots, t_{2,p_2}, \ldots, t_{n,1}, \ldots, t_{n.p_n}]$$

the polynomial ring in which the monomial ideal M_{pol} lives; here

$$p_i = \max\{c \,|\, x_i^c + 1 \text{ divides some monomial among } m_1, \dots, m_r\}.$$

Consider the sequence

$$\alpha = \{t_{j,i} - x_j \,|\, 1 \leq j \leq n, 1 \leq i \leq p_j\}\,.$$

Factoring S_{pol} by the ideal generated by α is called ***depolarization***; we can think of it as setting each variable $t_{j,i}$ to be equal to x_j. Note that

$$S_{\text{pol}}/(M_{\text{pol}} + (\alpha)) = S/M\,.$$

Example 21.8. Let $M = (x_1^3,\, x_2x_3^2x_4,\, x_3^3,\, x_1x_4)$. The polarization of x_1^3 is $x_1t_{1,1}t_{1,2}$. The polarization of $x_2x_3^2x_4$ is $x_2x_3t_{3,1}x_4$. The polarization of x_3^3 is $x_3t_{3,1}t_{3,2}$. The polarization of x_1x_4 is x_1x_4. Hence,

$$M_{\text{pol}} = (x_1t_{1,1}t_{1,2},\, x_2x_3t_{3,1}x_4,\, x_3t_{3,1}t_{3,2},\, x_1x_4)\,.$$

Construction 21.9. Let p_M be the maximal power of a variable appearing in $m_1, \dots, m_r$. Applying Construction 21.1 to M, we obtain a new monomial ideal M_1 with $p_{M_1} = p_M - 1$. The new variables added to S at this step are denoted $\{t_{j_i,1} \,|\, x_{j_i}^2$ divides some of $m_1, \dots, m_r\}$. We apply Construction 21.1 repeatedly, say s times, until we obtain a squarefree ideal M_s. After that applying Construction 21.1 will yield the same ideal. This squarefree ideal M_s is the polarization of M. Thus, polarization can be done by a sequence of partial polarizations.

The next result follows by Construction 21.9 and Lemma 21.3.

Theorem 21.10.

(1) *α is a regular sequence on $S_{\text{pol}}/M_{\text{pol}}$, and*

$$S_{\text{pol}}/(M_{\text{pol}} + (\alpha)) = S/M\,.$$

(2) *The minimal free resolution of S/M is obtained from the minimal free resolution of $S_{\text{pol}}/M_{\text{pol}}$ (over S_{pol}) by depolarization.*

(3) $\text{Hilb}_{S_{\text{pol}}/M_{\text{pol}}}(t) = (1-t)^a \text{Hilb}_{S/M}$, *where a is the number of the new variables (the t–variables) in S_{pol}.*

22 Deformations from Gröbner basis theory

We will discuss another application of Theorem 20.3. Gröbner basis theory can be used to reduce the study of some properties of graded ideals to properties of monomial ideals. The key point is that Gröbner basis theory provides deformations.

In this section we discuss how such deformations apply to resolutions using Theorem 20.3. Such deformations are constructed using weight vectors.

Definition 22.1. Set $\widetilde{R} = R \otimes k[t]$. Let $\widetilde{J}$ be an ideal in $\widetilde{R}$ such that $\widetilde{R}/\widetilde{J}$ is flat as a $k[t]$-module. For $\alpha \in k$, the quotient $\widetilde{R}/\widetilde{J} \otimes (k[t]/(t-\alpha))$ is denoted $(\widetilde{R}/\widetilde{J})_\alpha$ and is called the ***fiber over*** α. For any $\alpha, \beta \in k$ we say that the fibers $(\widetilde{R}/\widetilde{J})_\alpha$ and $(\widetilde{R}/\widetilde{J})_\beta$ are ***connected by a deformation over*** $\mathbf{A}^1_k$. We say that two ideals J and J' in R are connected by a deformation over $\mathbf{A}^1_k$ if R/J and R/J' are connected by a deformation over $\mathbf{A}^1_k$ (that is, if there exist $\widetilde{R}, \widetilde{J}$ and $\alpha, \beta \in k$ such that $R/J = (\widetilde{R}/\widetilde{J})_\alpha$ and $R/J' = (\widetilde{R}/\widetilde{J})_\beta$ are connected).

In this section we work over S and we consider some special deformations provided by Gröbner basis theory.

Definition 22.2. Given a weight vector $\mathbf{w} = (w_1, \ldots, w_n)$ with real coordinates define the ***weight order*** $\prec_\mathbf{w}$ on the monomials in S by

$$x_1^{\alpha_1} \ldots x_n^{\alpha_n} \succeq_w x_1^{\beta_1} \ldots x_n^{\beta_n} \quad \Longleftrightarrow \quad \sum_{i=1}^n w_i \alpha_i \geq \sum_{i=1}^n w_i \beta_i \,.$$

This is a partial order.

Theorem 22.3. *Let $\prec$ be a monomial order in S. Let J be an ideal in S and $\mathrm{in}_\prec(J)$ be the initial ideal of J with respect to the order $\prec$. There exists a weight vector $\mathbf{w}$ with strictly positive integer coordinates such that $\mathrm{in}_{\prec_\mathbf{w}}(J) = \mathrm{in}_\prec(J)$.*

The above result is proved in [Bayer]. It follows from [Eisenbud, Proposition 15.16] when we choose $\mathbf{w}$ so that $x_i >_\mathbf{w} 1$ for all i and so that $\mathbf{w}$ satisfies the conditions in [Eisenbud, Proposition 15.16].

Example 22.4. Consider the defining ideal

$$B = (ac - b^2,\ bc - ah,\ c^2 - bh)$$

of the twisted cubic curve in the polynomial ring $A = k[a, b, c, h]$. Straightforward computation shows that (ah, b^2, bh) is the initial ideal with respect to the lex order with $h > b > a > c$. We will find a weight vector $\mathbf{w} = (w_1, w_2, w_3, w_4)$ with the properties in the above theorem. By Corollary 39.7, it suffices to find a vector $\mathbf{w}$ such that $ah, b^2, bh \in \text{in}_{\prec_{\mathbf{w}}}(B)$. We have that

$$\begin{aligned}
\text{in}_{\prec_{\mathbf{w}}}(ah - bc) = ah &\quad\Longrightarrow\quad w_1 + w_4 > w_2 + w_3 \\
\text{in}_{\prec_{\mathbf{w}}}(b^2 - ac) = b^2 &\quad\Longrightarrow\quad 2w_2 > w_1 + w_3 \\
\text{in}_{\prec_{\mathbf{w}}}(bh - c^2) = bh &\quad\Longrightarrow\quad w_2 + w_4 > 2w_3\,.
\end{aligned}$$

Therefore, we need to find an integral solution of the system of linear inequalities

$$\begin{aligned}
w_1 - w_2 - w_3 + w_4 &> 0 \\
-w_1 + 2w_2 - w_3 \quad\quad &> 0 \\
w_2 - 2w_3 + w_4 &> 0\,.
\end{aligned}$$

The vector $\mathbf{w} = (1, 2, 1, 3)$ is a solution. Thus, (ah, b^2, bh) is the initial ideal with respect to the weight vector $\mathbf{w} = (1, 2, 1, 3)$.

Let $\mathbf{a} = (1, 1, 1, 1)$. For $p \in \mathbf{N}$, set

$$\mathbf{v}_p = \mathbf{w} + p\mathbf{a} = (1 + p, 2 + p, 1 + p, 3 + p)\,.$$

For every homogeneous polynomial $g \in B$ we have that $\text{in}_{\prec_{\mathbf{v}_p}}(g) = \text{in}_{\prec_{\mathbf{w}}}(g)$. Hence $\text{in}_{\prec_{\mathbf{v}_p}}(B) = \text{in}_{\prec_{\mathbf{w}}}(B) = \text{in}_{\prec}(B)$.

The following theorem is proved in [Mora-Robbiano].

Theorem 22.5. *There exists a fan in* $\mathbf{R}^n$ *so that* $\text{in}_{\prec_{\mathbf{w}}}(J) = \text{in}_{\prec_{\mathbf{v}}}(J)$ *if and only if* $\mathbf{w}$ *and* $\mathbf{v}$ *belong to the same face in the fan. Furthermore, if* J *is graded, then the fan is in the first quadrant.*

Construction 22.6. Given an integral weight vector $\mathbf{w} = (w_1, \dots, w_n)$, we define a weight function $\mathbf{w}$ on the monomials in S by

$$\mathbf{w}(x_1^{\gamma_1} \dots x_n^{\gamma_n}) = \mathbf{w} \cdot (\gamma_1, \dots, \gamma_n) = \sum_{i=1}^{n} w_i \gamma_i \,.$$

Consider the polynomial ring $\widetilde{S} = S[t]$ and the weight vector $\tilde{\mathbf{w}} = (w_1, \dots, w_n, 1)$. Let $f = \sum_i \alpha_i l_i \in S$, where $\alpha_i \in k \setminus 0$, and l_i is a monomial in S. Let l be a monomial in f such that $\mathbf{w}(l) = \max_i\{\mathbf{w}(l_i)\}$. Define $\tilde{f} = \sum_i \alpha_i t^{\mathbf{w}(l)-\mathbf{w}(l_i)} l_i$. If we grade $S[t]$ by $\deg(t) = 1$ and $\deg(x_i) = w_i$ for all i, then $\tilde{f}$ is homogeneous and is called the ***homogenization*** of the polynomial f. Now note that the image of $\tilde{f}$ in $S[t]/(t-1)$ is f, and its image in $S[t]/t$ is $\mathrm{in}_{\prec_{\mathbf{w}}}(f)$. The ideal

$$\widetilde{J} = (\tilde{f} \mid f \in J)$$

is called the ***homogenization*** of J. If $\mathrm{in}_{\prec_{\mathbf{w}}}(J)$ is a monomial ideal, then it is easy to prove that

$$\begin{aligned}\widetilde{J} &= (\tilde{f} \mid f \in J, \mathrm{in}_{\prec_{\mathbf{w}}}(f) \text{ is a minimal generator of } \mathrm{in}_{\prec_{\mathbf{w}}}(J)) \\ &= (\tilde{f} \mid f \text{ is in a fixed Gröbner basis of } J\,)\,.\end{aligned}$$

Thus, in order to obtain $\widetilde{J}$ it suffices to homogenize the elements in a fixed Gröbner basis. Denote by J_ν the image of $\widetilde{J}$ in $S[t]/(t-\nu)$ for $\nu \in k$. Clearly, $J_0 = \mathrm{in}_{\prec_{\mathbf{w}}}(J)$ and $J_1 = J$.

Example 22.7. Consider the defining ideal

$$B = (ac - b^2,\ bc - ah,\ c^2 - bh)$$

of the twisted cubic curve in the polynomial ring $A = k[a, b, c, h]$. It can be easily computed that

$$ac - b^2,\ bc - ah,\ c^2 - bh,\ b^3 - a^2h$$

is a Gröbner basis with respect to the weight order with $\mathbf{w} = (1, 2, 5, 1)$. Therefore, the homogenization of B is

$$\widetilde{B} = \big(ac - t^2b^2,\ bc - t^5ah,\ c^2 - t^7bh,\ b^3 - t^3a^2h\big)\,.$$

Set $\widetilde{A} = A[t]$. We have that $\widetilde{A}/\widetilde{B} \otimes \widetilde{A}/(t-1) = A/B$ and $\widetilde{A}/\widetilde{B} \otimes \widetilde{A}/(t) = A/\mathrm{in}_{\prec_{\mathbf{w}}}(B)$.

The following result is proved in [Eisenbud, Theorem 15.17].

Theorem 22.8. *Suppose that* $\mathrm{in}_{\prec_{\mathbf{w}}}(J)$ *is a monomial ideal. We have that* $\widetilde{S}/\widetilde{J}$ *is flat as a* $k[t]$*-module. In particular,* $t - \alpha$ *is a regular element on* $\widetilde{S}/\widetilde{J}$ *for every* $\alpha \in k$.

We say that $S[t]/\widetilde{J}$ is a ***flat family*** over $k[t]$ of quotients of S whose fiber over 0 is $S/\mathrm{in}_{\prec_{\mathbf{w}}}(J)$ and whose fiber over 1 is S/J. Thus, S/J and $S/\mathrm{in}_{\prec_{\mathbf{w}}}(J)$ are connected by a deformation over $\mathbf{A}^1_k$. This leads to the following upper bounds on Betti numbers.

Theorem 22.9. *Let* I *and* $B \supseteq I$ *be graded ideals in* S. *Fix a monomial order* $\prec$.

(1) *The graded Betti numbers of* $S/\mathrm{in}_{\prec}(B)$ *over the quotient ring* $S/\mathrm{in}_{\prec}(I)$ *are greater than or equal to those of* S/B *over the ring* S/I.
(2) *The graded Betti numbers of* S/B *over* S *are smaller or equal to those of* $S/\mathrm{in}_{\prec}(B)$.
(3) *The graded Betti numbers of* k *over* S/I *are smaller or equal to those of* k *over* $S/\mathrm{in}_{\prec}(I)$.

For the proof of the theorem we need the following exercise.

Exercise 22.10. *Let* $\mathbf{a} = (a_1, \ldots, a_n)$ *and* $\mathbf{c} = (c_1, \ldots, c_n)$ *be vectors with non-negative integer coordinates. Suppose that* $\mathbf{c}$ *has positive coordinates. We consider the following two gradings of* S:

- *the* $\mathbf{a}$*-grading with* $\deg(x_i) = a_i$ *for all* i.
- *the* $\mathbf{c}$*-grading with* $\deg(x_i) = c_i$ *for all* i.

(1) *Suppose that* V *is a finitely generated* S*-module which is both* $\mathbf{a}$*-graded and* $\mathbf{c}$*-graded. There exists a minimal free resolution of* V *which is both* $\mathbf{a}$*-graded and* $\mathbf{c}$*-graded.*
(2) *Let* I *and* $B \supseteq I$ *be ideals in* S *which are homogeneous with respect to both gradings. Then there exists a minimal free resolution of* S/B *over the ring* S/I *which is both* $\mathbf{a}$*-graded and* $\mathbf{c}$*-graded.*

We will prove Theorem 22.9.

Proof of 22.9. Note that (2) is a special case of (1) with $I = 0$. Also note that (3) is a special case of (1) with $B = (x_1, \ldots, x_n)$. We will prove (1).

Let $\mathbf{a} = (1, \ldots, 1)$. Note that I is graded with respect to the grading $\deg(x_i) = 1$ for all i.

Choose a weight vector $\mathbf{w}$ with positive integer coordinates such that $\text{in}_{\prec_\mathbf{w}}(I) = \text{in}_\prec(I)$ and $\text{in}_{\prec_\mathbf{w}}(B) = \text{in}_\prec(B)$. The ring $\widetilde{S}$ is graded by $\deg(x_i) = w_i$ for all i, and $\deg(t) = 1$. The ideal $\widetilde{I}$ is homogeneous with respect to this grading.

By Exercise 22.10, there exists a minimal free resolution $\widetilde{\mathbf{F}}_{\widetilde{S}/\widetilde{B}}$ of $\widetilde{S}/\widetilde{B}$ over the ring $\widetilde{S}/\widetilde{I}$ which is graded with respect to both gradings

- $\deg(x_i) = w_i$ for all i, and $\deg(t) = 1$
- $\deg(x_i) = 1$ for all i, and $\deg(t) = 0$.

The degrees in the former grading are strictly positive, and this assures that there exists a graded free resolution which is minimal. In the rest of the argument we use the latter grading.

We grade $\widetilde{S}$ by $\deg(x_i) = 1$ and $\deg(t) = 0$. We will apply Theorem 22.8 and Theorem 20.3. Let $\alpha \in k$ be such as in Theorem 22.8.

First, we consider the case $\alpha = 0$. We have that t is a homogeneous non-zero divisor by Theorem 22.8. Therefore, Theorem 20.3 shows that $\widetilde{\mathbf{F}}_{\widetilde{S}/\widetilde{B}} \otimes \widetilde{S}/(t)$ is a graded free resolution of $S/\text{in}_\prec(B)$ over the ring $S/\text{in}_\prec(I)$. The resolution is minimal, since the differential matrices in $\widetilde{\mathbf{F}}_{\widetilde{S}/\widetilde{B}}$ have entries in $(x_1, \ldots, x_n, t)$ and after we set $t = 0$ we get that the entries are in $(x_1, \ldots, x_n)$. Therefore, the i'th Betti number of $S/\text{in}_\prec(B)$ is equal to the rank of $\widetilde{F}_i$.

Now, we apply these theorems for $\alpha = 1$. We have that $t - 1$ is a homogeneous non-zero divisor by Theorem 22.8. Therefore, Theorem 20.3 shows that $\widetilde{\mathbf{F}}_{\widetilde{S}/\widetilde{B}} \otimes \widetilde{S}/(t-1)$ is a graded free resolution of S/B over the ring S/I. This resolution might be non-minimal because we have set $t = 1$ in the matrices of the differential. Therefore, the

i'th Betti number of S/B is less or equal to the rank of $\widetilde{F}_i$. □

Example 22.11. We will illustrate the construction in the above proof. We continue Example 22.7. We use non-graded free modules in order to simplify the notation. Computer computation shows that the minimal graded free resolution of $\widetilde{A}/\widetilde{B}$ over $\widetilde{A}$ is

$$\widetilde{\mathbf{F}}: \quad 0 \to \widetilde{A} \xrightarrow{\begin{pmatrix} c \\ -b \\ -t^2 \\ a \end{pmatrix}} \widetilde{A}^4 \xrightarrow{\begin{pmatrix} b & c & aht^3 & ht^5 \\ t^2 & 0 & c & 0 \\ -a & bt^2 & -b^2 & c \\ 0 & -a & 0 & -b \end{pmatrix}}$$

$$\widetilde{A}^4 \xrightarrow{\begin{pmatrix} ac-b^2t^2 & b^3-a^2ht^3 & bc-aht^5 & c^2-bht^7 \end{pmatrix}} \widetilde{A} \to \widetilde{A}/\widetilde{B}\,.$$

Tensoring $\widetilde{\mathbf{F}}$ by $\widetilde{A}/(t)$ means that we set $t = 0$, so we obtain the minimal graded free resolution

$$0 \to A \xrightarrow{\begin{pmatrix} c \\ -b \\ 0 \\ a \end{pmatrix}} A^4 \xrightarrow{\begin{pmatrix} b & c & 0 & 0 \\ 0 & 0 & c & 0 \\ -a & 0 & -b^2 & c \\ 0 & -a & 0 & -b \end{pmatrix}} A^4$$

$$\xrightarrow{\begin{pmatrix} ac & b^3 & bc & c^2 \end{pmatrix}} A \to A/\mathrm{in}(B)\,.$$

Tensoring $\widetilde{\mathbf{F}}$ by $\widetilde{A}/(t-1)$ means that we set $t = 1$, so we obtain the non-minimal graded free resolution

$$0 \to A \xrightarrow{\begin{pmatrix} c \\ -b \\ -1 \\ a \end{pmatrix}} A^4 \xrightarrow{\begin{pmatrix} b & c & ah & h \\ 1 & 0 & c & 0 \\ -a & b & -b^2 & c \\ 0 & -a & 0 & -b \end{pmatrix}} A^4$$

$$\xrightarrow{\begin{pmatrix} ac-b^2t^2 & b^3-a^2h & bc-ah & c^2-bh \end{pmatrix}} \widetilde{A} \to A/B\,.$$

We say that a sequence $q_{i,j}$ of numbers is obtained from a sequence $p_{i,j}$ by a ***consecutive cancellation*** if there exist indexes s

and r such that

$$q_{s,r} = p_{s,r} - 1, \quad q_{s+1,r} = p_{s+1,r} - 1,$$
$$q_{i,j} = p_{i,j} \text{ for all other values of } i, j\,.$$

Theorem 22.12. *Let I be a graded ideal. The sequence of graded Betti numbers of S/I is obtained from the sequence of graded Betti numbers of $S/\mathrm{in}(I)$ by consecutive cancellations.*

Proof. We use the notation in the proof of Theorem 22.9. We want to show that the sequence of graded Betti numbers of S/I is obtained from the sequence of graded Betti numbers of $\widetilde{S}/\widetilde{I}$ by consecutive cancellations. This follows from the fact that $\mathbf{F}_{\widetilde{S}/\widetilde{I}} \otimes S[t]/(t-1)$ is a graded free resolution of S/I and Theorem 7.5. Each trivial complex that is a summand in $\mathbf{F}_{\widetilde{S}/\widetilde{I}} \otimes S[t]/(t-1)$ contributes one consecutive cancellation. □

Corollary 22.13. *Let I be a graded ideal. If $\mathrm{in}_{\prec}(I)$ has a q-linear free resolution, then the graded Betti numbers of S/I are equal to those of $S/\mathrm{in}_{\prec}(I)$.*

Proof. Since the minimal graded free resolution of $\mathrm{in}_{\prec}(I)$ is q-linear, by Theorem 22.9(1) it follows that I has a q-linear resolution. By Corollary 17.6, the minimal free resolution of I is q-linear. Use that $S/\mathrm{in}_{\prec}(I)$ and S/I have the same Hilbert series and apply Theorem 17.11. □

23 Computing a graded free resolution

We will describe Schreyer's Algorithm for constructing a graded free resolution, cf. [Eisenbud, Theorem 15.10].

Construction 23.1. Let V be a finitely generated graded S-module. Let F be a free S-module with basis $f_1, \dots, f_v$. An element of the form mf_i, with m a monomial in S, is called a ***monomial*** in F; an element of the form $\alpha m f_i$, with $\alpha \in k$, is called a ***term***. Let $\prec$ be a monomial order in S. Define a monomial order on F by

$$\alpha m f_i \succ \alpha' m' f_j \quad \text{if } m \succ m', \text{or } m = m' \text{ and } i < j.$$

If $f = \sum_i \alpha_i m_i f_i \in F$, with $\alpha_i \in k$ and m_i monomial, then $\text{in}_{\prec}(f)$ denotes the greatest term. Let T be a submodule of F generated by homogeneous elements $g_1, \ldots, g_r$. Consider the free graded module $F' = \oplus_i S(-\deg(g_i))$ with basis denoted $\{\epsilon_i\}_i$ and the map

$$\begin{aligned} \varphi :\ \oplus_i S(-\deg(g_i)) &\to T \\ \epsilon_i &\mapsto g_i \,. \end{aligned}$$

We order the monomials in F' by

$$\begin{aligned} \alpha m \epsilon_i \succ \alpha' m' \epsilon_j \text{ if } &\text{in}_{\prec}(m g_i) \succ \text{in}_{\prec}(m' g_j), \\ &\text{or } \text{in}_{\prec}(m g_i) = \text{in}_{\prec}(m' g_j) \text{ up to a scalar and } i < j. \end{aligned}$$

Suppose that $g_1, \ldots, g_r$ form a Gröbner basis, cf. [Eisenbud, Chapter 15]. A pair of indices i, j such that $\text{in}_{\prec}(g_i)$ and $\text{in}_{\prec}(g_j)$ involve the same basis element of F is called an ***s-pair***. For each such pair we have: if $\text{in}_{\prec}(g_i) = \alpha m f$ and $\text{in}_{\prec}(g_j) = \alpha' m' f$ then

$$\alpha' \frac{m'}{\gcd(m, m')} g_i - \alpha \frac{m}{\gcd(m, m')} g_j = \sum_p l_p g_p \,,$$

where the right-hand side is the Gröbner reduction to zero of the s-pair g_i, g_j. We obtain the following element in the kernel of φ

$$\tau_{ij} = \alpha' \frac{m'}{\gcd(m, m')} \epsilon_i - \alpha \frac{m}{\gcd(m, m')} \epsilon_j - \sum_p l_p \epsilon_p \,.$$

By [Eisenbud, Theorem 15.10], the elements τ_{ij}, obtained in the above way, form a Gröbner basis for the module $\text{Ker}(\varphi)$.

Construction 23.2. We will construct a graded free resolution of V by induction on homological degree.

Step 0. We compute a Gröbner basis $g_1, \ldots, g_r$ of V following [Eisenbud, Section 15.4]. Then we apply Construction 23.1 (the map φ will be d_0) and we obtain a Gröbner basis of the module $\mathrm{Ker}(d_0)$.

Step i. Suppose that we have a Gröbner basis of the module $\mathrm{Ker}(d_{i-1})$. We apply Construction 23.1 (the map φ will be d_i) and we obtain a Gröbner basis of the module $\mathrm{Ker}(d_i)$.

Construction 23.3. We will construct a minimal graded free resolution of V. At each step of the above construction we perform the following: After we have computed a Gröbner basis of the module $\mathrm{Ker}(d_i)$, we obtain a minimal set of homogeneous generators of $\mathrm{Ker}(d_i)$ using the criterion in Theorem 2.12. That is, we choose preimages in $\mathrm{Ker}(d_i)$ of a basis of $\mathrm{Ker}(d_i)/\mathbf{m}\mathrm{Ker}(d_i)$.

These algorithms are implemented in several computer algebra systems.

24 Short resolutions

Exercise 24.1. *Each monomial ideal B in $A = k[a,b]$ can be written in the form*

$$B = (a^{\mu_i} b^{\nu_i} \,|\, 1 \le i \le r,\ 0 \le \mu_1 < \ldots < \mu_r,\ \nu_1 > \ldots \nu_r \ge 0)\,,$$

where r is the number of its minimal monomial generators. Note that the projective dimension of A/B is ≤ 2. The minimal free resolution of A/B is

$$0 \to \bigoplus_{i=1}^{r-1} A \to \bigoplus_{j=1}^{r} A \to A \to A/B \to 0\,,$$

where the matrix of d_1 has entries the monomials, and the matrix of d_2 can be described explicitly.

What happens in arbitrary resolutions of small length? The Hilbert-Burch Theorem, proved in cf. [Eisenbud, Theorem 20.15] (also see [Eisenbud 2]), gives the answer for length 2 resolutions.

Hilbert-Burch Theorem 24.2. *Let I be a graded ideal in S such* *that* $\mathrm{pd}_S(S/I) = 2$. *There exists a homogeneous non-zero divisor* u *and an* $r \times (r-1)$*-matrix* A *such that the minimal free resolution of* S/I *over* S *has the form*

$$0 \longrightarrow S^{r-1} \longrightarrow S^r \longrightarrow S\ ,$$

where the first differential is given by the matrix

$$u\big(\, det(A_1)\ \ldots\ (-1)^{i+1} det(A_i)\ \ldots\ (-1)^{r+1} det(A_r)\,\big)$$

and A_i *is the submatrix of* A *obtained by deleting the* i*'th row.*

The structure of length 3 minimal free resolutions can be quite complex.

25 Cohen-Macaulay and Gorenstein ideals

The results in this section are not used in the rest of the book; students, who lack background on Cohen-Macaulay and Gorenstein ideals, can skip the section.

Theorem 25.1. *If* V *is a graded finitely generated* S*-module, then* $\mathrm{pd}_S(V) = \mathrm{codim}(V)$ *if and only if* V *is Cohen-Macaulay.*

Proof. $\mathrm{pd}_S(V) = n - \mathrm{depth}(V) \geq n - \dim(V) = \mathrm{codim}(V)$. Equality holds if and only if V is Cohen-Macaulay. □

The following result is an immediate consequence of Theorem 20.3 and Proposition 20.1.

Theorem 25.2. *Suppose that the field* k *is infinite. If* I *is a graded Cohen-Macaulay ideal in* S, *then there exists an artinian graded ideal* J, *such that* S/I *and* S/J *have the same graded Betti numbers and* S/J *is obtained from* S/I *by factoring a regular sequence of linear forms.*

Proposition 25.3. *Let* I *be a graded ideal in* S. *Suppose that* S/I *is Cohen-Macaulay of dimension* q. *Let* $\mathbf{F}$ *be the minimal graded free resolution of* S/I *over* S. *The dual complex*

$$\mathbf{F}^* = \mathrm{Hom}(\mathbf{F}, S): \quad 0 \to F_0^* \to F_1^* \to \ldots \to F_{n-q}^*$$

is a minimal free resolution of $\operatorname{Ext}_S^{n-q}(S/I, S)$.

Proof. By Theorem 25.1, $\operatorname{pd}(S/I) = n - q$ holds. The dual complex $\mathbf{F}^*$ is acyclic since $\operatorname{Ext}_S^i(S/I, S) = 0$ for $i < n - q$. □

Proposition 25.4. *If a graded ideal I has a Cohen-Macaulay initial ideal, then I is Cohen-Macaulay.*

Proof. Let $\prec$ be a monomial order such that $S/\operatorname{in}_\prec(I)$ is Cohen-Macaulay. By Theorem 25.1, $\operatorname{pd}_S(S/\operatorname{in}_\prec(I)) = \operatorname{codim}(S/\operatorname{in}_\prec(I))$. Since we have

$$\operatorname{codim}(S/I) = \operatorname{codim}(S/\operatorname{in}_\prec(I))$$
$$\operatorname{pd}_S(S/I) \leq \operatorname{pd}_S(S/\operatorname{in}_\prec(I))$$

we conclude that $\operatorname{pd}_S(S/I) \leq \operatorname{codim}(S/I)$. Applying Theorem 25.1 again and Corollary 15.4, we conclude that S/I is Cohen-Macaulay.□

In the rest of the section we consider Gorenstein ideals.

Theorem 25.5. *Let I be a graded ideal in S. Suppose that S/I is artinian Gorenstein. Let*

$$q = \max\{i \mid (S/I)_i \neq 0\}.$$

Then

$$\dim_k (S/I)_i = \dim_k (S/I)_{q-i}.$$

Proof. Apply the argument in the proof of Theorem 25.6 to the graded Betti numbers, and combine it with Theorem 16.2. □

Theorem 25.6. *Let I be a graded ideal in S. If S/I is Gorenstein of dimension q, then*

$$b_i^S(S/I) = b_{n-q-i}^S(S/I) \quad \text{for } 0 \leq i \leq n - q.$$

Proof. Let $\mathbf{F}$ be the minimal graded free resolution of S/I over S. By Proposition 25.3, $\mathbf{F}^* = \operatorname{Hom}(\mathbf{F}, S)$ is a minimal free resolution of $\operatorname{Ext}_S^{n-q}(S/I, S) \cong S/I$. □

We have the following criterion for S/I to be Gorenstein, cf.

[Huneke, Proposition 3.2].

Theorem 25.7. *Let I be a graded ideal in S. Let* $\dim(S/I) = q$. *The quotient S/I is Gorenstein if and only if* $\mathrm{pd}(S/I) = n - q$ *and* $b^S_{n-q}(S/I) = 1$.

Proof. If S/I is Gorenstein, then apply Theorems 25.1 and 25.6.

Suppose that $\mathrm{pd}(S/I) = n - q$ and $b^S_{n-q}(S/I) = 1$. Let $\mathbf{F}$ be the minimal graded free resolution of S/I over S. By Theorem 25.1 we have that S/I is Cohen-Macaulay. By Proposition 25.3, the dual complex $\mathbf{F}^* = \mathrm{Hom}(\mathbf{F}, S)$ is a minimal free resolution of $\mathrm{Ext}^{n-q}_S(S/I, S)$. As $b^S_{n-q}(S/I) = 1$ we conclude that $\mathrm{Ext}^{n-q}_S(S/I, S) \cong S/J$ for some ideal J. Furthermore, note that I annihilates S/I so it annihilates $\mathrm{Ext}^{n-q}_S(S/I, S)$; therefore, $J \supseteq I$.

Now, we can apply the argument in the above paragraph to S/J and conclude that $I \supseteq J$.

Therefore, $\mathrm{Ext}^{n-q}_S(S/I, S) \cong S/I$ and we get that S/I is Gorenstein. □

26 Multigradings and Taylor's resolution

We consider a refined way to grade S, namely multigrading it. Multigradings are used throughout Chapters III and IV. In this section we also describe Taylor's free resolution which is simply structured and is very similar to the Koszul resolution.

Multigrading 26.1. The polynomial ring S is $\mathbf{N}^n$-graded by

$$\mathrm{mdeg}(x_i) = \text{the } i\text{'th standard vector in } \mathbf{N}^n\,,$$

where mdeg stands for multidegree. For every $\mathbf{a} = (a_1, \ldots, a_n) \in \mathbf{N}^n$ there exists a unique monomial of $\mathbf{N}^n$-degree $\mathbf{a}$, namely $\mathbf{x}^{\mathbf{a}} = x_1^{a_1} \cdots x_n^{a_n}$, and $\mathbf{a}$ is its ***exponent vector***. Usually we say that S is ***multigraded*** instead of $\mathbf{N}^n$-graded. Instead of "$\mathbf{N}^n$-degree $\mathbf{a}$", we can say "***multidegree*** $\mathbf{x}^{\mathbf{a}}$". Thus, S has a direct sum decomposition $S = \oplus_{m \text{ is a monomial}} S_m$ as a k-vector space. Note that $S_m S_{m'} = S_{mm'}$ for all monomials m, m'.

An S-module T is called ***multigraded***, if it has a direct sum decomposition $T = \oplus_{m \text{ is a monomial}} T_m$ as a k-vector space and $S_m T_{m'} \subseteq T_{mm'}$ for all monomials m, m'.

Denote by $S(\mathbf{x^a})$ the free S-module with one generator in multidegree $\mathbf{x^a}$.

A multigraded finitely generated module T has a ***multigraded Hilbert function***

$$\begin{aligned} h: \text{ monomials in } S &\to \mathbf{N} \\ m &\mapsto \dim_k T_m \end{aligned}$$

and a ***multigraded Hilbert series***

$$\mathrm{hilb}_T(x_1, \ldots, x_n) = \sum_m m \dim_k T_m \,,$$

where m runs over all monomials in S.

The Hilbert series is obtained from the multigraded Hilbert series by setting each $x_i = t$, that is

$$\mathrm{Hilb}_T(t) = \mathrm{hilb}_T(t, \ldots, t)\,.$$

Exercise 26.2. *The multigraded Hilbert series of S is*

$$\mathrm{hilb}_S(x_1, \ldots, x_n) = \frac{1}{(1 - x_1) \ldots (1 - x_n)}\,.$$

Exercise 26.3. *Every monomial ideal has a unique minimal system of monomial generators.*

Every monomial ideal is homogeneous with respect to the $\mathbf{N}^n$-grading. Note that it is also homogeneous with respect to the standard grading on the polynomial ring S.

Let M be a monomial ideal. Construction 4.2 works in the multigraded case. There exists a minimal free resolution $\mathbf{F}_M$ of S/M over S

which is $\mathbf{N}^n$-graded or ***multigraded***. We denote by d the differential in $\mathbf{F}_M$. We have ***multigraded Betti numbers***

$$b^S_{i,m}(S/M) \;=\; \dim_k \operatorname{Tor}^S_{i,m}(S/M,k) \quad \text{for } i \geq 0,\ m \text{ a monomial.}$$

Similarly to Construction 3.2, in the multigraded case the resolution can be written

$$0 \to \ldots \to \oplus_m\ S^{b_{3,m}}(m) \to \oplus_m\ S^{b_{2,m}}(m) \to \oplus_m S^{b_{1,m}}(m) \to S\,,$$

where the sums run over all monomials.

In the rest of this section, M stands for a monomial ideal minimally generated by monomials $m_1, \ldots, m_q$.

The next construction provides an analogue of Theorem 14.10 in the multigraded case.

Construction 26.4. We can compute the Betti numbers of M using the Koszul complex $\mathbf{K}$ that is the minimal free resolution of k over S. Let E be the exterior algebra over k on basis elements $e_1, \ldots, e_n$. The complex $\mathbf{K}$ equals $S \otimes E$ as an S-module and has differential

$$d(e_{j_1} \wedge \cdots \wedge e_{j_i}) \;=\; \sum\nolimits_{1 \leq p \leq i} (-1)^{p+1} \cdot x_{j_p} \cdot e_{j_1} \wedge \cdots \wedge \widehat{e}_{j_p} \wedge \cdots \wedge e_{j_i}\ ,$$

where $\widehat{e}_{j_p}$ means that e_{j_p} is omitted in the product. We have that

$$b^S_{i,m}(M) = \dim_k \operatorname{Tor}_i(M,k)_m = \dim \operatorname{H}_i(M \otimes \mathbf{K})_m\,.$$

The multidegree of e_i is x_i, for each i. The component of $\mathbf{K}$ in multidegree m has basis

$$\left\{ \frac{m}{x_{j_1} \ldots x_{j_i}} e_{j_1} \wedge \cdots \wedge e_{j_i} \;\middle|\; x_{j_p} \text{ divides } m \text{ for } 1 \leq p \leq i,\ 1 \leq j_1 < \ldots < j_i \leq n \right\}.$$

Hence, the component of $M \otimes \mathbf{K}$ in multidegree m has basis

$$\left\{ \frac{m}{x_{j_1} \dots x_{j_i}} e_{j_1} \wedge \dots \wedge e_{j_i} \;\middle|\; x_{j_p} \text{ divides } m \text{ for } 1 \le p \le i, \right.$$
$$\left. 1 \le j_1 < \dots < j_i \le n,\ \frac{m}{x_{j_1} \dots x_{j_i}} \in M \right\}.$$

Next we will describe Taylor's resolution, which resolves S/M. It is usually highly non-minimal. It is useful because of its simple structure, which is similar to that of the Koszul complex. Taylor's resolution was first constructed (using different terminology) by Taylor in her Ph.D. Thesis [Taylor]. The short formulation in Construction 26.5 is due to Eisenbud.

Construction 26.5. Let M be a monomial ideal minimally generated by monomials $m_1, \dots, m_q$. We will construct ***Taylor's resolution*** $\mathbf{T}_M$ of S/M in a way similar to the Koszul complex. We will use the notation from Construction 14.1. Let E be the exterior algebra over k on basis elements $e_1, \dots, e_q$. Denote by $\mathbf{T}_M$ the S-module $S \otimes E$ graded homologically by $\text{hdeg}(e_{j_1} \wedge \dots \wedge e_{j_i}) = i$ and equipped with the differential

$$d(e_{j_1} \wedge \dots \wedge e_{j_i})$$
$$= \sum_{1 \le p \le i} (-1)^{p-1} \frac{\text{lcm}(m_{j_1}, \dots, m_{j_i})}{\text{lcm}(m_{j_1}, \dots, \widehat{m}_{j_p}, \dots, m_{j_i})} e_{j_1} \wedge \dots \wedge \widehat{e}_{j_p} \wedge \dots \wedge e_{j_i},$$

where $\widehat{e}_{j_p}$ and $\widehat{m}_{j_p}$ mean that e_{j_p} and m_{j_p} are omitted respectively. The standard grading of $\mathbf{T}_M$ is given by

$$\deg(e_{j_1} \wedge \dots \wedge e_{j_i}) = \deg(\text{lcm}(m_{j_1}, \dots, m_{j_i})).$$

The multigrading of $\mathbf{T}_M$ is given by

$$\text{mdeg}(e_{j_1} \wedge \dots \wedge e_{j_i}) = \text{lcm}(m_{j_1}, \dots, m_{j_i}).$$

An argument similar to that in Construction 14.1 solves the following exercise.

Exercise 26.6. $\mathbf{T}_M$ *is a complex.*

Theorem 26.7. (Taylor) $\mathbf{T}_M$ *is a free resolution of* S/M.

Proof. Since $\mathbf{T}_M$ is multigraded it suffices to check exactness in each multidegree. Let m be a monomial. If $m \notin M$, then $(\mathbf{T}_M)_m = S_m \cong k$ where S_m is placed in homological degree 0. Suppose that $m \in M$. Consider the simplex Γ_m on vertexes $\{e_i \mid m_i \text{ divides } m\}$. Then

$$\frac{m}{\operatorname{lcm}(m_{j_1}, \dots, m_{j_i})} e_{j_1} \wedge \cdots \wedge e_{j_i} \mapsto \text{the face with vertices } e_{j_1}, \dots, e_{j_i}$$

gives an isomorphism of $(\mathbf{T}_M)_m$ to the chain complex computing the reduced homology of the simplex Γ_m. Hence, $(\mathbf{T}_M)_m$ is exact. □

Example 26.8. The free resolution

$$\mathbf{T}_Y : 0 \longrightarrow A \xrightarrow{\begin{pmatrix} y^2 \\ x \\ 1 \end{pmatrix}} A^3 \xrightarrow{\begin{pmatrix} -y & 0 & y^3 \\ x & -y^2 & 0 \\ 0 & x & -x^2 \end{pmatrix}} A^3 \xrightarrow{\begin{pmatrix} x^2 & xy & y^3 \end{pmatrix}} A$$

is the Taylor resolution of $k[x,y]/(x^2, xy, y^3)$ over the polynomial ring $A = k[x,y]$.

Corollary 26.9. *The entries in the differential maps in the multigraded minimal free resolution of* S/M *are scalar multiples of monomials which divide* $\operatorname{lcm}(m_1, \dots, m_q)$. *In particular, the degree of the monomial* $\operatorname{lcm}(m_1, \dots, m_q)$ *is an upper bound for the degree of each such entry.*

Proof. Since $\mathbf{F}_M$ is multigraded we have that each entry f in the matrices of the differentials in $\mathbf{F}_M$ is homogeneous. Let m be the multidegree of f. Since S_m is one dimensional, it follows that f is a scalar multiple of m. Note that m divides the multidegree of a homogeneous basis element in the resolution. Furthermore, the multidegree of every homogeneous basis element in $\mathbf{F}_M$ divides $\operatorname{lcm}(m_1, \dots, m_q)$, since $\mathbf{F}_M$ is a direct summand of Taylor's resolution by 7.5. □

A very useful corollary is that the Betti numbers of a squarefree monomial ideal are located in squarefree multidegrees.

Corollary 26.10. *Let M be a squarefree monomial ideal. If a monomial m is not squarefree, then $b^S_{i,m}(S/M) = 0$ for all $i \geq 0$.*

Corollary 26.11. (Hoa-Trung) *Suppose that the minimal generators $m_1, \ldots, m_q$ of M are ordered so that $\deg(m_1) \geq \ldots \geq \deg(m_q)$. Then*

$$\mathrm{reg}(S/M) \leq \deg(m_1) + \ldots + \deg(m_p) - p\,,$$

where $p = \min(n, q)$.

Proof. If $b^S_{i,m}(S/M) \neq 0$ then the monomial m has the form $m = \mathrm{lcm}(m_{j_1}, \ldots, m_{j_i})$ for some monomials $m_{j_1}, \ldots, m_{j_i}$. Hence

$$\begin{aligned}\deg(m) &\leq i + (\deg(m_{j_1}) + \ldots + \deg(m_{j_i}) - i) \\ &\leq i + (\deg(m_1) + \ldots + \deg(m_p) - p).\end{aligned}$$

□

Corollary 26.12. (Hoa-Trung) *Let $u = \mathrm{lcm}(m_1, \ldots, m_q)$. For $s \geq 1$ we have that*

$$\mathrm{reg}(S/M^s) \leq s\deg(u)\,.$$

Proof. Corollary 26.9 implies that for any monomial ideal J we have that $\mathrm{reg}(S/J)$ is bounded above by the degree of the lcm of its minimal monomial generators. The degree of the lcm of the minimal monomial generators of M^s is less or equal to $s\deg(u)$. □

27 Mapping cones

Construction 27.1. Let I be a graded ideal in S, and $R = S/I$. Let

$$\varphi: \ (\mathbf{U}, d) \to (\mathbf{U}', d')$$

be a map of complexes of finitely generated R-modules. We say that φ is a ***comparison map***. The ***mapping cone*** of φ is the complex

W with differential ∂, defined as follows:

$$\begin{aligned} W_i &= U_{i-1} \oplus U'_i \text{ as a module} \\ \partial\,|_{\,U_{i-1}} &= -d + \varphi :\ U_{i-1} \to U_{i-2} \oplus U'_{i-1} \\ \partial\,|_{\,U'_i} &= d' :\ U'_i \to U'_{i-1} \end{aligned}$$

for each i. We have the diagram

$$\begin{array}{ccc} W_i & \xrightarrow{\partial_i} & W_{i-1} \\ \| & & \| \\ U'_i & \xrightarrow{d'_i} & U'_{i-1} \\ \oplus & \nearrow_{\varphi_i} & \oplus \\ U_{i-1} & \xrightarrow[-d_{i-1}]{} & U_{i-2} \end{array}$$

Exercise 27.2. **W** *is a complex.*

Clearly, $\mathbf{U}'$ is a subcomplex of **W**. The quotient $\mathbf{W}/\mathbf{U}'$ is isomorphic to $\mathbf{U}[-1]$ (where $\mathbf{U}[-1]$ means the complex **U** with homological grading shifted one step, that is $\mathbf{U}[-1]_i = U_{i-1}$). We have the short exact sequence of complexes

$$0 \to \mathbf{U}' \to \mathbf{W} \to \mathbf{U}[-1] \to 0\,.$$

By Theorem 13.2, it yields the long exact sequence in homology

$$\ldots \to \mathrm{H}_i(\mathbf{U}') \to \mathrm{H}_i(\mathbf{W}) \to \mathrm{H}_i(\mathbf{U}[-1]) \to \mathrm{H}_{i-1}(\mathbf{U}') \to \ldots\ ,$$

so

$$\ldots \to \mathrm{H}_i(\mathbf{U}') \to \mathrm{H}_i(\mathbf{W}) \to \mathrm{H}_{i-1}(\mathbf{U}) \to \mathrm{H}_{i-1}(\mathbf{U}') \to \ldots\ .$$

The following straightforward computation following 13.1 (and using the notation in 13.1) shows that the connecting homomorphism $\mathrm{H}_{i-1}(\mathbf{U}) \to \mathrm{H}_{i-1}(\mathbf{U}')$ is the map on homology induced by φ: choose an $x \in \mathbf{U}[-1]_i = U_{i-1}$, then choose the preimage $y = (x, 0) \in W_i$, then

$$\partial(y) = (-d(x), \varphi(x)) = (0, \varphi(x))\,,$$

then choose $z = \varphi(x)$.

Now, suppose that $\mathbf{U}$ and $\mathbf{U}'$ are free resolutions of finitely generated R-modules V and V' respectively, and suppose that $\varphi : V \to V'$ is an injective homomorphism of modules. There exists a lifting $\varphi : \mathbf{U} \to \mathbf{U}'$ that induces $\varphi : V \to V'$. The long exact homology sequence above implies that $\mathrm{H}_i(\mathbf{W}) = 0$ for $i \geq 2$ and

$$0 \to \mathrm{H}_1(\mathbf{W}) \to V \to V' \to \mathrm{H}_0(\mathbf{W}) \to 0$$

is exact. Since $\varphi : V \to V'$ is injective, we conclude that $\mathrm{H}_1(\mathbf{W}) = 0$. It follows that $\mathbf{W}$ is a free resolution of $V'/\varphi(V)$. If in addition the modules V and V' are graded and φ has degree 0, then $\mathbf{W}$ is a graded free resolution of $V'/\varphi(V)$. Note, that it is possible that the graded free resolutions $\mathbf{U}$ and $\mathbf{U}'$ are minimal but the mapping cone $\mathbf{W}$ is not.

We will apply the mapping cone to monomial ideals.

Construction 27.3. Let M be an ideal minimally generated by monomials $m_1, \dots, m_r$. Set $M_i = (m_1, \dots, m_i)$ for $1 \leq i \leq r$. Thus, $M = M_r$. For each $i \geq 1$, we have the short exact sequence

$$0 \to \ S/(M_i : m_{i+1}) \xrightarrow{m_{i+1}} S/M_i \ \to \ S/M_{i+1} \to 0\,.$$

The comparison map is multiplication by m_{i+1}. Therefore, in order to make the comparison map of degree 0, we shift the left module $S/(M_i : m_{i+1})$ in multidegree by m_{i+1}. Then, we have the graded short exact sequence

$$0 \to \ S/(M_i : m_{i+1})\,(m_{i+1}) \xrightarrow{m_{i+1}} S/M_i \ \to \ S/M_{i+1} \to 0\,.$$

Assuming that a multigraded free resolution $\mathbf{F}_i$ of S/M_i is already known, and that a multigraded free resolution $\mathbf{G}_i$ of $S/(M_i : m_{i+1})$ is also known, we can construct the mapping cone and obtain a multigraded free resolution $\mathbf{F}_{i+1}$ of S/M_{i+1}. We say that the multigraded free resolution $\mathbf{F}_r$ of S/M obtained in this way, is obtained by ***iterated mapping cones***.

Example 27.4. Let $B = (x^2, y^3, xy)$ in the ring $A = k[x,y]$. We will apply Construction 27.3. We have that $B_1 = (x^2)$, $B_2 = (x^2, y^3)$

and $B_3 = B = (x^2, y^3, xy)$. The second short exact sequence is

$$0 \to A/(B_2 : xy) = A/(x, y^2) \xrightarrow{xy} A/(x^2, y^3) \to A/B \to 0\,.$$

Furthermore, $A/(x, y^2)$ is resolved minimally by the Koszul resolution on x, y^2, and $A/(x^2, y^3)$ is resolved minimally by the Koszul resolution on x^2, y^3. The comparison map for the mapping cone is

$$\begin{array}{ccccccc}
0 \to A(x^2y^3) & \xrightarrow{\begin{pmatrix} y^2 \\ -x \end{pmatrix}} & A(x^2y) \oplus A(xy^3) & \xrightarrow{(x\ y^2)} & A(xy) & \to & \frac{A}{(x,y^2)}(xy) \\
\downarrow 1 & & \downarrow \begin{pmatrix} y & 0 \\ 0 & x \end{pmatrix} & & \downarrow xy & & \downarrow xy \\
0 \to A(x^2y^3) & \xrightarrow{\begin{pmatrix} y^3 \\ -x^2 \end{pmatrix}} & A(x^2) \oplus A(y^3) & \xrightarrow{(x^2\ y^3)} & A & \to & A/(x^2, y^3)\,.
\end{array}$$

Note that the minimal free resolution in the top row is shifted in multidegree by xy in order to make the comparison map of degree 0. Then, we obtain the mapping cone

$$0 \to A(x^2y^3) \xrightarrow{\begin{pmatrix} 1 \\ -y^2 \\ x \end{pmatrix}} A(x^2y^3) \oplus A(x^2y) \oplus A(xy^3) \xrightarrow{\begin{pmatrix} y^3 & y & 0 \\ -x^2 & 0 & x \\ 0 & -x & -y^2 \end{pmatrix}}$$

$$A(x^2) \oplus A(y^3) \oplus A(xy) \xrightarrow{(x^2\ y^3\ xy)} A \to A/(x^2, y^3, xy) \to 0\,.$$

Exercise 27.5. *Prove Theorem* 26.7 *using iterated mapping cones.*

28 The Eliahou-Kervaire resolution

In this section we discuss Borel ideals and describe their minimal free resolutions. This class of ideals has many applications.

A monomial ideal M in S is called ***Borel*** if it satisfies the ***Borel property***: whenever $i < j$ and g is a monomial such that $gx_j \in M$, we have $gx_i \in M$ as well.

Exercise 28.1. *A monomial ideal M is Borel if and only if whenever $i < j$ and g is a monomial such that gx_j is a minimal monomial generator of M, we have $gx_i \in M$ as well.*

A monomial $m' \in S$ is said to be in the ***big shadow*** of a monomial $m \in S$ if $m' = \dfrac{x_i m}{x_j}$ for some x_j dividing m and some $i \leq j$. Thus, a monomial ideal M is Borel if and only if it contains all monomials in the big shadows of its minimal monomial generators.

The interest in studying such special monomial ideals comes from generic initial ideals.

Construction 28.2. The general linear group $\mathrm{GL}(n, k)$ of invertible $(n \times n)$-matrices acts as a group of algebra automorphisms on S by acting on the variables (such an action is a linear change of the variables, and is also called a linear change of coordinates). Denote by $\mathrm{B}(n, k)$ the *Borel subgroup* consisting of upper triangular invertible matrices. The following result of Galligo, Bayer, and Stillman is proved in [Eisenbud, Theorem 15.18].

Theorem 28.3. *Let J be a graded ideal. Fix a monomial order $>$ such that $x_1 > \ldots > x_n$. There exists a nonempty Zariski open set $\mathcal{B}$ in $GL(n, k)$, such that $\mathrm{in}_>(\varphi(J))$ does not depend on $\varphi \in \mathcal{B}$. Furthermore, $\mathcal{B} \cap B(n, k) \neq \emptyset$.*

For every $\varphi \in \mathcal{B}$, the monomial ideal $\mathrm{in}_>(\varphi(J))$ is called the ***generic initial ideal*** of J (in ***generic coordinates***), and is denoted $\mathrm{gin}(J)$. The importance of this ideal comes from Theorem 18.10. Furthermore, the structure of generic initial ideals is given in the following result, cf. [Eisenbud, 15.20 and 15.23].

Theorem 28.4. (Galligo) *If* $\mathrm{char}(k) = 0$ *then the generic initial ideal of a graded ideal is Borel.*

For a monomial m denote

$$\max(m) = \max\{i \mid x_i \text{ divides } m\}$$
$$\min(m) = \min\{i \mid x_i \text{ divides } m\}.$$

Exercise 28.5. *Let M be a Borel ideal. If $w \in M$ is a monomial, then there is a unique decomposition $w = uv$, such that u is a minimal monomial generator of M and* $\max(u) \le \min(v)$.

In the notation of the above exercise, we set $b(w) = u$ and call it the ***beginning*** of the monomial w, and we set $e(w) = v$ and call it the ***end*** of w.

Construction 28.6. [Eliahou-Kervaire] Let M be a Borel ideal. Denote by $m_1, \dots, m_r$ the set of minimal monomial generators of M. For each m_i and for each sequence $1 \le j_1 < \dots < j_p < \max(m_i)$ of strictly increasing natural numbers, we consider the free S-module $S(m_i x_{j_1} \dots x_{j_p})$ with one generator, denoted $(m_i; j_1, \dots, j_p)$, in homological degree $p+1$ and multidegree $m_i x_{j_1} \dots x_{j_p}$. The ***Eliahou-Kervaire resolution*** $\mathbf{E}_M$ of S/M has basis denoted

$$\mathcal{B} = \{1\} \cup \Big\{ (m_i; j_1, \dots, j_p) \,\Big|\, 1 \le j_1 < \dots < j_p < \max(m_i),\ 1 \le i \le r \Big\}.$$

The element 1 is the basis of S in homological degree 0. The basis in homological degree 1 consists of the elements $(m_1; \emptyset), \dots, (m_r; \emptyset)$. Define maps ∂ and μ by:

$$\partial(m_i; j_1, \dots, j_p) = \sum_{q=1}^{p} (-1)^q \, x_{j_q} (m_i; j_1, \dots, \hat{j}_q, \dots, j_p)$$

$$\mu(m_i; j_1, \dots, j_p) = \sum_{q=1}^{p-1} (-1)^q \, \frac{m_i x_{j_q}}{b(m_i x_{j_q})} \left(b(m_i x_{j_q}); j_1, \dots, \hat{j}_q, \dots, j_p \right),$$

where $b(m_i x_{j_q})$ is the beginning of $m_i x_{j_q}$. Note that the coefficient

$$\frac{m_i x_{j_q}}{b(m_i x_{j_q})} = e(m_i x_{j_q})$$

is the end of $m_i x_{j_q}$. Note also that $\left(b(m_i x_{j_q}); j_1, \dots, \hat{j}_q, \dots, j_p \right)$ is considered zero if $j_p \ge \max(b(m_i x_{j_q}))$. The differential in $\mathbf{E}_M$ is defined by

$$d = \partial - \mu \,.$$

The differential is multihomogeneous.

Example 28.7. The ideal (x^2, xy, xz, y^3) is Borel in the ring $A = k[x, y, z]$. The basis of the Eliahou-Kervaire resolution is

$$\begin{array}{rl} 1 & \text{in homological degree } 0 \\ (x^2;\emptyset),\ (xy;\emptyset),\ (xz;\emptyset),\ (y^3;\emptyset) & \text{in homological degree } 1 \\ (xy;1),\ (xz;1),\ (xz;2),\ (y^3;1) & \text{in homological degree } 2 \\ (xz;1,2) & \text{in homological degree } 3\,. \end{array}$$

The Eliahou-Kervaire resolution is

$$0 \to A(x^2yz) \xrightarrow{\begin{pmatrix} -z \\ y \\ -x \\ 0 \end{pmatrix}} A(x^2y) \oplus A(x^2z) \oplus A(xyz) \oplus A(xy^3)$$

$$\xrightarrow{\begin{pmatrix} y & z & 0 & 0 \\ -x & 0 & z & y^2 \\ 0 & -x & -y & 0 \\ 0 & 0 & 0 & -x \end{pmatrix}} A(x^2) \oplus A(xy) \oplus A(xz) \oplus A(y^3)$$

$$\xrightarrow{(x^2\ xy\ \ xz\ y^3)} A\,.$$

Exercise 28.8. *If M is Borel, then $\mathbf{E}_M$ is a complex.*

Theorem 28.9. [Eliahou-Kervaire] *If M is Borel, then $\mathbf{E}_M$ is the minimal free resolution of S/M.*

We will prove the theorem using the mapping cone Construction 27.3. First, we need a lemma.

Lemma 28.10. *Suppose that the minimal monomial generators m_1, $\ldots, m_r$ of M are ordered so that $s < q$ if either the degree of m_s is* *less than the degree of m_q, or the degrees are equal and $m_s \succ_{rlex} m_q$. For $i \geq 1$ we have*

$$(m_1, \ldots, m_i) : m_{i+1} = (x_1, \ldots, x_{\max(m_{i+1})-1}) .$$

Proof. Note that the ideal $(m_1, \ldots, m_i)$ is Borel.

The inclusion $\supseteq$ follows immediately from the Borel property. Assume that the equality does not hold. There exists a monomial w such that $wm_{i+1} \in (m_1, \ldots, m_i)$ and $\min(w) \geq \max(m_{i+1})$. Hence, $b(wm_{i+1}) = m_{i+1}$. Since for $1 \leq t \leq i$ the degree of m_t is less or equal to the degree of m_{i+1}, it follows that m_{i+1} is divisible by some monomial among $m_1, \ldots, m_i$. This is a contradiction. □

There are several different proofs of the Eliahou-Kervaire resolution. In [Eliahou-Kervaire] and [Green] the proofs use Gröbner basis theory. In [Aramova-Herzog] the proof uses the homology of the Koszul complex. G. Evans and M. Stillman (unpublished) realized that mapping cones can be used for computing the Betti numbers, cf. [Charalambous-Evans]. We give the proof from [Peeva-Stillman 2].

Proof of Theorem 28.9. Order $m_1, \ldots, m_r$ as in Lemma 28.10. The proof is by induction on the number r of minimal monomial generators of M. Let $i \geq 1$. Set $J = (m_1, \ldots, m_i)$ and $m = m_{i+1}$. Denote by $\mathbf{E_J}$ the minimal free Eliahou-Kervaire resolution of S/J.

Denote by $\mathbf{K}$ be the Koszul complex resolving $S/(J : m) = S/(x_1, \ldots, x_{\max(m_{i+1})-1})$ and multigraded so that the free module in homological degree 0 is $S(m)$ with one generator of multidegree m. We denote the basis of $\mathbf{K}$ by

$$\{ (m; j_1, \ldots, j_p) \,|\, 1 \leq j_1 < \ldots < j_p < \max(m) \} ,$$

where $(m; j_1, \ldots, j_p)$ has multidegree $mx_{j_1} \ldots x_{j_p}$ and homological degree p. Thus, $(m; j_1, \ldots, j_p)$ stands for $e_{j_1} \wedge \ldots \wedge e_{j_p}$ in the notation of Construction 14.1. The element $(m, \emptyset)$ is the basis in homological degree 0. The differential in $\mathbf{K}$ is $-\partial$ constructed in Construction 28.6.

Define $\mu(m, \emptyset) = -m$, and recall the map μ in Construction 28.6. We will show that the map $-\mu : \mathbf{K} \to \mathbf{E}_J$ is a multigraded map of complexes of degree 0 and that it lifts the homomorphism

$$S/(J:m) \xrightarrow{m} S/J\,.$$

It is clear, that the map $-\mu$ is multigraded of degree 0. In order to show that $-\mu$ is a map of complexes, we have to verify that $-\mu(-\partial(m;j_1,\dots,j_p)) = d(-\mu(m;j_1,\dots,j_p))$. Indeed,

$$\begin{aligned}\mu\partial = -\partial^2 + \mu\partial &= (-\partial+\mu)\partial = -d\partial = -d(\partial-\mu+\mu) = -d^2 - d\mu \\ &= -d\mu\,.\end{aligned}$$

In the above computation, we used Exercise 28.8.

By Construction 27.3, the comparison map $\mu :\ \mathbf{K} \to \mathbf{E}_J$ yields a mapping cone, which we denote by $\mathbf{E}$, and which is a multigraded free resolution of $S/(J+(m))$. Note that $\mathbf{E}$ has the basis of an Eliahou-Kervaire resolution. The differential on $\mathbf{E}$ is:

- d on $\mathbf{E}_J$
- $\partial - \mu$ on $\mathbf{K}$,

so the differential on $\mathbf{E}$ is d, as desired. Thus, $\mathbf{E}$ is the Eliahou-Kervaire resolution of $S/(J+(m))$.

The resolution $\mathbf{E}$ is minimal, since $d(\mathbf{E}) \subseteq (x_1,\dots,x_n)\mathbf{E}$. □

Example 28.11. Consider again Example 28.7. The ideal $B = (x^2, xy, xz, y^3)$ is Borel in $A = k[x,y,z]$. Set $J = (x^2, xy)$ and $P = (x^2, xy, xz)$.

The first step of the iterated mapping cones construction provides the Eliahou-Kervaire resolution of A/P. Since $(J : xz) = (x,y)$, we have the short exact sequence

$$0 \to\ A/(J:xz) = A/(x,y) \xrightarrow{xz} A/J \ \to\ A/P \to 0\,.$$

This yields the mapping cone

$$\begin{array}{ccccccc}
0 \to A(x^2yz) & \xrightarrow{\begin{pmatrix} -y \\ x \end{pmatrix}} & A(x^2z)\oplus A(xyz) & \xrightarrow{(x\ y)} & A(xz) & \to & A/(x,y)\,(xz) \to 0 \\
\downarrow {\scriptstyle -z} & & \downarrow \begin{pmatrix} z & 0 \\ 0 & z \end{pmatrix} & & \downarrow {\scriptstyle xz} & & \downarrow {\scriptstyle xz} \\
0 \to A(x^2y) & \xrightarrow{\begin{pmatrix} y \\ -x \end{pmatrix}} & A(x^2)\oplus A(xy) & \xrightarrow{(x^2\ xy)} & A & \to & A/J \to 0\,.
\end{array}$$

From this mapping cone we obtain the following Eliahou-Kervaire resolution of A/P:

$$0 \rightarrow A(x^2yz) \xrightarrow{\begin{pmatrix} -z \\ y \\ -x \end{pmatrix}} A(x^2y) \oplus A(x^2z) \oplus A(xyz)$$

$$\xrightarrow{\begin{pmatrix} y & z & 0 \\ -x & 0 & z \\ 0 & -x & -y \end{pmatrix}} A(x^2) \oplus A(xy) \oplus A(xz) \xrightarrow{(x^2\ xy\ xz)} A\,.$$

The next step of the iterated mapping cone construction provides the Eliahou-Kervaire resolution of $A/(x^2, xy, xz, y^3)$. Since $(P : y^3) = (x)$, we have the short exact sequence

$$0 \rightarrow\ A/(P : y^3) = A/(x) \xrightarrow{y^3} A/P\ \rightarrow\ A/B \rightarrow 0\,.$$

This yields the mapping cone

$$\begin{array}{ccccc} 0 \rightarrow & A(xy^3) & \xrightarrow{x} & A(y^3) \rightarrow \frac{A}{(x)}(y^3) \rightarrow 0 & \\ & \downarrow \begin{pmatrix} 0 \\ y^2 \\ 0 \end{pmatrix} & & \downarrow y^3 \qquad \downarrow y^3 & \end{array}$$

$$\cdots \xrightarrow{\begin{pmatrix} y & z & 0 \\ -x & 0 & z \\ 0 & -x & -y \end{pmatrix}} A(x^2) \oplus A(xy) \oplus A(xz) \xrightarrow{(x^2\ xy\ xz)} A \rightarrow A/P \rightarrow 0\,.$$

From this mapping cone we obtain the Eliahou-Kervaire resolution of

A/B as follows

$$0 \to A(x^2yz) \xrightarrow{\begin{pmatrix} -z \\ y \\ -x \\ 0 \end{pmatrix}} A(x^2y) \oplus A(x^2z) \oplus A(xyz) \oplus A(xy^3)$$

$$\xrightarrow{\begin{pmatrix} y & z & 0 & 0 \\ -x & 0 & z & y^2 \\ 0 & -x & -y & 0 \\ 0 & 0 & 0 & -x \end{pmatrix}} A(x^2) \oplus A(xy) \oplus A(xz) \oplus A(y^3)$$

$$\xrightarrow{(x^2 \ xy \ \ xz \ y^3)} A\,.$$

It coincides with the resolution described in Example 28.7.

Corollary 28.12. [Eliahou-Kervaire] *Let M be a Borel ideal minimally generated by monomials $m_1, \ldots, m_r$. If P is an associated prime of M, then $P = (x_1, \ldots, x_q)$ for some $q \geq 1$. We can express the codimension, regularity, the projective dimension, and the Betti numbers as follows:*

$$\operatorname{codim}(M) = \max\{j \mid \text{a power of } x_j \text{ is in } M\}$$

$$\operatorname{reg}(M) = \text{highest degree of a minimal monomial generator of } M$$

$$\operatorname{pd}(M) = \max\{\max(m_i) - 1 \mid 1 \leq i \leq r\}$$

$$b^S_{p,p+q}(M) = \sum_{\deg(m_i)=q,\ 1 \leq i \leq r} \binom{\max(m_i) - 1}{p}$$

$$b^S_p(M) = \sum_{i=1}^{r} \binom{\max(m_i) - 1}{p}.$$

Proof. Let P be an associated prime of M. Since M is a monomial ideal, P is generated by some of the variables. Let $q = \max\{j|x_j \in P\}$. Since P is an associated prime, we have that $P = (M : g)$ for some polynomial g. As $x_q g \in M$, the Borel property implies that $x_j g \in M$ for every $j \leq q$. Hence $x_j \in P$. Therefore, $P = (x_1, \ldots, x_q)$.

If a monomial $w \in M$, then by the Borel property it follows that a power of $x_{\min(w)}$ is in M. Hence $x_{\min(w)}$ is contained in every prime ideal that contains M. Therefore, the minimal prime ideal containing M is

$$(\{x_j \mid \text{ a power of } x_j \text{ is in } M\}) = (x_1, \ldots, x_t),$$

where $t = \max\{j \mid \text{a power of } x_j \text{ is in } M\}$. Hence, $\dim(S/M) = n - t$.

In order to prove the remaining formulas, note that the Eliahou-Kervaire minimal free resolution of M has basis

$$\left\{ (m_i; j_1, \ldots, j_p) \mid 1 \le j_1 < \ldots < j_p < \max(m_i),\ 1 \le i \le r \right\}$$

in homological degree p, by Construction 28.6. For a fixed m_i, there are $\binom{\max(m_i)-1}{p}$ choices for the sequence $j_1, \ldots, j_p$. □

Corollary 28.13. *If M is a Borel ideal generated by monomials of degree q, then its minimal graded free resolution is q-linear.*

In the rest of this section we discuss the class of p-Borel ideals which arises from the study of generic initial ideals when the characteristic of the ground field k is $p > 0$.

Borel ideals are also called 0-***Borel fixed*** ideals (or just 0-***Borel***).

Let p be a prime number. For two natural numbers a and b, we define that $a \prec_p b$ if each digit in the base-p expansion of a is $\le$ the corresponding digit in the base-p expansion of b. A monomial ideal T is called p-***Borel fixed*** (or just p-***Borel***) if for each minimal monomial generator m of T and each x_j that divides m, the following property holds: if x_j^t is the highest power of x_j that divides m, then $\left(\frac{x_i}{x_j}\right)^s m \in T$ for each $s \prec_p t$ and $i \le j$.

Example 28.14. For any prime number p, the ideal $(x_1^p, \ldots, x_n^p)$ is p-Borel.

The interest in studying such special monomial ideals comes

from the following result, cf. [Eisenbud, 15.20 and 15.23] and Theorem 18.10.

Theorem 28.15. *Let p be a prime number, and* $\operatorname{char}(k) = p$. *The generic initial ideal (with respect to any monomial order) of a graded ideal J in S is p-Borel. The regularity of J is equal to the regularity of its generic initial ideal with respect to revlex order.*

The following problems are open.

Problems 28.16. (folklore)

(1) *What is the regularity of a p-Borel ideal?*
(2) *Can the Betti numbers of a p-Borel ideal depend on the characteristic of k?*
(3) *Find upper (and lower) bounds on the regularity of a p-Borel ideal.*
(4) *Find upper (and lower) bounds on the Betti numbers of a p-Borel ideal.*
(5) *Describe the minimal free resolution of a p-Borel ideal for some classes of ideals.*

The following example shows that the mapping cone construction and the Eliahou-Kervaire resolution are not applicable.

Example 28.17. [Peeva-Stillman 2] Let $A = k[x, y, z]$. The ideals

$$M = (x^{11},\, x^{10}y,\, x^9y^2,\, x^8y^3,\, x^7y^4,\, x^6y^5,\, x^9z^2,\, x^5y^4z^2,\, x^7z^4)$$
$$N = (M,\, x^6yz^4)$$

are 2-Borel. If the mapping cone construction or the Eliahou-Kervaire resolution worked, then the Betti numbers of A/N would have been greater than those of A/M. However,

$$b_{2,15}(A/M) = 2 \quad \text{and} \quad b_{2,15}(A/N) = 1\,.$$

29 Applications of Eliahou-Kervaire's resolution

In this section we give applications of the Eliahou-Kervaire resolution.

The regularity of a graded ideal is equal to the regularity of its generic initial ideal with respect to the revlex order by Theorem 18.10. Combining this with Theorem 28.4 and Corollary 28.12 yields the following useful result.

Theorem 29.1. *Suppose that* $\mathrm{char}(k) = 0$. *The regularity of a graded ideal is equal to the highest degree of a minimal monomial generator of its generic initial ideal with respect to the revlex order.*

We will prove that the Regularity Conjecture 18.14 holds for Cohen-Macaulay ideals.

Theorem 29.2. *Suppose that* $\mathrm{char}(k) = 0$. *If* $P \subseteq \mathbf{m}^2$ *is a Cohen-Macaulay graded ideal in* S, *then*

$$\mathrm{reg}(P) \leq \mathrm{mult}(S/P) - \mathrm{codim}(P) + 1\,,$$

where $\mathrm{mult}(S/P)$ *is the multiplicity of* S/P.

The Cohen-Macaulay case is proved in [Eisenbud-Goto]. We present a different proof from [Peeva-Stillman 2].

Proof. First, we reduce to the case when the ideal is Borel. Denote by J the generic initial ideal of P with respect to the revlex order. By Theorem 29.1 it follows that $\mathrm{reg}(P)$ equals to the highest degree of a minimal monomial generator of J.

By Proposition 20.1 and since the coordinates (variables) are generic, it follows that we can choose a regular sequence on S/P that consists of some of the variables. Theorem 15.5 implies that the last $\dim(S/J)$ variables form a regular sequence in S/J. Hence the ideal J is Cohen-Macaulay. Therefore, in order to prove the desired inequality it suffices to show that

$$\mathrm{reg}(J) \leq \mathrm{mult}(S/J) - \mathrm{codim}(J) + 1$$

for a Cohen-Macaulay Borel ideal J.

Next, we reduce to the case when the ideal is artinian Borel. Set

$$i = \max\{j \mid \text{ a power of } x_j \text{ is in } J\}\,.$$

We have that $\operatorname{codim}(J) = i$ by Corollary 28.12. Since the last $n-i$ variables form a regular sequence in S/J, it follows that none of the variables $x_{i+1}, \ldots, x_n$ appears in the minimal monomial generators of J. Set $\widetilde{S} = S/(x_{i+1}, \ldots, x_n)$ and $\widetilde{J} = J \otimes \widetilde{S}$. The ideal $\widetilde{J}$ is artinian and Borel. The multiplicity of J is equal to $\operatorname{length}(\widetilde{S}/\widetilde{J})$.

Let x_i^p be a minimal monomial generator of J. It follows that p is the highest degree of a minimal monomial generator of J. Hence, $\operatorname{reg}(J) = p$ by Corollary 28.12.

Therefore, the inequality

$$\operatorname{reg}(J) \leq \operatorname{mult}(S/J) - \operatorname{codim}(J) + 1$$

is equivalent to $p \leq \operatorname{length}(\widetilde{S}/\widetilde{J}) - i + 1$, and to

$$p + i - 1 \leq \operatorname{length}(\widetilde{S}/\widetilde{J})\,.$$

We will prove that the above inequality holds. For $0 \leq j < p$, the monomial x_i^j is not in $\widetilde{J}_j$. In addition, the monomials $x_1, \ldots, x_{i-1}$ are not in $\widetilde{J}_1$. Hence,

$$p + (i-1) \leq \operatorname{length}(\widetilde{S}/\widetilde{J})\,.$$

□

Recall from Section 19 that for a $p \geq 0$ we can consider the truncation $I_{\geq p} = \oplus_{i \geq p} I_i$. We present a second proof of Theorem 19.7 in the case when $\operatorname{char}(k) = 0$.

Proof of Theorem 19.7. Denote by J the generic initial ideal of I with respect to the revlex order. Then $J_{\geq p}$ is the generic initial ideal of $I_{\geq p}$. By Theorem 29.1 it follows that $\operatorname{reg}(I_{\geq p})$ equals to the highest degree of a minimal monomial generator of $J_{\geq p}$.

By Theorem 29.1 it also follows that J has no minimal monomial generators of degree higher than $\operatorname{reg}(I)$. Therefore, all the minimal monomial generators of $J_{\geq p}$ are in degree p.

Hence, $\operatorname{reg}(I_{\geq p}) = p$. □

Next, we give a useful Gröbner basis criterion.

Green's Crystallization Principle 29.3. [Green] *Let J be a graded ideal and $\prec$ be a monomial order such that the initial ideal $\text{in}_\prec(J)$ is Borel. Suppose that J is generated in degrees $\leq p$. Let $q \geq p$ be such that $\text{in}_\prec(J)$ has no minimal monomial generators in degree $q+1$. Then $\text{in}_\prec(J)$ is generated in degrees $\leq q$.*

Proof. Let $\mathcal{G}$ be a set of homogeneous polynomials in J such that $\text{in}_\prec(\mathcal{G})$ is the set of minimal monomial generators of $\text{in}_\prec(J)$ of degrees $\leq q$. We will show that $\mathcal{G}$ is a Gröbner basis.

We have to check all s-pairs of polynomials in $\mathcal{G}$ that stem from first syzygies of $\left(\text{in}_\prec(J)\right)_{\leq q}$. By Corollary 28.12 it follows that every such s-pair has degree $\leq q+1$. Since $\text{in}_\prec(J)$ has no minimal monomial generators in degree $q+1$, it follows that all s-pairs reduce to zero. Therefore $\mathcal{G}$ is a Gröbner basis. □

Our last application is about Stillman's Problem 15.8. Caviglia showed that the problem is equivalent to the following problem on regularity.

Problem 29.4. *Assume* $\text{char}(k) = 0$. *Fix a sequence of natural numbers $a_1, \ldots, a_s$. Does there exist a number q, such that* $\text{reg}_T(T/J) \leq q$ *if T is a polynomial ring and J is a graded ideal with a minimal system of homogeneous generators of degrees $a_1, \ldots, a_s$? Note that the number of variables in the polynomial ring T is not fixed.*

Theorem 29.5. (Caviglia) *Suppose* $\text{char}(k) = 0$. *Fix a sequence of natural numbers $a_1 \leq \ldots \leq a_s$. The following are equivalent.*

(1) *There exists a number p, such that* $\text{pd}_T(T/J) \leq p$ *if T is a polynomial ring and J is a graded ideal with a minimal system of homogeneous generators of degrees $a_1, \ldots, a_s$.*
(2) *There exists a number q, such that* $\text{reg}_T(T/J) \leq q$ *if T is a polynomial ring and J is a graded ideal with a minimal system of homogeneous generators of degrees $a_1, \ldots, a_s$.*

Proof. First, we will show that (1) implies (2). Let J be an ideal with a minimal system of homogeneous generators of degrees $a_1 \leq \ldots \leq a_s$. Suppose that J is an ideal in a polynomial ring T with u variables.

By the Auslander-Buchsbaum Formula, we have that

$$\operatorname{depth}(T/J) = u - \operatorname{pd}(T/J) \geq u - p\,.$$

By Proposition 20.1, there exist $u - p$ linear forms $g_1, \ldots, g_{u-p}$ that are a regular sequence. By Theorem 20.3,

$$\operatorname{reg}(T/J) = \operatorname{reg}\big(T/(J + (g_1, \ldots, g_{u-p}))\big) = \operatorname{reg}(Q/J_Q),$$

where $Q = T/(g_1, \ldots, g_{u-p})$ is a polynomial ring in p variables and J_Q is the image of J in Q. Since the number of variables is fixed, we can apply Theorem 18.11 which says that there exists an upper bound on $\operatorname{reg}(Q/J_Q)$.

Next, we will show that (2) implies (1). Suppose that we have an upper bound q such that $\operatorname{reg}(T/J) \leq q$ for every ideal J generated by forms $f_1, \ldots, f_s$ of degrees $a_1, \ldots, a_s$. Consider the generic initial ideal $\operatorname{gin}(J)$ of J with respect to the revlex order. We have the following inequalities

$$\begin{aligned}
\operatorname{pd}(S/J) &\leq \operatorname{pd}(S/\operatorname{gin}(J)) \quad \textit{by Theorem } 22.9 \\
&= \big(\text{the number of variables appearing in the minimal monomial} \\
&\qquad \text{generators of } \operatorname{gin}(J)\big) \quad \textit{by Theorem } 28.12 \\
&\leq \big(\text{the sum of the degrees of the minimal monomial} \\
&\qquad \text{generators of } \operatorname{gin}(J)\big) \\
&\leq \big(\text{the number of the minimal monomial generators of } \operatorname{gin}(J)\big) \\
&\qquad \cdot \big(\text{maximal degree of a minimal monomial generator of } \operatorname{gin}(J)\big) \\
&= \big(\text{the number of the minimal monomial generators of } \operatorname{gin}(J)\big) \\
&\qquad \cdot \operatorname{reg}(\operatorname{gin}(J)) \\
&= \big(\text{the number of the minimal monomial generators of } \operatorname{gin}(J)\big) \\
&\qquad \cdot \operatorname{reg}(J) \qquad \textit{by Theorem } 18.10 \\
&\leq \big(\text{the number of the minimal monomial generators of } \operatorname{gin}(J)\big)\, q\,.
\end{aligned}$$

So, it suffices to prove that the number of the minimal monomial generators of $\mathrm{gin}(J)$ is bounded above. We will show that the cardinality of a minimal generic Gröbner basis of J with respect to revlex is bounded above in terms of s and q.

Suppose that we work in generic coordinates. Consider the steps in the process of constructing a minimal Gröbner basis of J. We start with the generators $f_1, \dots, f_s$ of J. At each step, we add at most one remainder for each s-pair of elements from the previous step. At the first step we add $v \leq \binom{s}{2}$ new elements. At the next step we add at most

$$\binom{s+v}{2} - v = \binom{v}{2} + vs$$

elements. At any step the number of elements, that we adjoin, is bounded by a polynomial function of s (note that the function does not depend on the number of variables). Furthermore, note that the degree of the remainder cannot be equal to the degree of one of the elements in the s-pair since we are constructing a minimal Gröbner basis. Hence, the set is enlarged only if the remainder has strictly bigger degree than each of the elements in the s-pair. Since q is an upper bound on the degree of the elements that we adjoin, it follows that there exists an upper bound (in terms of s and q) of the number of the elements that we adjoin. □

30 Double complexes

A ***double complex*** of finitely generated R-modules is a doubly indexed module $\mathbf{G} = \oplus_{p,q} G_{p,q}$ with ***horizontal differential*** d' and ***vertical differential*** d'', that are homomorphisms

$$d'_{pq} : G_{p,q} \to G_{p-1,q} \qquad d''_{pq} : G_{p,q} \to G_{p,q-1}$$

such that

$$d''d'' = 0, \qquad d'd'' = d''d', \qquad d''d'' = 0\,.$$

The following commutative diagram is helpful:

$$
\begin{array}{lccccccc}
 & & & p\text{'th column} & & & & \\
 & & & \vdots & & \vdots & & \\
 & & & \downarrow & & \downarrow & & \\
q\text{'th row} & \dots & \longrightarrow & G_{p,q} & \xrightarrow{d'} & G_{p-1,q} & \longrightarrow & \dots \\
 & & & d'' \downarrow & & \downarrow d'' & & \\
 & \dots & \longrightarrow & G_{p,q-1} & \xrightarrow{d'} & G_{p-1,q-1} & \longrightarrow & \dots \\
 & & & \downarrow & & \downarrow & & \\
 & & & \vdots & & \vdots & &
\end{array}
$$

The q'th row in the diagram is a complex denoted $\mathbf{G}_{*,q}$, and the p'th column is a complex denoted $\mathbf{G}_{p,*}$.

The ***total complex*** $\mathbf{W}$ of $\mathbf{G}$ is defined as follows: for $i \in \mathbf{Z}$ we set $W_i = \oplus_{p+q=i} G_{p,q}$ and the differential d acts as $d(x) = d'(x) + (-1)^p d''(x)$ for $x \in G_{p,q}$.

Exercise 30.1. *Show that* $\mathbf{W}$ *is a complex.*

31 DG algebras

We will discuss how to consider a free resolution as an algebra. The existence of a differential graded algebra structure on a minimal free resolution is usually a helpful tool for studying the properties of the resolution. The paper [Miller] is an overview of this topic.

Definition 31.1. Let I be a graded ideal in S. Let $(\mathbf{A}, d)$ be a graded resolution of the S-module S/I and let $A_0 = S$. Suppose that $\mathbf{A}$ is a graded k-algebra with multiplication $* : \mathbf{A} \times \mathbf{A} \to \mathbf{A}$ respecting the S-module structure (that is $(s_1 a_1) * (s_2 a_2) = s_1 s_2 (a_1 * a_2)$ for any $s_1, s_2 \in S$ and $a_1, a_2 \in \mathbf{A}$). Then $(\mathbf{A}, d, *)$ is a ***commutative associative differential graded*** S-***algebra*** (or a DG-algebra) if the following properties hold for any elements $\alpha, \beta, \gamma \in \mathbf{A}$ homogeneous with respect to the homological grading (denoted by hdeg) and internal grading (denoted by deg):

(1) the multiplication respects the homological grading:

$$\mathrm{hdeg}(\alpha * \beta) = \mathrm{hdeg}(\alpha) + \mathrm{hdeg}(\beta)$$

(2) the multiplication respects the grading:

$$\deg(\alpha * \beta) = \deg(\alpha) + \deg(\beta)$$

(3) the product is skew commutative:

$$\alpha * \beta = (-1)^{\mathrm{hdeg}(\alpha)\mathrm{hdeg}(\beta)} \beta * \alpha$$

(4) the product is associative:

$$(\alpha * \beta) * \gamma = \alpha * (\beta * \gamma)$$

(5) the Leibniz rule holds:

$$d(\alpha * \beta) = d(\alpha) * \beta + (-1)^{\mathrm{hdeg}(\alpha)} \alpha * d(\beta).$$

In this case we say that **A** has a ***graded DG-algebra structure*** or that **A** is an ***associative commutative differential graded algebra.*** In [Buchsbaum-Eisenbud] it was addressed that the existence of a graded DG-algebra structure on a minimal free resolution is a helpful tool for investigating the properties of the resolution. This paper inspired further study of graded DG-algebra structures.

Theorem 31.2. *The Koszul minimal free resolution* **K** *of* k *over* S *is an associative commutative differential graded algebra.*

Proof. Consider the multiplication induced by the underlying exterior algebra E and denoted by $\wedge$. The multiplication respects the homological and internal gradings, is skew commutative, and is associative. We will show that the Leibniz rule holds. It suffices to check that the Leibniz rule holds for two elements of the form $u = e_{j_1} \wedge \cdots \wedge e_{j_i}$ and $v = e_{q_1} \wedge \cdots \wedge e_{q_r}$. We have that

$$\begin{aligned} d(u \wedge v) = & \sum\nolimits_{1 \le p \le i} (-1)^{p+1} \cdot x_{j_p} \cdot e_{j_1} \wedge \cdots \wedge \widehat{e}_{j_p} \wedge \cdots \wedge e_{j_i} \wedge v \\ & + \sum\nolimits_{1 \le p \le r} (-1)^{\mathrm{hdeg}(u)+p+1} \cdot x_{q_p} \cdot u \wedge e_{q_1} \wedge \cdots \wedge \widehat{e}_{q_p} \wedge \cdots \wedge e_{q_r} \\ & = d(u) \wedge v + (-1)^{\mathrm{hdeg}(u)} u \wedge d(v)\,. \end{aligned}$$

Thus, $\mathbf{K}$ is a DG-algebra. □

In the same way, we have that the Koszul complex $\mathbf{K}(f_1, \dots, f_r)$ is a DG-algebra for any sequence of homogeneous elements $f_1, \dots, f_r$.

Let I be a graded ideal in S. The complex $\mathrm{Tor}_*^S(S/I, k) \cong \mathrm{H}_*(S/I \otimes \mathbf{K})$ has a graded DG-algebra structure induced by the multiplication on $\mathbf{K}$ discussed in Theorem 31.2. We call it the ***Tor-algebra*** of S/I.

Proposition 31.3. *Taylor's resolution is an associative commutative differential graded algebra.*

Proof. Let M be a monomial ideal. Use the notation in Construction 26.5. For $\tau = \{j_1 < \dots < j_p\}$ set $e_\tau = e_{j_1} \wedge \dots \wedge e_{j_p}$. Furthermore, set $\mathrm{mdeg}(\tau) = \mathrm{lcm}(m_{j_1}, \dots, m_{j_p})$ (here we use mdeg to denote multidegree). We define multiplication on Taylor's resolution $\mathbf{T}_M$ by

$$e_\tau * e_{\tau'} = \begin{cases} \mathrm{sign}(\tau, \tau') \dfrac{\mathrm{mdeg}(\tau)\mathrm{mdeg}(\tau')}{\mathrm{mdeg}(\tau \cup \tau')} \, e_{\tau \cup \tau'} & \text{if } \tau \cap \tau' = \emptyset \\ 0 & \text{otherwise.} \end{cases} .$$

Here, if $\tau = \{i_1 < \dots < i_q\}$ and $\tau' = \{j_1 < \dots < j_p\}$, then $\mathrm{sign}(\tau, \tau')$ is the sign of the permutation which makes $i_1, \dots, i_q, j_1, \dots, j_p$ an increasing sequence. Straightforward verification shows that this product is graded, associative, skew commutative, and satisfies Leibniz's rule. □

Open Problems 31.4.

(1) (Buchsbaum-Eisenbud) *If a graded DG-algebra structure exists on the graded minimal free resolution of S/I, then is it unique?* (The answer is negative over a local ring.)

(2) (folklore) *Find minimal graded free resolutions which have a DG-algebra structure.*

Srinivasan has constructed a DG-algebra structure on the minimal free resolution of $S/(x_1, \dots, x_n)^p$ for $p \geq 2$. Peeva has constructed in a different way a DG-algebra structure on the Eliahou-Kervaire resolution. It is not known if the DG-algebra structures constructed by Peeva and Srinivasan coincide.

Counterexample 31.5. By [Buchsbaum-Eisenbud, Proposition 1.1], any resolution of a cyclic module admits a commutative differential graded S-algebra structure. The problem is that such structures are rarely associative. In [Avramov 2] a homological obstruction is provided for the existence of a graded DG-algebra structure and there exist examples in which the obstruction does not vanish. If $M = (x^2, xy^2z, y^2z^2, yz^2w, w^2)$, then no graded DG-algebra structure exists on the minimal free resolution of $k[x,y,z,w]/M$ (Backelin, see [Avramov 2, 5.2.3]). Also, by [Avramov 2, Theorem 2.3.1] if $M = (x_1^2,\ x_1x_2,\ x_2x_3,\ x_3x_4,\ x_4^2)$, then the minimal free resolution of $k[x_1,\dots,x_4]/M$ does not admit a DG-algebra structure.

32 Two resolutions

Let I be a graded ideal in S. In this short section we draw attention to two core resolutions which are closely related to the properties of S/I.

The first resolution which one can consider is the minimal graded free resolution $\mathbf{F}_{S/I}$ of S/I over S. This resolution starts with

$$\mathbf{F}_{S/I}: \quad \ldots \to S^p \xrightarrow{(f_1\, f_2 \ldots f_p)} S \to S/I \to 0\,,$$

where $f_1, f_2, \dots, f_p$ is a minimal system of homogeneous generators of I. This resolution is finite. Sometimes, in the literature, the minimal free resolution $\mathbf{F}_I$ of I is considered. Note that $\mathbf{F}_I$ coincides with $(\mathbf{F}_{S/I})_{\geq 1}$ (but the homological degree is shifted), so it starts with

$$\ldots \to S^p \to I \to 0\,.$$

The Betti numbers of S/I and those of I are related by

$$\mathrm{P}^S_{S/I}(t,z) = 1 + t\,\mathrm{P}^S_I(t,z)$$
$$\mathrm{reg}(I) = \mathrm{reg}(S/I) + 1\,.$$

Another well-studied resolution is the minimal graded free resolution $\mathbf{F}_k$ of k over the ring S/I. By Theorem 15.10 that resolution is infinite unless I is generated by linear forms. It starts with

$$\mathbf{F}_k : \quad \ldots \to (S/I)^n \xrightarrow{(x_1\, x_2\, \ldots\, x_n)} S/I \to k \to 0\,.$$

The resolutions $\mathbf{F}_{S/I}$ and $\mathbf{F}_k$ are related via a spectral sequence, cf. [Avramov, Proposition 3.2.4]. We present a beautiful structure theorem for $\mathbf{F}_k$ in a special case.

Theorem 32.1. [Tate] *Suppose that* $\mathrm{char}(k) = 0$. *Let I be generated by a homogeneous regular sequence $f_1, \ldots, f_q$. Write $f_i = \sum_{1\le j}^{n} g_{ij}x_j$ for $1 \le i \le q$ with some $g_{ij} \in S$. Let*

$$\mathbf{F}_k = S/I\,\langle e_1, \ldots, e_n, y_1, \ldots, y_q\rangle$$

be the skew-commutative algebra generated by exterior variables $e_1, \ldots,$ e_n of homological degree 1 and polynomial variables $y_1, \ldots, y_q$ of homological degree 2; skew-commutative means that

$$uv = (-1)^{\deg(u)\deg(v)}vu$$

$$w^2 = 0 \quad \textit{if } \deg(w) \textit{ is odd.}$$

The module $\mathbf{F}_k$ is equipped with differential d which acts by

$$d(e_i) = x_i \quad \textit{for } 1 \le i \le n$$

$$d(y_i) = \sum_{1\le j\le n} g_{ij}e_j \quad \textit{for } 1 \le i \le q$$

$$d(uv) = d(u)v + (-1)^{\deg(u)}ud(v)\,.$$

The minimal free resolution of k over S/I is $\mathbf{F}_k$. In particular, the Poincarè series of k is

$$\mathrm{P}_k^{S/I}(t) = \frac{(1+t)^n}{(1-t^2)^q}\,.$$

There exists a simply structured free resolution of k over R, but usually it is highly non-minimal. We describe that resolution in the next construction.

The Bar resolution 32.2. We will follow [Mac Lane, Ch. 10, Sec. 2] to construct and prove the Bar resolution (note that the construction in [Mac Lane] is much more general). Let $\widetilde{R}$ be the cokernel of the inclusion of vector spaces $k \to R$. For $i \geq 0$ set $B_i = R \otimes \widetilde{R} \otimes \ldots \otimes \widetilde{R}$, where we have i factors $\widetilde{R}$. The left factor R gives B_i a structure of a free R-module. Fix a basis Λ of R over k such that $1 \in \Lambda$. Let $\lambda \in R$ and $\lambda_1, \ldots, \lambda_i \in \Lambda$. We denote by $\lambda[\lambda_1 \mid \ldots \mid \lambda_i]$ the element in B_i replacing $\otimes$ by a vertical bar; in particular, the elements of B_0 are written as $\lambda[\,]$. Note that $\lambda[\lambda_1 \mid \ldots \mid \lambda_i] = 0$ if some $\lambda_j \in k$. Consider the sequence

$$\mathbf{B}: \quad \ldots \to B_i \to B_{i-1} \to \ldots \to B_0 = R \to k \to 0$$

with differential d defined by

$$d_i(\lambda[\lambda_1 \mid \ldots \mid \lambda_i]) = \lambda\lambda_1[\lambda_2 \mid \ldots \mid \lambda_i] + \sum_{1 \leq j \leq i-1} (-1)^j \lambda[\lambda_1 \mid \ldots \mid \lambda_j\lambda_{j+1} \mid \ldots \mid \lambda_i].$$

The differential is well defined since if $\lambda_j = 1$ for some j, then the terms

$$(-1)^j \lambda[\lambda_1 \mid \ldots \mid \lambda_j\lambda_{j+1} \mid \ldots \mid \lambda_i]$$
$$(-1)^{j-1} \lambda[\lambda_1 \mid \ldots \mid \lambda_{j-1}\lambda_j \mid \ldots \mid \lambda_i]$$

cancell and all other terms vanish.

We will show that $\mathbf{B}$ is a free resolution of k over R. Straightforward computation shows that $d^2 = 0$, so $\mathbf{B}$ is a complex.

We construct a map $h : \mathbf{B} \to \mathbf{B}$ by

$$\begin{aligned} h_i : B_i &\to B_{i+1} \\ \lambda[\lambda_1 \mid \ldots \mid \lambda_i] &\mapsto 1[\lambda\lambda_1 \mid \ldots \mid \lambda_i], \\ h_{-1} : k &\to B_0 \\ 1 &\mapsto 1\,[\,]. \end{aligned}$$

Straightforward computation shows that $dh + hd = \mathrm{id}$, so h is a k-homotopy. By Thoerem 6.4 we conclude that $\mathbf{B}$ is a free resolution of k. The resolution $\mathbf{B}$ is called the ***Bar resolution*** of k.

33 Betti numbers of infinite free resolutions

In this short section we provide a glimpse into the world of *infinite* graded free resolutions. Expository lectures in this area are given in [Avramov].

First, we give a simple example of an *infinite* free resolution:

$$\ldots \longrightarrow A \xrightarrow{x} A \xrightarrow{x} A \xrightarrow{x} A \qquad \text{for } A = k[x]/(x^2).$$

The graded form of this resolution is

$$\ldots \longrightarrow A(-3) \xrightarrow{x} A(-2) \xrightarrow{x} A(-1) \xrightarrow{x} A\,.$$

The structure of infinite free resolutions can be quite complex. For example, by [Anick] there exist examples of rings R such that the Poincarè series $\sum_{i\geq 0} b_i^R(k)t^i$ is irrational. It is natural to ask some basic questions about the growth of the Betti numbers in an infinite minimal free resolution. We list some such questions, cf. also [Avramov].

Problem 33.1. (Avramov) *What type of growth can the sequence of Betti numbers of a graded finitely generated R-module have? Are polynomial and exponential growth the only possibilities?*

Problem 33.2. (Avramov) *Is it true that the Betti numbers of every finitely generated graded R-module are eventually non-decreasing?*

Problem 33.3. (Ramras) *Is it true that if the Betti numbers of a graded finitely generated R-module are bounded, then they are eventually constant?*

A finitely generated graded R-module U is ***periodic*** if there exists a p such that $\mathrm{Syz}_p^R(U) \cong U$.

Problem 33.4. (Avramov) *Does there exist a periodic module with non-constant Betti numbers?*

Open-Ended Problem 33.5. (folklore) *Find classes of graded rings over which every finitely generated graded module has a rational Poincarè series. What can be said about the denominator in such cases?*

Complete intersections are the main class of quotient rings over which the answers are known.

Theorem 33.6. [Avramov-Gasharov-Peeva] and [Eisenbud 3] *Let I be generated by a homogeneous regular sequence. Let U be a graded finitely generated module over $R = S/I$. Then*

(1) *The Poincarè series $\sum_{i\geq 0} b_i^R(U)t^i$ is rational.*
(2) *The Betti numbers of U are eventually non-decreasing.*
(3) *There exist two polynomials $f, g \in \mathbf{Q}[t]$ of the same degree and with the same leading coefficient such that $b_{2i}^R(U) = f(i)$ and $b_{2i+1}^R(U) = g(i)$ for all $i \gg 0$.*
(4) *If the Betti numbers of U are bounded, then they are eventually constant.*
(5) *If the Betti numbers of U are bounded then U is eventually periodic of period 2, that is, there exists a j such that $\mathrm{Syz}_j^R(U) \cong \mathrm{Syz}_{j+2}^R(U)$.*

34 Koszul rings

Since infinite minimal free resolutions are often very complicated, it makes sense to study some special classes of rings over which resolutions are nice. Koszul rings are such a class and they appear in many interesting and important situations in several mathematical fields. The book [Polishchuk-Positselski] is focused on Koszul rings.

Throughout this section I is a graded ideal in S. Recall that $R = S/I$ is called Koszul if the minimal free resolution of k over R is linear, that is, the entries in the matrices of the differential are linear forms (sometimes we say that I is Koszul in this case). This happens if and only if $b_{i,j}^R(k)$ vanishes for $i \neq j$.

Example 34.1. This is a simple example of a Koszul ring. The ring

$A = k[x]/x^2$ is Koszul, since k has the linear free resolution

$$\ldots \to A \xrightarrow{x} A \to \ldots \xrightarrow{x} A \to k \to 0\,.$$

Exercise 34.2. *If S/I is Koszul, then I is generated by quadrics and linear forms.*

Using the methods in Section 16, one can prove the following result.

Theorem 34.3. *Suppose that the ring S/I is Koszul. The Poincaré series of k is expressed in terms of the Hilbert series of S/I as follows*

$$\sum_{i\geq 0} b_i^{S/I}(k)t^i = \frac{1}{\sum_{i\geq 0} \dim_k (S/I)_i\,(-t)^i}\,.$$

Exercise 34.4. *Prove the above theorem.*

Suppose that S/I is Koszul. The minimal free resolution of k is the generalized Koszul complex, introduced in [Priddy], cf. [Eisenbud, 17.22].

Koszul algebras have extraordinary homological properties and appear in many interesting contexts. The following problem is of interest.

Open-Ended Problem 34.5. (folklore) *Find classes of Koszul rings.*

The following nice recult is proved in [Avramov-Eisenbud].

Theorem 34.6. *If S/I is Koszul, then every graded finitely generated S/I-module has finite regularity over the quotient ring S/I.*

Example 34.7. [Roos] It would have been very useful if one could recognize whether a ring is Koszul or not by just looking at the beginning of the infinite minimal free resolution of k. Unfortunately, this is impossible. Roos constructed and proved the following example. Let

$A = k[x,y,z,u,v,w]$. Choose a $2 \le q \in \mathbf{N}$. Let B be the ideal

$$B = \Big(x^2,\ xy,\ yz,\ z^2,\ zu,\ u^2,\ uv,\ vw,\ w^2,$$
$$xz + qzw - uw,\ zw + xu + (q-2)uw\Big).$$

Then

$$b^{A/B}_{i,j}(k) = 0 \quad \text{for } j \ne i \text{ and } i \le q$$
$$b^{A/B}_{q+1,q+2}(k) \ne 0.$$

The following technique using filtrations provides a way of proving that a ring is Koszul.

Definition 34.8. Fix a graded ideal I in S. Let $\mathcal{K}$ be a set of tuples $(L;l)$, where L is a linear ideal (that is, L is generated by linear forms) in $R = S/I$ and l is a linear form in L. Denote by $\bar{\mathcal{K}}$ the set of linear ideals appearing in the tuples in $\mathcal{K}$. A ***Koszul filtration*** of R is a set $\mathcal{K}$ such that the following conditions are satisfied:

(1) $(x_1,\dots,x_n) \in \bar{\mathcal{K}}$.

(2) If $(L;l) \in \mathcal{K}$ and $L \ne 0$, then there exists a proper subideal $M \subset L$ such that $L = (M,l)$, and $(M : l) \in \bar{\mathcal{K}}$, and $M \in \bar{\mathcal{K}}$.

Note that $\bar{\mathcal{K}}$ necessarily contains (0). In practice, (see the example below), it is more convenient to give a Koszul filtration by giving a list of the ideals $(M : l)$; given such a list one can easily recover the Koszul filtration since $(M : l)$ corresponds to $(L;l) = ((M,l);l)$.

Note that we do not assume that I is generated by quadrics.

Example 34.9. [Aramova-Herzog-Hibi] The computations in this example are made by computer. Let $S = k[x_1,\dots,x_8]$, and I be the kernel of the map

$$k[x_1,\dots,x_8] \longrightarrow k[t_1t_2, t_1t_3, t_1t_4, t_1t_5, t_2t_3, t_2t_4, t_2t_5, t_3t_4]$$

sending the variables $x_1,\dots,x_n$ to the monomials $t_1t_2, t_1t_3, t_1t_4, t_1t_5,$

$t_2t_3, t_2t_4, t_2t_5, t_3t_4$, respectively. The following is a Koszul filtration.

$$
\begin{aligned}
&(x_1,\dots,x_7) : x_8 = (x_1,\dots,x_7)\\
&(x_1,\dots,x_6) : x_7 = (x_1,\dots,x_6)\\
&(x_1,\dots,x_5) : x_6 = (x_1,\dots,x_5)\\
&(x_1,\dots,x_4) : x_5 = (x_1,\dots,x_4)\\
&(x_1,x_2,x_3) : x_4 = (x_1,x_2,x_3,x_5,x_6)
\end{aligned}
$$

$$
\begin{aligned}
&(x_1,x_2) : x_3 = (x_1,x_2,x_5)\\
&(x_1) : x_2 = (x_1,x_6)\\
&(0) : x_1 = 0\\
&(x_1,x_2,x_3,x_5) : x_6 = (x_1,\dots,x_5)\\
&(x_1,x_2,x_3) : x_5 = (x_1,\dots,x_4)\\
&(x_1,x_2) : x_5 = (x_1,\dots,x_4)\\
&(x_1) : x_6 = (x_1,x_2)\,.
\end{aligned}
$$

Proofs of Koszulness or rate involving filtrations have been used in various forms, for example by Eisenbud and Herzog, cf. [Eisenbud-Reeves-Totaro]. Later, the method was formally introduced in [Conca-Trung-Valla] with the name "Koszul filtration".

Theorem 34.10. *If there exists a Koszul filtration of R, then R is Koszul.*

Proof. We use the notation in 34.8. The short exact sequence

$$0 \to \frac{R}{(M:l)}(-1) \to R/M \to R/(M,l) = R/L \to 0$$

of homomorphisms of degree 0 yields the long exact sequence

$$\dots \to \operatorname{Tor}_i^R(R/M,k)_j \to \operatorname{Tor}_i^R(R/L,k)_j \to \operatorname{Tor}_{i-1}^R(R/(M:l),k)_{j-1} \\ \to \dots$$

for each j.

The proof is by induction on homological degree and induction on the number of minimal generators of the linear ideal L. By induction hypothesis, we have that $\operatorname{Tor}_i^R(R/M, k)_j = 0$ for $i \neq j$ since the ideal M has fewer minimal generators than L. On the other hand, $\operatorname{Tor}_{i-1}^R(R/(M : l), k)_{j-1} = 0$ for $i \neq j$, by induction on homological degree. Hence $\operatorname{Tor}_i^R(R/L, k)_j = 0$ for $i \neq j$. □

We call a Koszul filtration ***simple*** if the following two conditions are satisfied:

(1) All linear ideals appearing in the filtration are generated by variables.
(2) All linear forms appearing in the filtration are variables.

Every tuple in a simple filtration has the form

$$\big((x_{i_1}, \ldots, x_{i_j}) \, ; \, x_{i_j} \big) .$$

The filtration in Example 34.9 is simple.

Theorem 34.11. (Fröberg) *If I is generated by quadratic monomials, then R is Koszul.*

Proof. Take an ideal P in R generated by some variables, and let p be a variable. The ideal

$$(P : p)_R = \{ m \in R \,|\, mp \in P \}$$

is again generated by variables. Therefore, the set

$$\mathcal{K} = \{ \text{all ideals in } R \text{ generated by variables} \}$$

is a simple Koszul filtration. By Theorem 34.10, it follows that R is Koszul. □

This leads to the following criterion using Gröbner basis.

Theorem 34.12. *If I has a quadratic Gröbner basis then R is Koszul.*

Proof. The result follows from Theorem 34.11 and Theorem 22.9. □

Proposition 34.13. *Suppose that I is generated by quadrics, and $\prec$ is a monomial order such that there is no cubic in the Gröbner basis* *with respect to $\prec$. Then I has a quadratic Gröbner basis with respect to $\prec$, and S/I is Koszul.*

Proof. Let $\mathcal{G}$ be a set of quadrics that generate I. Every s-pair of two polynomials in $\mathcal{G}$ either has degree ≤ 3 or the two initial terms are relatively prime. In the latter case the s-pair reduces to zero. In the former case, the s-pair reduces to zero since there is no cubic in the Gröbner basis with respect to $\prec$. Hence $\mathcal{G}$ is a Gröbner basis. □

Exercise 34.14. *If h is a homogeneous non-zero divisor of degree* 1 *or* 2*, then either both R and R/h are Koszul or both R and R/h are not Koszul.* (Use Theorem 20.2 and Theorem 20.6.)

35 Rate

If the regularity of a minimal free resolution is infinite, then a meaningful numerical invariant is rate.

Definition 35.1. [Backelin 2] Let I be a graded ideal in S. Define

$$\begin{aligned} \mathrm{rate}_{S/I}(k) &= \sup\Big\{ \frac{p_i - 1}{i-1} \,\Big|\, i \geq 2 \Big\}, \\ \text{where } p_i &= \max\{\, j \mid b_{i,j}^{S/I}(k) \neq 0 \text{ or } j = i\}, \end{aligned}$$

called the ***rate*** of of k over S/I (or the rate of S/I). Note the following property.

Proposition 35.2. *S/I is Koszul if and only if* $\mathrm{rate}_{S/I}(k) = 1$.

Example 35.3. Let $A/B = k[x]/x^3$. The graded minimal free resolution of k is

$$\ldots \to A/B(-4) \xrightarrow{x} A/B(-3) \xrightarrow{x^2} A/B(-1) \xrightarrow{x} A/B \to k \to 0 .$$

Hence $b_{2j,3j}^{A/B}(k) = 1$ and $b_{2j+1,3j+1}^{A/B}(k) = 1$ for $j \geq 0$. Therefore, $\mathrm{reg}_{A/B}(k) = \infty$ and

$$\mathrm{rate}_{A/B}(k) = \sup\left\{ \frac{3j-1}{2j-1}, \frac{3j}{2j} \,\middle|\, j \geq 1 \right\} = 2\,.$$

Similarly to Koszul filtrations, described in Section 34, we consider s-filtrations.

Definition 35.4. Let I be a graded ideal in S. Let $\mathcal{K}$ be a set of tuples $(L; l)$, where L is an ideal in $R = S/I$ generated by homogeneous polynomials of degree $\leq s$, and l is a form of degree $\leq s$ in L. Denote by $\bar{\mathcal{K}}$ the set of ideals appearing in the tuples in $\mathcal{K}$. An ***s-filtration*** of R is a set $\mathcal{K}$ such that:

(1) $(x_1, \ldots, x_n) \in \bar{\mathcal{K}}$.

(2) If $(L; l) \in \mathcal{K}$ and $L \neq 0$, then there exists a proper subideal $M \subset L$ such that $L = (M, l)$, and $(M : l) \in \bar{\mathcal{K}}$, and $M \in \bar{\mathcal{K}}$.

In practice, (as for Koszul filtrations), it is more convenient to give a filtration by giving a list of the ideals $(M : l)$; given such a list one can easily recover the filtration since $(M : l)$ corresponds to $(L; l) = ((M, l); l)$. Note that we do not assume any bound on the degrees of the generators of I.

A minor modification in the argument in the proof of Theorem 34.10 leads to the next result.

Theorem 35.5. *If there exists an s-filtration of R, then* $\mathrm{rate}_R(k) \leq s$.

Theorem 35.6. [Eisenbud-Reeves-Totaro] *If M is an ideal minimally generated by monomials $m_1, \ldots, m_r$, then*

$$\mathrm{rate}_{S/M}(k) \leq \max\{\deg(m_i) \,|\, 1 \leq i \leq r\} - 1\,.$$

Proof. Set $s = \max\{\deg(m_i) \,|\, 1 \leq i \leq r\}$. Take a monomial ideal P in S/M generated by monomials of degrees $\leq s - 1$, and let p be a monomial of degree $\leq s - 1$. The ideal

$$(P : p)_{S/M} = \{m \in S/M \,|\, mp \in P\}$$

is generated by monomials of degrees $\leq s - 1$. Therefore, the set

$$\mathcal{K} = \{\text{ all monomial ideals in } S/M \text{ generated by monomials of degrees } \leq s - 1\,\}$$

is an $(s-1)$-filtration. By the above theorem, we obtain the desired inequality $\text{rate}_{S/M}(k) \le s-1$. □

This leads to the following criterion using Gröbner basis.

Theorem 35.7. *If a graded ideal I in S has an initial ideal generated up to degree s, then* $\text{rate}_{S/I}(k) \le s-1$.

Proof. Apply the above theorem and Theorem 22.9. □

Corollary 35.8. *If I is a graded ideal in S, then*

$$\text{rate}_{S/I}(k) < \infty\,.$$

Remark 35.9. Let I be a graded ideal in S. In view of the notation used in Betti diagrams (see Section 12) it makes sense to consider a modified version of rate. We consider the minimal slope of a diagonal line that goes through the slot $(0,0)$ and such that all Betti numbers below it vanish. For example, if S/I is Koszul and we consider the minimal free resolution of k, then the considered line is horizontal since all Betti numbers are placed in the top row in the Betti diagram. Define

$$\text{slope}_{S/I}(k) = \sup\Big\{\, \frac{p_i}{i} \,\Big|\, i \ge 1\Big\},$$

$$\text{where } p_i \;=\; \max\{\, j \mid b^{S/I}_{i,i+j}(k) \ne 0 \text{ or } j = 0\},$$

called the ***slope*** of k over S/I. Note that

$$S/I \text{ is Koszul} \iff \text{slope}_{S/I}(k) = 0\,.$$

36 Topological tools

In several situations (for example, for monomial and toric ideals; see Chapters III and IV) the Betti numbers can be computed using homology of simplicial complexes. In this section we review some techniques from Algebraic Topology on computing homology of simplicial complexes. We consider only finite simplicial complexes. Most of the material in this section is from [Björner].

Background on simplicial complexes 36.1. We consider only finite simplicial complexes. A ***simplicial complex*** Δ on ***vertices*** $\{v_1, \ldots, v_p\}$ is a collection of subsets, called ***faces***, such that $\tau \in \Delta$ whenever $\tau \subset \sigma \in \Delta$. We make no distinction between an abstract simplicial complex Δ and an arbitrary geometric realization of Δ. The maximal faces of Δ are called ***facets***. Note that Δ is determined by the list of its facets. We say that Δ is a ***simplex*** if it has one facet (that is, $\{v_1, \ldots, v_p\}$ is the facet).

In this book, usually the vertex set of a simplicial complex is either $\{1, \ldots, n\}$, or $\{x_1, \ldots, x_n\}$, or $\{m_1, \ldots, m_r\}$ where $m_1, \ldots, m_r$ are monomials in S.

The ***dimension*** of a face σ is $|\sigma| - 1$. The dimension of Δ is the maximum of the dimensions of its faces, or $-\infty$ if Δ is the ***void complex*** that has no faces. By convention, $\emptyset$ (that is, the simplicial complex with one face $\emptyset$) has dimension -1. A simplicial complex is called ***pure*** if all of its facets have the same dimension.

Throughout this section, Δ stands for a finite simplicial complex.

Background on posets 36.2. A ***poset*** P is a partially ordered set. A totally ordered subset is called a ***chain***. A ***bottom*** element x is an element such that $x \leq z$ for every $z \in P$; if the bottom element exists, then it is sometimes denoted $\hat{0}$. A ***top*** element y is an element such that $y \geq z$ for every $z \in P$; if the top element exists, then it is sometimes denoted $\hat{1}$. The poset P is ***bounded*** if it has a bottom element and a top element.

Let $x, y \in P$. An ***upper bound*** of x and y is an element $z \in P$ such that $z \geq x$ and $z \geq y$. A ***least upper bound*** of x and y is an upper bound w such that every upper bound z of x and y satisfies $z \geq w$; if a least upper bound exists, then it is unique and is denoted $x \vee y$ and is called the ***join*** of x and y. A ***lower bound*** of x and y is an element $v \in P$ such that $v \leq x$ and $v \leq y$. A ***greatest lower bound*** of x and y is a lower bound u such that every lower bound v of x and y satisfies $u \geq v$; if a greatest lower bound exists, then it is unique and is denoted $x \wedge y$ and is called the ***meet*** of x and y. A subset Q of P has an ***upper bound*** if there exists an upper bound in

P of all of its elements; Q has a ***lower bound*** if there exists a lower bound in P of all of its elements. We say that Q is ***bounded*** if it has either an upper bound or a lower bound in P. If the least upper bound of the elements in Q exists, then it is denoted $\vee Q$ and is called the ***join***. If the greatest lower bound of the elements in Q exists, then it is denoted $\wedge Q$ and is called the ***meet***.

For $x, y \in P$, we say that y ***covers*** x if $y > x$ and there exists no $z \in P$ such that $y > z > x$. The ***Hasse diagram*** of P is the graph with vertices the elements in P so that if y covers x then y is placed higher than x and they are connected with an edge.

The ***order complex*** $O(P)$ of P is the abstract simplicial complex whose vertices are the elements of P and whose faces are the chains in the poset. We implicitly think of a poset P as a topological space by considering its order complex $O(P)$.

For a simplicial complex Δ, the ***face poset*** $F(\Delta)$ has elements the non-empty faces of Δ ordered by inclusion.

Example 36.3. Consider the poset in Figure 1. Its order complex is also shown in Figure 1. Its facets are the triangles $\{a, d, e\}$, $\{b, d, e\}$, and $\{c, d, e\}$.

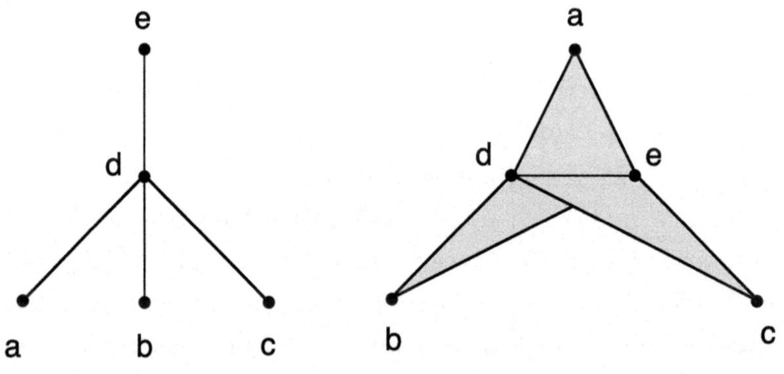

Figure 1.

Consider the simplicial complex in Figure 2. Its face poset is shown in Figure 2.

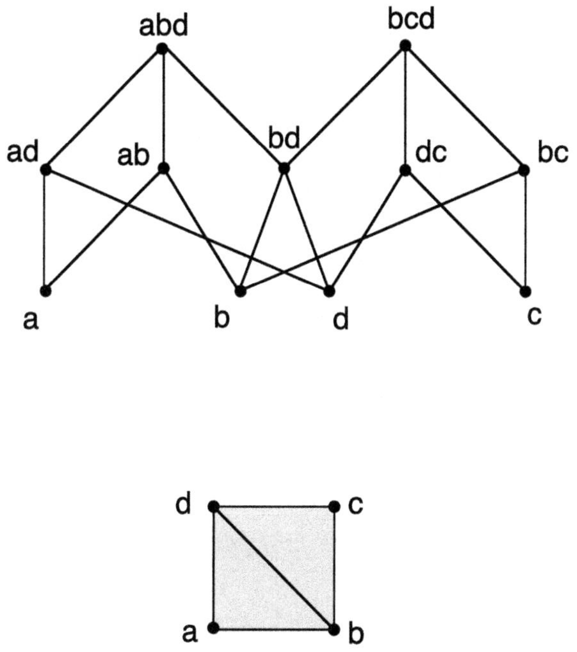

Figure 2.

A poset map $f : P \to Q$ is ***order-preserving*** if $f(x) \leq f(y)$ whenever $x \leq y$. The map is ***order-reversing*** if $f(x) \geq f(y)$ whenever $x \leq y$.

If Δ_1 and Δ_2 are two simplicial complexes, then a ***simplicial map*** $f : \Delta_1 \to \Delta_2$ maps the vertices $v_1, \ldots, v_n$ of Δ_1 to vertices of Δ_2 so that if $\{v_{i_1}, \ldots, v_{i_p}\}$ is a face of Δ_1 then $\{f(v_{i_1}), \ldots, f(v_{i_p})\}$ is a face of Δ_2. Such an f induces the order-preserving poset map $f : F(\Delta_1) \to F(\Delta_2)$, and then induces a simplicial map $f : O(F(\Delta_1)) \to O(F(\Delta_2))$.

Exercise 36.4. *Let Δ be a finite simplical complex and P be a finite poset. A simplicial map $f : \Delta \to O(P)$ sends the vertices of Δ to elements of P so that each face of Δ is mapped to a chain in P.*

Exercise 36.5. *An order-preserving or order-reversing map of finite posets $f : P \to Q$ induces a simplicial map $O(P) \to O(Q)$.*

We present without proofs the following two theorems.

Theorem 36.6. *Let $f : P \to P$ be an order-preserving map of finite posets.*

(1) *If $f(x) \geq x$ for all $x \in P$, then $O(P)$ and $O(f(P))$ are homotopic.*
(2) *If $f^2(x) = f(x)$ for all $x \in P$, then $O(P)$ and $O(f(P))$ are homotopic. In this case, f is called a* ***closure operator***.

Fiber Theorem 36.7. *Let Δ be a finite simplicial complex, P a finite poset, and $f : \Delta \to O(P)$ a simplicial map. By $P_{\geq x}$ we denote the poset $\{y \in P | \, y \geq x\}$. If for every $x \in P$ the fiber $f^{-1}O(P_{\geq x})$ is contractible, then Δ and $O(P)$ are homotopic.*

Exercise 36.8. *$O(F(\Delta))$ is the first barycentric subdivision of Δ.*

Corollary 36.9. *Δ and $O(F(\Delta))$ are homotopic.*

Construction 36.10. The ***nerve*** of a finite set of finite simplicial complexes $\{\Lambda_i\}_{i \in \mathcal{A}}$ is the simplicial complex N on vertex set $\mathcal{A}$ and with faces

$$\{\sigma \subseteq \mathcal{A} | \, \cap_{i \in \sigma} \Lambda_i \neq \emptyset\} .$$

Nerve Theorem 36.11. *Let Δ be a finite simplicial complex and $\{\Lambda_i\}_{i \in \mathcal{A}}$ be a finite cover (that is, a set of subcomplexes such that $\cup_{i \in \mathcal{A}} \Lambda_i = \Delta$). Suppose that every non-empty intersection $\cap_{i \in \sigma} \Lambda_i$ is contractible. Then Δ and the nerve N are homotopic.*

Proof. Define an order-reversing poset map

$$\begin{aligned} f : F(\Delta) &\to F(N) \\ \sigma &\mapsto \{i \in \mathcal{A} \, | \, \sigma \in \Lambda_i\} . \end{aligned}$$

It induces a simplicial map $f : O(F(\Delta)) \to O(F(N))$.

We will apply the Fiber Theorem 36.7 to the simplicial complex $O(F(\Delta))$ and the poset $F(N)$. Let $x \in F(N)$. Consider the fiber $f^{-1}O(F(N)_{\geq x})$. Set $\sigma_x = \cap_{i \in x} \Lambda_i \in F(\Delta)$. Then

$$f^{-1}\Big(F(N)_{\geq x}\Big) = f^{-1}\Big(\big(F(N) \cap \mathrm{Im}(f)\big)_{\geq x}\Big)$$
$$= f^{-1}\Big(\big(F(N) \cap \mathrm{Im}(f)\big)_{\geq f(\sigma_x)}\Big).$$

Since $f : F(\Delta) \to F(N)$ is order-reversing and since $f(\sigma_x)$ is the bottom element of $\big(F(N) \cap \mathrm{Im}(f)\big)_{\geq f(\sigma_x)}$, we get that σ_x is the top element in $f^{-1}\big(F(N) \cap \mathrm{Im}(f)\big)_{\geq f(\sigma_x)}$. Hence, $O\big(f^{-1}\big(F(N) \cap \mathrm{Im}(f)\big)_{\geq x}\big)$ is contractible. By the Fiber Theorem 36.7, $O(F(\Delta))$ and $O(F(N))$ are homotopic. By Corollary 36.9, $O(F(N))$ and N are homotopic, and also $O(F(\Delta))$ and Δ are homotopic. □

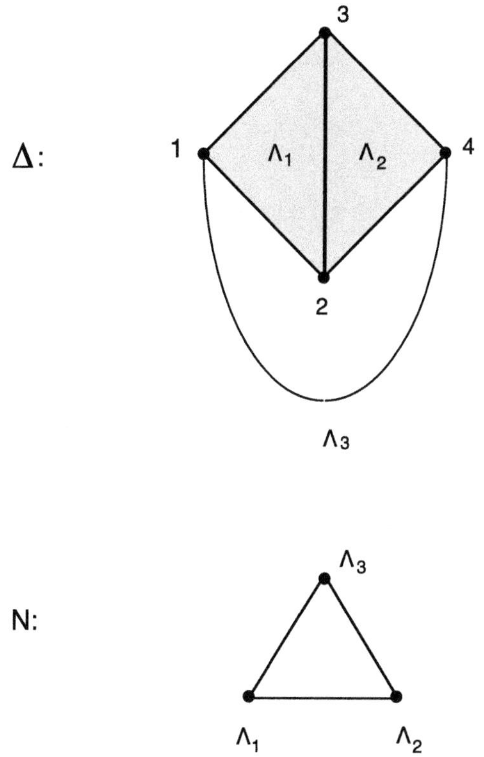

Figure 3.

Example 36.12. Let Δ be the simplicial complex on vertices $\{1, 2, 3, 4\}$, whose facets are the two triangles $\{1, 2, 3\}$ and $\{2, 3, 4\}$ and the edge $\{1, 4\}$. Denote by $\Lambda_1, \Lambda_2, \Lambda_3$ the facets.

The nerve N is the empty triangle on vertices $\Lambda_1, \Lambda_2, \Lambda_3$. See Figure 3 above.

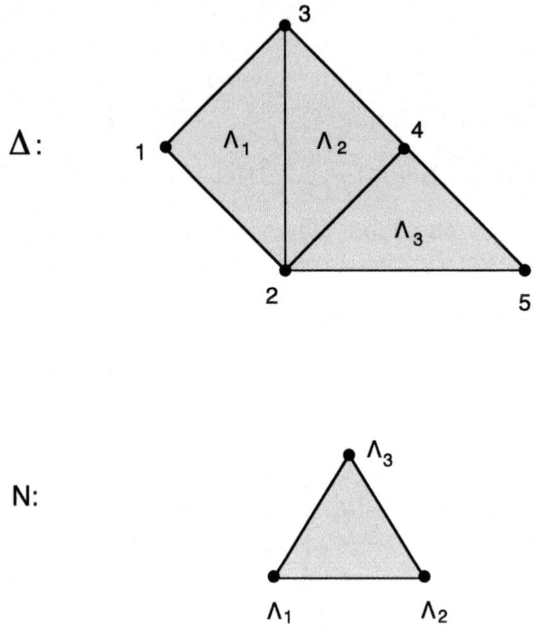

Figure 4.

Let Δ be the simplicial complex on vertices labeled $\{1, 2, 3, 4, 5\}$, whose facets are the three triangles $\{1, 2, 3\}$, $\{2, 3, 4\}$, and $\{2, 4, 5\}$. Denote by $\Lambda_1, \Lambda_2, \Lambda_3$ the facets.

The nerve N is the simplex on vertices $\Lambda_1, \Lambda_2, \Lambda_3$. See Figure 4.

Corollary 36.13. *Let Δ be a finite simplicial complex and $F(\Delta)$ be its face poset. Let $T(\Delta)$ be the poset with elements the non-empty intersections of the facets of Δ. Then, $O(F(\Delta))$ is homotopic to $O(T(\Delta))$.*

Proof. Consider the cover of Δ that consists of all the facets $\Lambda_1, \dots \Lambda_s$

(ordered arbitrarily). Every non-empty intersection of facets is a simplex, so it is contractible. Let N be the nurve.

Now, note that $T(\Delta)$ is the face poset of the simplicial complex N. By Corollary 36.9, it follows that $O(T(\Delta))$ and N are homotopic. By Theorem 36.11, we have that N is homotopic to Δ. Finally, by Corollary 36.9, Δ is homotopic to $O(F(\Delta))$. $\square$

Construction 36.14. Let P be a finite poset. A subset $D \subset P$ is called a ***crosscut*** if it consists of non-comparable elements such that the following conditions are satisified:

(1) For every chain $\sigma \in P$ there exits a $c_\sigma \in D$ which is comparable to each element in the chain σ.

(2) If $A \subseteq D$ is bounded, then either the join $\vee A$ or the meet $\wedge A$ exists in P.

The ***crosscut complex*** C is the simplicial complex on vertices in D and with faces the bounded subsets of D.

Example 36.15. Consider the poset P in Figure 5 below. The elements a, b, c form a crosscut. The crosscut complex C is the simplicial complex on vertices a, b, c with two edges $\{a, b\}$ and $\{b, c\}$. See Figure 5.

Crosscut Theorem 36.16. *Let P be a finite poset with a bottom element $\emptyset$, and D be a crosscut. The crosscut complex C and $O(P \setminus \emptyset)$ are homotopic.*

Proof. We will apply the Nerve Theorem 36.11. For every $x \in D$, set $\Lambda_x = O(P_{\geq x} \cup (P_{\leq x} \setminus \emptyset))$. Let $\sigma \neq \emptyset$ be a chain in P. By (1), there exits a $c_\sigma \in D$ which is comparable to each element in σ. Hence $\sigma \in \Lambda_{c_\sigma}$. Therefore, $\{O(\Lambda_x)\}_{x \in D}$ is a cover of $O(P \setminus \emptyset)$.

For a subset A of D, suppose that $y \in \cap_{x \in A} \Lambda_x$. If there exist $x_1, x_2 \in A$ such that $y > x_1$ and $y < x_2$, then it follows that $x_2 > x_1$ contradicting to the fact that x_1 and x_2 are incomparable. Hence, y is either an upper bound for A or a lower bound for A. By condition (2) above, either the join $\vee A$ or the meet $\wedge A$ exists in P. Consider the following properties.

$$(*) \qquad \begin{aligned} &(1)\ \cap_{x\in A}\Lambda_x \neq \emptyset \\ &(2)\ A \text{ is bounded} \\ &(3)\ \text{either } \vee A \text{ or } \wedge A \text{ exists, and is in } \cap_{x\in A}\Lambda_x\,. \end{aligned}$$

The argument above shows that (1) implies (3). Clearly, (2) implies (1). Also, (2) and (3) are equivalent.

If $\cap_{x\in A} O(\Lambda_x) \neq \emptyset$ then $\cap_{x\in A}\Lambda_x \neq \emptyset$, so $\cap_{x\in A} O(\Lambda_x)$ is a cone since every maximal face of this intersection contains either $\vee A$ or $\wedge A$, so the intersection is contractible.

By the Nerve Theorem 36.11, it follows that $O(P \setminus \emptyset)$ and the nerve N are homotopic. By $(*)$, the nerve coincides with the crosscut complex. $\square$

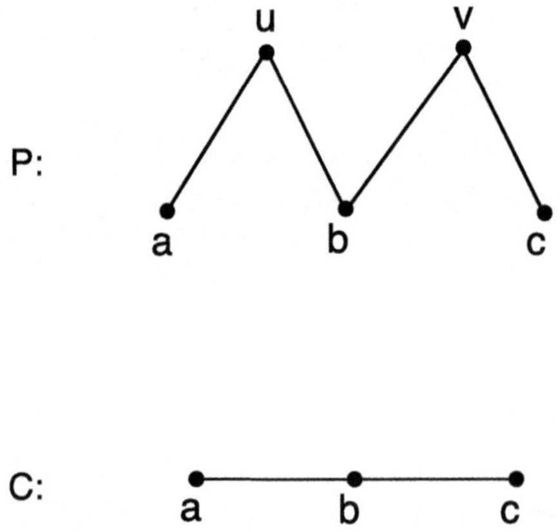

Figure 5.

Construction 36.17. The ***link*** of a face τ in a simplicial complex Δ is

$$\text{link}_\Delta(\tau) = \{\sigma \in \Delta \,|\, \sigma \cup \tau \in \Delta \text{ and } \sigma \cap \tau = \emptyset\}\,,$$

and the ***star*** of τ is

$$\text{star}_\Delta(\tau) = \{\sigma \in \Delta \,|\, \sigma \cup \tau \in \Delta\}\,.$$

The ***restriction*** of Δ on τ is denoted by Δ_τ and is the maximal subcomplex of Δ on the vertices of τ.

Another helpful technique is Discrete Morse Theory, developed in [Forman].

Construction 36.18. Let Δ be a simplicial complex, and $p \in \mathbf{N}$. A ***p-matching*** on Δ is a set

$$\mathcal{M} = \{ (\sigma, \tau) \mid \sigma \subset \tau \in \Delta, \ \dim \tau = \dim \sigma + 1 \},$$

such that the following two conditions are satisfied:

1) every face of Δ of dimension $\leq p$ is contained in exactly one pair.
2) a face of Δ of dimension $> p$ is contained in at most one pair.

We construct an oriented graph $G_{\mathcal{M}}$ on vertices labeled by the faces of Δ as follows: Let σ and τ be two vertices, such that $\dim \tau = \dim \sigma + 1$. We have an edge from σ to τ if $(\sigma, \tau) \in \mathcal{M}$. We have an edge from τ to σ if $(\sigma, \tau) \notin \mathcal{M}$.

We say that the p-matching $\mathcal{M}$ is ***acyclic*** if the graph $G_{\mathcal{M}}$ contains no oriented cycle.

Theorem 36.19. [Forman] *Let Δ be a finite simplicial complex, and $p \in \mathbf{N}$. If there exists an acyclic p-matching on Δ, then $\widetilde{\mathrm{H}}_i(\Delta; k) = 0$ for $i \leq p$.*

Example 36.20. Let Δ be the empty triangle on vertices $\{a, b, c\}$, so it has edges $\{a, b\}$, $\{b, c\}$, $\{a, c\}$. Then

$$\{ (\emptyset, a), \ (b, \{b, c\}), \ (c, \{a, c\}) \}$$

is a 0-acyclic matching, so $\widetilde{\mathrm{H}}_0(\Delta; k) = 0$. See Figure 6 below.

Let Δ' be the triangle on vertices $\{a, b, c\}$, so it has edges $\{a, b\}$, $\{b, c\}$, $\{a, c\}$ and the facet $\{a, b, c\}$. Then

$$\{ (\emptyset, a), \ (b, \{b, c\}), \ (c, \{a, c\}), \ (\{a, b\}, \ \{a, b, c\}) \}$$

is a 1-acyclic matching, so $\widetilde{\mathrm{H}}_0(\Delta'; k) = 0$ and $\widetilde{\mathrm{H}}_1(\Delta'; k) = 0$. See Figure 7 below.

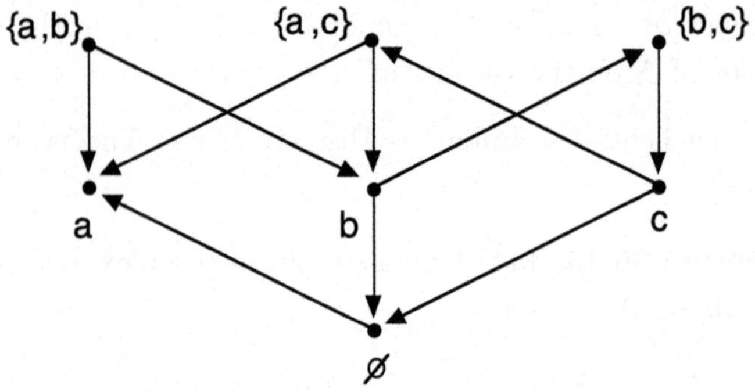

Figure 6.

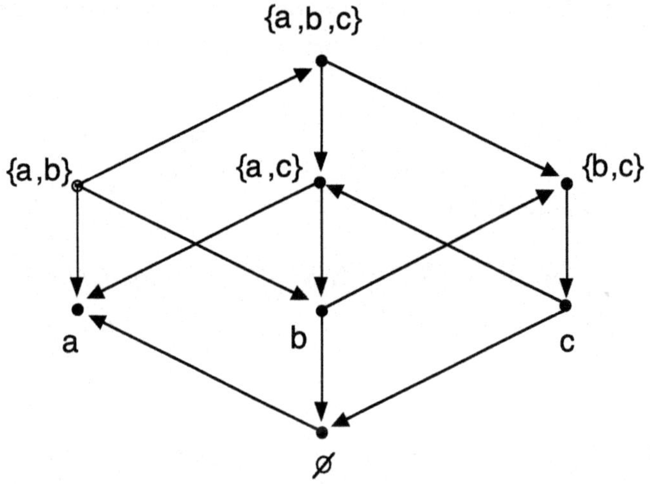

Figure 7.

Another helpful tool is the reduced Mayer-Vietoris exact sequence in the next theorem.

Theorem 36.21. *Let Δ_1 and Δ_2 be simplicial complexes in $\mathbf{R}^s$ and suppose that $\Delta_1 \cup \Delta_2$ is a simplicial complex in $\mathbf{R}^s$ as well. We have the long exact sequence*

$$\ldots \to \widetilde{\mathrm{H}}_i(\Delta_1 \cap \Delta_2; k) \longrightarrow \widetilde{\mathrm{H}}_i(\Delta_1; k) \oplus \widetilde{\mathrm{H}}_i(\Delta_2; k) \longrightarrow \widetilde{\mathrm{H}}_i(\Delta_1 \cup \Delta_2; k)$$

$$\longrightarrow \ldots$$

$$\to \widetilde{\mathrm{H}}_0(\Delta_1 \cap \Delta_2; k) \longrightarrow \widetilde{\mathrm{H}}_0(\Delta_1; k) \oplus \widetilde{\mathrm{H}}_0(\Delta_2; k) \longrightarrow \widetilde{\mathrm{H}}_0(\Delta_1 \cup \Delta_2; k) \to$$

$$\widetilde{\mathrm{H}}_{-1}(\Delta_1 \cap \Delta_2; k) \longrightarrow \widetilde{\mathrm{H}}_{-1}(\Delta_1; k) \oplus \widetilde{\mathrm{H}}_{-1}(\Delta_2; k) \longrightarrow \widetilde{\mathrm{H}}_{-1}(\Delta_1 \cup \Delta_2; k)$$

$$\longrightarrow 0\,.$$

Sometimes, homology can be computed using a shelling.

Definition 36.22. If τ is a facet of Δ then we define the closed interval

$$[\emptyset : \tau] = \{\sigma \in \Delta \,|\, \sigma \subseteq \tau\}\,.$$

A pure simplicial complex Δ is ***shellable*** if its facets can be ordered as $\tau_1,\ \tau_2, \ldots$ so that for each $i \geq 1$ we have that the intersection

$$[\emptyset : \tau_{i+1}] \cap \left(\cup_{1 \leq j \leq i} [\emptyset : \tau_j]\right)$$

is pure of dimension $\dim(\Delta) - 1$, that is, the intersection consists of elements in the boundary of τ_{i+1}.

Theorem 36.23. *Suppose that Δ is a shellable pure simplicial complex. Then $\widetilde{\mathrm{H}}_i(\Delta; k) = 0$ for $i \neq \dim(\Delta)$. Furthermore, $\widetilde{\mathrm{H}}_{\dim(\Delta)}(\Delta; k)$ is equal to the number of facets τ_{i+1} such that the intersection*

$$[\emptyset : \tau_{i+1}] \cap (\cup_{1 \leq j \leq i} [\emptyset : \tau_j]$$

contains all subfaces of τ_{i+1} of dimension $\dim(\Delta) - 1$ (that is, the intersection is the entire boundary of τ_{i+1}).

Proof. The proof is by induction on i. Set $\Delta_i = \cup_{j=1}^{i} [\emptyset : \tau_j]$ and use the reduced Mayer-Vietoris exact sequence 36.21. We get

$$\ldots \to \widetilde{\mathrm{H}}_p(\Delta_i; k) \oplus \widetilde{\mathrm{H}}_p(\tau_{i+1}; k) \to \widetilde{\mathrm{H}}_p(\Delta_i \cup \tau_{i+1}; k)$$

$$\to \widetilde{\mathrm{H}}_{p-1}(\Delta_i \cap \tau_{i+1}; k) \to \ldots\,.$$

We have that τ_{i+1} has no homology, Δ_i is shellable, and the intersection $\Delta_i \cap \tau_{i+1}$ either has no homology or is equal to the entire boundary of τ_{i+1}. □

Proposition 36.24. *Let Δ' be a subcomplex of the simplcial complex Δ. The reduced relative simplicial homology with coefficients in k of the pair (Δ, Δ') appears in the following long exact sequence:*

$$\ldots \to \widetilde{\mathrm{H}}_{i-1}(\Delta; k) \to \widetilde{\mathrm{H}}_{i-1}(\Delta, \Delta'; k)$$
$$\to \widetilde{\mathrm{H}}_{i-2}(\Delta'; k) \to \widetilde{\mathrm{H}}_{i-2}(\Delta; k) \to \ldots$$

37 Appendix: Tools from homological algebra

The following lemma is proved in [Northcott, Section 4.4].

Lemma 37.1. *Let*

$$\mathbf{A}: \ A_2 \to A_1 \to A_0$$
$$\mathbf{B}: \ B_2 \to B_1 \to B_0$$
$$\mathbf{C}: \ C_2 \to C_1 \to C_0$$

be three complexes of finitely generated R-modules, and

$$\mathbf{A} \to \mathbf{B} \to \mathbf{C}$$

be homomorphisms of complexes. Suppose that we have a commutative diagram

$$\begin{array}{ccccccc} & A_2 & \to & B_2 & \to & C_2 & \to 0 \\ & \downarrow & & \downarrow & & \downarrow & \\ & A_1 & \to & B_1 & \to & C_1 & \\ & \downarrow & & \downarrow & & \downarrow & \\ 0 \to & A_0 & \to & B_0 & \to & C_0 & \end{array}$$

of R-modules such that
(1) $B_2 \to C_2$ *is surjective*
(2) $A_1 \to B_1 \to C_1$ *is exact*
(3) $A_0 \to B_0$ *is injective.*
Then $\mathrm{H}_1(\mathbf{A}) \to \mathrm{H}_1(\mathbf{B}) \to \mathrm{H}_1(\mathbf{C})$ *is exact.*

Corollary 37.2. *Let*

$$\begin{array}{ccccc} A & \to & B & \to & C \\ \downarrow \alpha & & \downarrow \beta & & \downarrow \gamma \\ A' & \to & B' & \to & C' \end{array}$$

be a commutative diagram of finitely generated R-modules with exact rows.

(1) *If $A' \to B'$ is injective, then* $\mathrm{Ker}(\alpha) \to \mathrm{Ker}(\beta) \to \mathrm{Ker}(\gamma)$ *is exact.*

(2) *If $B \to C$ is surjective, then* $\mathrm{Coker}(\alpha) \to \mathrm{Coker}(\beta) \to \mathrm{Coker}(\gamma)$ *is exact.*

Proof. To prove (1) apply Lemma 37.1 to the diagram

$$\begin{array}{ccccccc} & & 0 & \to & 0 & \to & 0 & \to 0 \\ & & \downarrow & & \downarrow & & \downarrow & \\ & & A & \to & B & \to & C & \\ & & \downarrow & & \downarrow & & \downarrow & \\ 0 & \to & A' & \to & B' & \to & C' & \end{array}$$

and use that $\mathrm{H}(0 \to A \to A') = \mathrm{Ker}(\alpha)$, $\mathrm{H}(0 \to B \to B') = \mathrm{Ker}(\beta)$, and $\mathrm{H}(0 \to C \to C') = \mathrm{Ker}(\gamma)$.

To prove (2) apply Lemma 37.1 to the diagram

$$\begin{array}{ccccccc} & & A & \to & B & \to & C & \to 0 \\ & & \downarrow & & \downarrow & & \downarrow & \\ & & A' & \to & B' & \to & C' & \\ & & \downarrow & & \downarrow & & \downarrow & \\ 0 & \to & 0 & \to & 0 & \to & 0 & \end{array}$$

and use that $\mathrm{H}(A \to A' \to 0) = \mathrm{Coker}(\alpha)$, $\mathrm{H}(B \to B' \to 0) = \mathrm{Coker}(\beta)$, and $\mathrm{H}(C \to C' \to 0) = \mathrm{Coker}(\gamma)$. □

The Snake Lemma 37.3. *Let*

$$\begin{array}{ccccccc} & & A & \to & B & \to & C & \to 0 \\ & & \downarrow \alpha & & \downarrow \beta & & \downarrow \gamma & \\ 0 & \to & A' & \to & B' & \to & C' & \end{array}$$

be a commutative diagram of finitely generated R-modules with exact rows. The sequence

$$\mathrm{Ker}(\alpha) \to \mathrm{Ker}(\beta) \to \mathrm{Ker}(\gamma) \xrightarrow{\tau} \mathrm{Coker}(\alpha) \to \mathrm{Coker}(\beta) \to \mathrm{Coker}(\gamma)$$

is exact, where τ is the connecting homomorphism.

If in addition $A \to B$ is injective and $B' \to C'$ is surjective, then *the sequence*

$$0 \to \operatorname{Ker}(\alpha) \to \operatorname{Ker}(\beta) \to \operatorname{Ker}(\gamma) \xrightarrow{\tau} \operatorname{Coker}(\alpha) \to \operatorname{Coker}(\beta) \to \operatorname{Coker}(\gamma) \to 0$$

is exact.

By Corollary 37.2 we have that the sequences

$$\operatorname{Ker}(\alpha) \to \operatorname{Ker}(\beta) \to \operatorname{Ker}(\gamma)$$
$$\operatorname{Coker}(\alpha) \to \operatorname{Coker}(\beta) \to \operatorname{Coker}(\gamma)$$

are exact. The rest of the proof of the Snake Lemma can be found in cf. [Northcott, Section 4.5].

38 Appendix: Tor and Ext

Let U and W be finitely generated R-modules. Let $\mathbf{F}$ be a free resolution of U. Consider the complex

$$\mathbf{F} \otimes_R W: \quad \ldots \longrightarrow F_i \otimes_R W \xrightarrow{d_i \otimes W} F_{i-1} \otimes_R W \to \ldots \to F_0 \otimes_R W \to 0\,.$$

The R-module $\mathrm{H}_i(\mathbf{F} \otimes_R W)$ is denoted $\operatorname{Tor}_i^R(U,W)$.

Set $\operatorname{Tor}_*^R(U,W) = \oplus_{i \geq 0} \operatorname{Tor}_i^R(U,W)$.

Theorem 38.1.

(1) $\operatorname{Tor}_0^R(U,W) = U \otimes_R W$.

(2) $\operatorname{Tor}_*^R(U,W)$ *is defined uniquely up to an isomorphism, that is, it does not depend on the choice of the resolution.*

(3) $\operatorname{Tor}_*^R(U,W) \cong \operatorname{Tor}_*^R(W,U)$.

Now consider the complex

$$\operatorname{Hom}_R(\mathbf{F},W): \ 0 \to \operatorname{Hom}_R(F_0,W) \xrightarrow{\operatorname{Hom}_R(d_1,W)} \operatorname{Hom}_R(F_1,W) \to \ldots\,.$$

The R-module $\mathrm{H}_i(\operatorname{Hom}_R(\mathbf{F},W))$ is denoted $\operatorname{Ext}_R^i(U,W)$.

Set $\operatorname{Ext}_R^*(U,W) = \oplus_{i \geq 0} \operatorname{Ext}_R^i(U,W)$.

Theorem 38.2.
(1) $\mathrm{Ext}^0_R(U,W) = \mathrm{Hom}_R(U,W)$.
(2) $\mathrm{Ext}^*_R(U,W)$ *is defined uniquely up to an isomorphism, that is, it does not depend on the choice of the resolution.*

Theorem 38.3. *Suppose that* $0 \to U \to U' \to U'' \to 0$ *is a short exact sequence of finitely generated R-modules. Then we have the long exact sequence*

$$\begin{aligned} \ldots \to \mathrm{Tor}^R_i(U,W) &\to \mathrm{Tor}^R_i(U',W) \to \mathrm{Tor}^R_i(U'',W) \to \\ \to \mathrm{Tor}^R_{i-1}(U,W) &\to \mathrm{Tor}^R_{i-1}(U',W) \to \mathrm{Tor}^R_{i-1}(U'',W) \to \\ \ldots \to \mathrm{Tor}^R_0(U,W) &\to \mathrm{Tor}^R_0(U',W) \to \mathrm{Tor}^R_0(U'',W) \to 0 \end{aligned}$$

and also the long exact sequence

$$\begin{aligned} 0 \to \mathrm{Ext}^0_R(U'',W) &\to \mathrm{Ext}^0_R(U',W) \to \mathrm{Ext}^0_R(U,W) \to \\ \ldots \to \mathrm{Ext}^{i-1}_R(U'',W) &\to \mathrm{Ext}^{i-1}_R(U',W) \to \mathrm{Ext}^{i-1}_R(U,W) \to \\ \to \mathrm{Ext}^i_R(U'',W) &\to \mathrm{Ext}^i_R(U',W) \to \mathrm{Ext}^i_R(U,W) \to \ldots\ . \end{aligned}$$

Suppose that U and W are graded, and that $\mathbf{F}$ is a graded resolution. Therefore, $\mathbf{F} \otimes_R W$ is a graded complex as well. So its homology is graded. Thus, in this case we obtain the bigraded module

$$\mathrm{Tor}^R_*(U,W) = \oplus_{i,j} \mathrm{Tor}^R_i(U,W)_j$$

where i is the homological degree and j is the internal degree. Furthermore, we would like to obtain a graded version of the Ext-module. The complex $\mathrm{Hom}(\mathbf{F},W)$ is bigraded since for every i the module $\mathrm{Hom}(F_i,W)$ is equal to the graded hom $\mathcal{H}(F_i,W)$ in the notation of 2.7 (the equality holds by Proposition 2.7 because F_i is finitely generated over R), so it is graded internaly. Hence, its homology is graded. Thus, we obtain the bigraded module

$$\mathrm{Ext}^*_R(U,W) = \oplus_{i,j} \mathrm{Ext}^i_R(U,W)_j\,,$$

where i is the homological degree and j is the internal degree.

39 Appendix: Gröbner basis

Sometimes monomial ideals are easier to study. Using Gröbner basis theory one can reduce the study of some properties of a given ideal to the study of properties of a monomial ideal. We will discuss Gröbner basis theory of graded ideals in S. For a more detailed exposition, see [Eisenbud, Chapter 15]. For Gröbner basis theory over exterior algebras, see [Green], [Aramova-Herzog-Hibi 2]. For Gröbner basis theory of modules over S, see [Eisenbud, Chapter 15].

Definition 39.1. A ***monomial order*** on S is a total order $\prec$ on the monomials in S, such that the following two properties hold:

(1) $w \succ 1$ for every monomial $w \neq 1$.

(2) if $m' \succ m$, then $wm' \succ wm$ for every monomials m, m' and $w \neq 1$.

Clearly, 1 is the smallest monomial.

Construction 39.2. We will describe some monomial orders which are widely used. Fix the order of the variables to be

$$x_1 > \ldots > x_n \, .$$

Let $\alpha = (\alpha_1, \ldots, \alpha_n)$ and $\beta = (\beta_1, \ldots, \beta_n)$ be in $\mathbf{N}^n$. Consider the monomials $\mathbf{x}^\alpha = x_1^{\alpha_1} \ldots x_n^{\alpha_n}$ and $\mathbf{x}^\beta = x_1^{\beta_1} \ldots x_n^{\beta_n}$.

(1) The ***lex order*** (or ***lexicographic order***, or ***pure-lex order***), denoted by $\prec_{lex}$, is the monomial order defined by

$$x_1^{\alpha_1} \ldots x_n^{\alpha_n} \succ_{lex} x_1^{\beta_1} \ldots x_n^{\beta_n} \quad \text{if } \alpha_i > \beta_i \text{ for the first index } i \text{ such that } \alpha_i \neq \beta_i \, .$$

For example, $x_1^2 x_2^3 x_5 \succ_{lex} x_1^2 x_3^2 x_4^5$.

(2) The ***degree-lex order***, denoted $\prec_{dlex}$, is the monomial order defined

$$\mathbf{x}^\alpha \succ_{dlex} \mathbf{x}^\beta \quad \text{if } \deg(\mathbf{x}^\alpha) > \deg(\mathbf{x}^\beta) \, , \quad \text{or } \deg(\mathbf{x}^\alpha) = \deg(\mathbf{x}^\beta) \text{ and } \mathbf{x}^\alpha \succ_{lex} \mathbf{x}^\beta \, .$$

For example, $x_1^2x_3^2x_4^5 \prec_{dlex} x_1^2x_2^3x_5$ and $x_1^2x_2^3x_5 \succ_{dlex} x_1^2x_3^2x_4^2$. Note that if I is graded, then $\text{in}_{lex}(I) = \text{in}_{dlex}(I)$.

(3) The ***revlex order*** (or ***reverse lex***), denoted $\prec_{rlex}$, is the monomial order defined by

$$\begin{aligned}\mathbf{x}^\alpha \succ_{rlex} \mathbf{x}^\beta \text{ if } &\deg(\mathbf{x}^\alpha) > \deg(\mathbf{x}^\beta)\,,\\ &\text{or } \deg(\mathbf{x}^\alpha) = \deg(\mathbf{x}^\beta) \text{ and } \alpha_i < \beta_i \text{ for the last index}\\ &i \text{ with } \alpha_i \neq \beta_i\,.\end{aligned}$$

For example, $x_1^2x_2^4x_5 \succ_{rlex} x_1^2x_3^2x_4^2x_5$.

(4) Let $<$ be a partial order on the monomials in S that satisfies (1) and (2) in Definition 39.1. Let $\lhd$ be a monomial order. We say that $\prec$ is the order $<$ ***refined*** by $\lhd$ if

$$\begin{aligned}\mathbf{x}^\alpha \succ \mathbf{x}^\beta \quad &\text{if either } \mathbf{x}^\alpha > \mathbf{x}^\beta\,,\\ &\text{or } \mathbf{x}^\alpha \text{ and } \mathbf{x}^\beta \text{ are equal with respect to the order } > \text{ but}\\ &\mathbf{x}^\alpha \rhd \mathbf{x}^\beta\,.\end{aligned}$$

For example, if $<$ is the partial order on the monomials in S by degree and if $\lhd$ is the lex-order, then the order $<$ refined by $\lhd$ is the degree-lex order.

Definition 39.3. The elements of k are called ***scalars***. A scalar multiple of a monomial is called a ***term***. If f is a polynomial in S, then we denote by $\text{in}_\prec(f)$ the greatest term of f with respect to the order $\prec$ and call it the ***initial term*** of f. Let J be an ideal in S. The ideal

$$\text{in}_\prec(J) = \big(\text{in}_\prec(f) \mid f \in J\big)$$

is called the ***initial ideal*** of J with respect to $\prec$. We remark that if $m \in \text{in}_\prec(J)$ is a monomial, then there exists an $f \in J$ such that $\text{in}_\prec(f) = m$. A subset $\mathcal{G}$ of J is called a ***Gröbner basis*** if $\{\text{in}_\prec(f) \mid f \in \mathcal{G}\}$ generate $\text{in}_\prec(J)$. We say that $\mathcal{G}$ is a ***reduced Gröbner basis*** if for every $f, g \in \mathcal{G}$ we have that $\text{in}_\prec(f)$ divides no term of g.

For simplicity, instead of $\text{in}_{\prec}$ we write "in" when we consider a fixed monomial order.

Example 39.4. Consider the defining ideal $B = (ac-b^2,\ bc-ah,\ c^2-bh)$ of the twisted cubic curve in the polynomial ring $A = k[a,b,c,h]$. Straightforward computation shows that

$$\text{in}_{lex}(B) = \text{in}_{dlex}(B) = (ac,\ ah,\ bh)$$
$$\text{in}_{rlex}(B) = (b^2,\ bc,\ c^2)\,.$$

Computer computation shows that the list of all monomial initial ideals of B is

$$(bh,ah,ac),\ (bh,b^2,ah),\ (bh,bc,b^2,ah^2),\ (c^3,bh,bc,b^2),$$
$$(c^2,bc,b^2),\ (c^2,bc,b^3,ac),\ (c^2,bc,ac,a^2h),\ (c^2,ah,ac)\,.$$

Theorem 39.5. *A Gröbner basis $\mathcal{G}$ of J generates J.*

Proof. Suppose that $J \neq (\mathcal{G})$. Let $f \in J$ be an element such that $f \notin (\mathcal{G})$ and f has a minimal initial term among the elements with this property. Since $\text{in}_{\prec}(f) \in \text{in}_{\prec}(J) = (\text{in}_{\prec}(\mathcal{G}))$, there exists a $g \in (\mathcal{G})$ with $\text{in}_{\prec}(g) = \text{in}_{\prec}(f)$. Hence $f - g \in J$, $f - g \notin (\mathcal{G})$, and it has a smaller initial term than f. This is a contradiction. Therefore $J = (\mathcal{G})$. □

Theorem 39.6. (Macaulay) *Let J be an ideal in S. The monomials not in $\text{in}_{\prec}(J)$ form a basis of the k-vector space S/J.*

Proof. Let $m_1, \ldots, m_p$ be monomials that are not in $\text{in}_{\prec}(J)$. We will prove that they are linearly independent. Assume the opposite. This means that there exist $\alpha_1, \ldots, \alpha_p \in k\backslash 0$ such that $\alpha_1 m_1 + \ldots + \alpha_p m_p \in J$. Therefore, $\text{in}_{\prec}(\alpha_1 m_1 + \ldots + \alpha_p m_p) \in \text{in}_{\prec}(J)$. But the initial term is a scalar multiple of one of the monomials, so it is not in $\text{in}_{\prec}(J)$. This is a contradiction. Hence $m_1, \ldots, m_p$ are linearly independent.

Now, we want to show that the monomials not in $\text{in}_{\prec}(J)$ span S/J. Denote by T the span of the monomials not in $\text{in}_{\prec}(J)$. We will prove that $\{T, J\}$ span S. Assume the opposite. Choose an $f \in S$ such that $f \notin \operatorname{span}\{T, J\}$ and it has a minimal initial term among those polynomials not in $\operatorname{span}\{T, J\}$. Set $m = \operatorname{in}_\prec(f)$. If $m \notin \operatorname{in}_\prec(J)$, then $f - m \notin \operatorname{span}\{T, J\}$ has a smaller initial term than f, which is a contradiction. Hence $m \in \operatorname{in}_\prec(J)$. Choose a $g \in J$ such that $\operatorname{in}_\prec(g) = m$. Then $f - g \notin \operatorname{span}\{T, J\}$ has a smaller initial term than f, which is a contradiction. Therefore, T spans S/J. □

Corollary 39.7. *Let I be a graded ideal in S.*

(1) *S/I and $S/\operatorname{in}_\prec(I)$ have the same Hilbert function.*
(2) *Let $\prec$ and $<$ be monomial orders in I. If $\operatorname{in}_\prec(I) \subseteq \operatorname{in}_<(I)$, then $\operatorname{in}_\prec(I) = \operatorname{in}_<(I)$.*
(3) *If $M \subseteq \operatorname{in}_\prec(I)$ is a monomial ideal with the same Hilbert function as I, then $M = \operatorname{in}_\prec(I)$.*

Next, we will outline Buchberger's Algorithm for computing a Gröbner basis with respect to a fixed monomial order $\prec$. Fix a finite subset T of J. Let $f \in J$. A first reduction of f is obtained by subtracting a multiple $\tilde{g}$ of an element g in T so that $\operatorname{in}_\prec(f)$ and $\operatorname{in}_\prec(\tilde{g})$ cancel. A remainder $r(f)$ is obtained by repeatedly reducing f as many times as possible. Note that $r(f) \in J$. The reductions and the remainder are not uniquely defined. We say that f ***reduces to zero*** (with respect to T) if we can get $r(f) = 0$.

For two polynomials f, g set

$$\tau_{f,g} = \frac{\operatorname{in}(g)}{\gcd(\operatorname{in}(f), \operatorname{in}(g))} f - \frac{\operatorname{in}(f)}{\gcd(\operatorname{in}(f), \operatorname{in}(g))} g \, .$$

This difference is formed with the goal to cancel the initial terms of f and g in the most efficient way. If $f, g \in J$, then $\tau_{f,g} \in J$, and hence $r(\tau_{f,g}) \in J$.

Buchberger's Criterion 39.8. *Fix a monomial order $\prec$. Let $\mathcal{G}$ be a subset of J such that $(\mathcal{G}) = J$. The set $\mathcal{G}$ is a Gröbner basis of J if and only if $\tau_{f,g}$ reduces to zero for each $f, g \in \mathcal{G}$.*

Buchberger's Algorithm 39.9. *Fix a monomial order $\prec$. Let $\mathcal{G}_0$ be a finite system of generators of J. Define inductively*

$$\mathcal{G}_i = \mathcal{G}_{i-1} \cup \{\textit{a remainder of } \tau_{f,g} \,|\, f, g \in \mathcal{G}_{i-1},\ \tau_{f,g} \textit{ does not reduce to zero}\}.$$

This process terminates after finitely many steps. The obtained set is a Gröbner basis of J.

For convenience, instead of $r(\tau_{f,g})$ we may sometimes add $-r(\tau_{f,g})$ to the set $\mathcal{G}_{i-1}$.

Proof. As noted above, if $f, g \in J$ then $r(\tau_{f,g}) \in J$. Therefore, $\mathcal{G}_i \subset J$ for each i. Hence $(\text{in}(\mathcal{G}_i))$ is a monomial ideal contained in $\text{in}(J)$. If $\mathcal{G}_i \neq \mathcal{G}_{i-1}$, then $\mathcal{G}_{i-1} \subset \mathcal{G}_i$. We cannot have an infinite increasing sequence of monomial ideals contained in the monomial ideal $\text{in}(J)$. Hence, the process terminates after finitely many steps. We obtain a set $\mathcal{G}$ such that $r(\tau_{f,g}) = 0$ for all $f, g \in \mathcal{G}$. Buchberger's Criterion says that $\mathcal{G}$ is a Gröbner basis. ☐

The following theorem is proved in [Mora-Robbiano].

Theorem 39.10. *There exists a finite subset $\mathcal{G}$ of J such that $\mathcal{G}$ is a Gröbner basis of J with respect to every monomial order.*

Chapter II

Hilbert Functions

Abstract. A well-studied and important numerical invariant of a graded ideal over a graded polynomial ring S is the Hilbert function. It gives the sizes of the graded components of the ideal.

The Hilbert function encodes important information (for example, dimension and multiplicity). Hilbert's insight was that it is determined by finitely many of its values.

In many recent papers and books, Hilbert functions are studied using clever computations with binomials; we mention the binomial-approach briefly and avoid such computations whenever possible. Instead our arguments are founded upon Macaulay's key idea in 1927: *There exist highly structured monomial ideals - lex ideals - which attain all Hilbert functions. Lex ideals play an important role in many results on Hilbert functions.* The pivotal property is that a lex ideal grows as slowly as possible.

Another exciting direction of research is to *parametrize all graded ideals in S with a fixed Hilbert function, and then study their (common) properties and the structure of the parameter space.* Lex ideals play crucial role in Hartshorne's Theorem that Grothendieck's Hilbert scheme is connected.

40 Notation

Let W be a graded finitely generated R-module. It decomposes as a direct sum of its components $W = \oplus_{q \geq 0} W_q$. Its Hilbert function is defined by $q \to \dim_k W_q$. We denote

$$|W_q| = \dim_k(W_q)\,.$$

Recall that the Hilbert series of W is

$$\mathrm{Hilb}_W(t) = \sum_{q \geq 0} \dim_k(W_q)\, t^q\,.$$

I. Peeva, *Graded Syzygies*, Algebra and Applications 14,
DOI 10.1007/978-0-85729-177-6_2, © Springer-Verlag London Limited 2011

Throughout this chapter V stands for a graded finitely generated S-module.

41 Lex ideals

Macaulay's Theorem 41.7 characterizes the Hilbert functions of graded ideals in S. The theorem is well-known and has many applications. The key idea is that each Hilbert function is attained by a lex ideal. Lex ideals are highly structured: they are defined combinatorially and it is easy to derive the inequalities characterizing their Hilbert functions. They play other important roles; for example,

- Hartshorne's [Hartshorne 2] proof that the Hilbert scheme is connected uses lex ideals in an essential way.
- The homological properties of lex ideals are combinatorially tractable by Theorem 41.9. This leads to results in Section 47, showing that the lex ideals have greatest Betti numbers.

Notation and Definition 41.1. Recall that S_q is the k-vector space spanned by all monomials in S of degree q. So, S_1 is the k-vector space spanned by the variables. We order the variables lexicographically by $x_1 > \ldots > x_n$. We denote by $\succ_{lex}$ the degree-lex order on the monomials, that is, $m \succ_{lex} m'$ if either $\deg(m) > \deg(m')$ or $\deg(m) = \deg(m')$ and m is lex-greater than m'. Sometimes we say lex-last instead of lex-smallest.

We say that A_q is an ***S_q-monomial space*** if it can be spanned by monomials of degree q. We denote by $\{A_q\}$ the set of monomials (non-zero monomials in S_q) contained in A_q. The cardinality of this set is $|A_q| = \dim_k A_q$. By S_1A_q we mean the k-vector subspace $(A_q)_{q+1}$ of S_{q+1}, (where (A_q) is the ideal generated by the elements in A_q).

The ***lex-segment*** $M_{q,p}$ of length p in degree q is defined as the k-vector space spanned by the lex-greatest p monomials in S_q. An S_q-monomial space M_q is ***lex*** in S_q if there exists a p such that $M_q = M_{q,p}$. The monomial space 0 is lex in S_q by convention. For a monomial space A_q, we say that $M_{q,|A_q|}$ is its ***S_q-lexification.***

For an S_q-monomial space A_q sometimes we say for simplicity that A_q is a monomial space in S_q or a monomial space; in the latter case the index q indicates that $A_q \subseteq S_q$.

An S_q-monomial space T_q is ***greater lexicographically*** than an S_q-monomial space A_q if when we order the monomials in $\{T_q\}$ and $\{A_q\}$ lexicographically, and then compare the two ordered sets lexicographically, we get that the first ordered set is greater.

Proposition 41.2. *If a monomial space M_q is lex in S_q, then $S_1 M_q$ is lex in S_{q+1}.*

Proof. Let $m \in M_q$ be a monomial and let $u \succ_{lex} x_i m$ be a monomial in S_{q+1}. We have to show that $u \in S_1 M_q$. Write $x_i m = m'z$, where z is the lex-last variable that divides $x_i m$, and $m' = \frac{x_i m}{z}$. It follows that $m' \succ_{lex} m$, so $m' \in M_q$.

Similarly write $u = u'y$, where where y is the lex-last variable that divides u, and $u' = \frac{u}{y}$. Since $u'y = u \succ_{lex} x_i m = m'z$, it follows that $u' \succ_{lex} m'$. As $m' \in M_q$ and M_q is lex, we get that $u' \in M_q$. Therefore, $u = yu' \in S_1 M_q$. □

Proposition 41.3. *Let L be a monomial ideal in S. The following conditions are equivalent.*

1. *For each $q \geq 0$, we have that L_q is lex.*
2. *If m is a monomial, such that $m \succ_{lex} m'$ and $\deg(m) = \deg(m')$ for some monomial $m' \in L$, then $m \in L$.*
3. *Let p be a number, such that L has no minimal monomial generators in degrees $> p$. For each $q \leq p$, we have that L_q is lex.*
4. *Let L be minimally generated by the monomials $l_1, \dots, l_r$. If m is a monomial, $m \succ_{lex} l_i$ and $\deg(m) = \deg(l_i)$ for some $1 \leq i \leq r$, then $m \in L$.*

Proof. (1) $\iff$ (2) and (3) $\Longrightarrow$ (4) by the definition of lex-segment.

We will show that (4) $\Longrightarrow$ (3) by induction on the degree q. Suppose that L_q is lex; we will prove that L_{q+1} is lex as well.

If L has no minimal monomial generators of degree $q+1$, then by Proposition 41.2 it follows that L_{q+1} is lex.

If u is the lex-last minimal monomial generator of L of degree

$q+1$, then by Proposition 41.2 and (4) it follows that L_{q+1} is the lex monomial space in S_q whose end (that is, whose lex-last monomial) is u.

(1) $\Longrightarrow$ (3). By 41.2 it follows that (3) implies (1). □

Definition 41.4. A monomial ideal L is ***lex*** (or ***lexicographic***) if it satisfies the equivalent conditions in Proposition 41.3.

We usually use (4) in order to show that a given ideal is lex. On the other hand, (1) is the condition usually used in proofs.

Example 41.5. By (4), the ideal $(x_1^2, x_1x_2, x_1x_3, x_2^5, x_2^4x_3, x_2^3x_3^2, x_2^2x_3^3, x_2x_3^6, x_3^9)$ is lex in $k[x_1, x_2, x_3]$.

We are ready to discuss Macaulay's Theorem 41.7, which characterizes the Hilbert functions of graded ideals in S.

Proposition 41.6. *The following properties are equivalent.*

(1) *Let A_q be an S_q-monomial space and L_q be its lexification in S_q. Then $|\, S_1L_q \,| \leq |\, S_1A_q \,|$.*
(2) *For every graded ideal J in S there exists a lex ideal L with the same Hilbert function.*

The key property of lex ideals is expressed in (1) above: among all subspaces of the same dimension, the lex monomial space generates as little as possible in the next degree.

Proof. We will prove that (1) and (2) are equivalent. (2) implies (1). Assume that (1) holds. We will prove (2). We can assume that J is a monomial ideal by Gröbner basis theory. For each $q \geq 0$, let L_q be the lexification of J_q. By (1), it follows that $L = \oplus_{q\geq 0} L_q$ is an ideal. By construction, it is a lex-ideal and has the same Hilbert function as J in all degrees. □

In Section 45, we will prove that (1) holds which will establish Macaulay's Theorem.

Macaulay's Theorem 41.7. *The equivalent properties in Proposition* 41.6 *hold.*

space with disjoint union

$$\{A_q\} = \coprod_{0 \le j \le q} x_i^{q-j}\{U_j\}$$

where each U_j is a monomial space in $(S/x_i)_j$. For each $0 \le j \le q$, let L_j be the lexification of the space U_j in $(S/x_i)_j$. The S_q-monomial space C_q defined by

$$\{C_q\} = \coprod_{0 \le j \le q} x_i^{q-j}\{L_j\}$$

is the ***i-compression*** of A_q. Clearly, $|C_q| = |A_q|$.

Example 42.7. Let A_2 be the S_2-monomial space spanned by $\{x_1^2, x_2x_3, x_2^2, x_3x_4\}$. We have the disjoint union

$$\{A_2\} = x_2^2\{1\} \coprod x_2\{x_3\} \coprod 1\{x_1^2, x_3x_4\}$$

so U_2 is spanned by $\{x_1^2, x_3x_4\}$, U_1 is spanned by $\{x_3\}$, and U_0 is spanned by $\{1\}$. Therefore L_2 is spanned by $\{x_1^2, x_1x_3\}$, L_1 is spanned by $\{x_1\}$, and L_0 is spanned by $\{1\}$. The 2-compression of A_2 is

$$\{C_2\} = x_2^2\{1\} \coprod x_2\{x_1\} \coprod 1\{x_1^2, x_1x_3\}.$$

Lemma 42.8. *Let A_q be an S_q-monomial space. Fix an $1 \le i \le n$. Let C_q be the i-compression of A_q. We have that $|C_q| = |A_q|$ and $|S_1C_q| \le |S_1A_q|$.*

Proof. We use induction on the number of variables, and assume that Theorem 41.7(1) holds for $n-1$ variables.

Suppose that A_q is not i-compressed. Set $z = x_i$ and $\mathbf{n} = S_1/z$. Use the notation in Construction 42.6. We have the disjoint unions

$$\{S_1A_q\} = \coprod_{0 \le j \le q+1} z^{q-j+1}\{U_j + \mathbf{n}U_{j-1}\}$$

$$\{S_1C_q\} = \coprod_{0 \le j \le q+1} z^{q-j+1}\{L_j + \mathbf{n}L_{j-1}\}.$$

We will show that

$$|L_j + \mathbf{n}L_{j-1}| = \max\Big\{ |L_j|, |\mathbf{n}L_{j-1}| \Big\}$$
$$\le \max\Big\{ |U_j|, |\mathbf{n}U_{j-1}| \Big\} \le |U_j + \mathbf{n}U_{j-1}| .$$

The first equality above holds because both L_j and $\mathbf{n}L_{j-1}$ are lex $(S/z)_j$-monomial spaces, so $L_j + \mathbf{n}L_{j-1}$ is the longer of these two lex monomial spaces. The last inequality is obvious. The middle inequality holds since: by construction L_{j-1} is the lexification of U_{j-1}, so $|L_{j-1}| = |U_{j-1}|$ and by induction on the number of variables we can apply Macaulay's Theorem 41.7 to the ring S/z.

Thus, $|L_j + \mathbf{n}L_{j-1}| \le |U_j + \mathbf{n}U_{j-1}|$ for each j. This implies the desired inequality $|S_1 C_q| \le |S_1 A_q|$. □

Compression Lemma 42.9. (Clements-Lindström) *Let A_q be an S_q-monomial space. There exists a compressed monomial space T_q in S_q such that $|T_q| = |A_q|$ and $|S_1 T_q| \le |S_1 A_q|$.*

Proof. Suppose that A_q is not i-compressed for some $1 \le i \le n$. Let C_q be the i-compression of A_q. By the above lemma, we have that $|C_q| = |A_q|$ and $|S_1 C_q| \le |S_1 A_q|$.

Note that $\{C_q\}$ is greater lexicographically than $\{A_q\}$. If C_q is not compressed, we can apply the argument above. After finitely many steps in this way, the process must terminate because at each step we construct a lexicographically greater S_q-monomial space. Thus, after finitely many steps, we reach a compressed monomial space. □

43 Multicompression

In this section we describe a multigraded version of the technique of compression.

Let $\mathcal{A} \subset \{x_1, \ldots, x_n\}$; its ***complement*** is $\mathcal{A}^c = \{x_1, \ldots, x_n\} \setminus \mathcal{A}$. Denote by $\oplus_m$ the direct sum over all monomials m in the variables in $\mathcal{A}^c$. An S_q-monomial space C_q can be written uniquely in the form

$$C_q = \bigoplus_m m\, V_m ,$$

where V_m is a monomial space in the ring $k[\mathcal{A}] = k[x_i \,|\, x_i \in \mathcal{A}]$.

We say that C_q is ***$\mathcal{A}$-multicompressed*** if each V_m is lex in $k[\mathcal{A}]$.

Furthermore, we say that C_q is ***(j)-multicompressed*** if it is $\mathcal{A}$-multicompressed for every set $\mathcal{A}$ of size j. We say that C_q is ***multicompressed*** if it is $\mathcal{A}$-multicompressed for every set $\mathcal{A}$.

A monomial ideal P is ***$\mathcal{A}$-multicompressed*** if P_q is $\mathcal{A}$-multicompressed for all $q \geq 0$. The ideal is ***(j)-multicompressed*** if P_q is (j)-multicompressed for all $q \geq 0$.

Example 43.1. Let $\mathcal{A} = \{x_1, x_3\} \subset \{x_1, x_2, x_3, x_4\}$ and C_2 be spanned by the monomials

$$x_2^2, x_1x_2, x_1^2, x_1x_3, x_4^2, x_1x_4, x_2x_4 .$$

We have the decomposition

$$\{C_2\} = x_2^2\{1\} \coprod x_2\{x_1\} \coprod 1\{x_1^2, x_1x_3\} \coprod x_4^2\{1\} \\ \coprod x_4\{x_1\} \coprod x_2x_4\{1\} .$$

We see that

$$\{V_{x_2^2}\} = \{1\}, \qquad \{V_{x_2}\} = \{x_1\},$$
$$\{V_{x_4^2}\} = \{1\}, \qquad \{V_1\} = \{x_1^2, x_1x_3\},$$
$$\{V_{x_2x_4}\} = \{1\}, \qquad \{V_{x_4}\} = \{x_1\}$$

are all lex, so C_2 is $\{x_1, x_3\}$-compressed.

Exercise 43.2. *If C_q is $\mathcal{A}$-multicompressed in S_q, then S_1C_q is $\mathcal{A}$-multicompressed in S_{q+1}.*

Exercise 43.3. *Let P be a monomial ideal and p be a number, such that P has no minimal monomial generators in degrees $> p$. If P_q is $\mathcal{A}$-multicompressed for every $0 \leq q \leq p$, then P is $\mathcal{A}$-multicompressed.*

Exercise 43.4. *If L_q is lex, then it is $\mathcal{A}$-multicompressed for every set $\mathcal{A}$.*

Exercise 43.5. *If $\mathcal{A}'$ is a subset of $\mathcal{A}$ and C_q is $\mathcal{A}$-multicompressed in S_q, then C_q is $\mathcal{A}'$-multicompressed.*

Exercise 43.6. *If C_q is (j)-multicompressed, then it is (i)-multicompressed for every $i \le j$.*

Structure Theorem 43.7. [Mermin]

(1) *A monomial space C_q is Borel if and only if it is (2)-multicompressed.*

(2) *A monomial space C_q is lex if and only if it is (3)-multicompressed.*

Proof. First, we prove (1).

Let C_q be (2)-multicompressed. We will prove that it is Borel. Let $x_j m' \in C_q$ be a monomial and fix an $1 \le i < j$. Set $\mathcal{A} = \{x_i, x_j\}$. Write $x_j m' = x_i^s x_j^t m$ so that m is not divisible by either x_i or x_j. Hence $x_i^s x_j^t \in \{V_m\}$. The monomial $x_i^{s+1} x_j^{t-1}$ is lex-greater than $x_i^s x_j^t$. Since V_m is lex, it follows that $x_i^{s+1} x_j^{t-1} \in \{V_m\}$. Hence $x_i m' \in C_q$.

Let C_q be a Borel monomial space. We will prove that it is (2)-multicompressed. Fix a set $\mathcal{A} = \{x_i, x_j\}$ with $1 \le i < j$. We will show that each V_m is lex. Let $x_i^s x_j^t \in V_m$. Let $x_i^{s+h} x_j^{t-h}$ be a monomial that is lex-greater than $x_i^s x_j^t$. Since $x_i^s x_j^t m \in C_q$ and C_q is Borel, it follows that $x_i^{s+h} x_j^{t-h} m \in C_q$. Hence $x_i^{s+h} x_j^{t-h} \in V_m$. Therefore, V_m is lex.

Now, we prove (2). If C_q is lex then it is (3)-multicompressed by Exercise 43.4. Suppose that C_q is (3)-multicompressed. We will show that it is lex. By (1) and Exercise 43.6, it follows that C_q is Borel.

Let $u = x_1^{a_1} \dots x_n^{a_n}$ be a monomial in C_q. Let $v = x_1^{c_1} \dots x_n^{c_n}$ be a monomial that is lex-greater than u. We will show that $v \in C_q$. Let i be minimal so that $a_i \neq c_i$. Then $a_i < c_i$ since v is lex-greater than u. Set $w = x_1^{a_1} \dots x_{i-1}^{a_{i-1}}$ and $e = \deg(x_{i+1}^{a_{i+1}} \dots x_n^{a_n}) = a_{i+1} + \dots + a_n$.

Since $u \in C_q$, we can use that C_q is Borel in order to conclude that $wx_i^{a_i}x_{i+1}^e \in C_q$. Set $\mathcal{A} = \{x_i, x_{i+1}, x_n\}$. Then $x_i^{a_i}x_{i+1}^e \in V_w$. As C_q is $\{x_i, x_{i+1}, x_n\}$-multicompressed, it follows that V_w is lex. The monomial $x_i^{a_i+1}x_n^{e-1}$ is lex-greater than $x_i^{a_i}x_{i+1}^e$, so $x_i^{a_i+1}x_n^{e-1} \in V_w$. Hence $wx_i^{a_i+1}x_n^{e-1} \in C_q$. As C_q is Borel it follows that $v \in C_q$. □

The following is an immediate corollary.

Structure Theorem 43.8. [Mermin]

(1) *If $n < 3$, then every monomial space is multicompressed.*

(2) *If $n = 3$, then the multicompressed monomial spaces are exactly the Borel spaces.*

(3) *If $n > 3$ then the multicompressed monomial spaces are exactly the lex spaces.*

The following lemma is proved similarly to the Compression Lemma 42.9.

Lemma 43.9. *Let $\mathcal{A} \subset \{x_1, \dots, x_n\}$. Let A_q be an S_q-monomial space. There exists an $\mathcal{A}$-compressed monomial space T_q in S_q such that $|T_q| = |A_q|$ and $|S_1T_q| \le |S_1A_q|$.*

Lemma 43.10. *Fix a $1 \le j \le n-1$. Let A_q be an S_q-monomial space. There exists a (j)-compressed monomial space C_q in S_q such that $|C_q| = |A_q|$ and $|S_1C_q| \le |S_1A_q|$.*

Proof. Apply Lemma 43.9 repeatedly if necessary. □

44 Green's Theorem

Green's Theorem describes the change in the Hilbert function when we factor out a generic form.

For a monomial m define

$$\max(m) = \max\{i \mid x_i \text{ divides } m\}$$
$$\min(m) = \min\{i \mid x_i \text{ divides } m\}.$$

For an S_q-monomial space A_q set

$$r_{i,j}(A_q) = \Big|\{m \in \{A_q\} \mid \max(m) \leq i \text{ and } x_i^j \text{ does not divide } m\}\Big|$$

$$t_i(A_q) = \Big|\{m \in \{A_q\} \mid \max(m) \leq i\}\Big|.$$

Lemma 44.1. (Bigatti) *If an S_q-monomial space B_q is Borel, then $\{S_1 B_q\}$ is the set*

$$\mathcal{B} = \coprod_{i=1}^{n} x_i \{m \in \{B_q\} \mid \max(m) \leq i\}$$

and

$$\Big|\{S_1 B_q\}\Big| = \sum_{i=1}^{n} t_i(B_q).$$

Proof. Let $w \in \{B_q\}$. For $j \geq \max(w)$ we have that $x_j w \in \mathcal{B}$. Let $j < \max(w)$. Then $v = x_j \dfrac{w}{x_{\max(w)}} \in B_q$. So, $x_j w = x_{\max(w)} v \in \mathcal{B}$. $\square$

Lemma 44.2. *Let A_q be a Borel S_q-monomial space. Its n-compression C_q is Borel.*

Proof. We use the notation in Construction 42.6. Consider the disjoint unions

$$\{A_q\} = \coprod_{0 \leq j \leq q} x_n^{q-j} \{U_j\}$$

$$\{C_q\} = \coprod_{0 \leq j \leq q} x_n^{q-j} \{L_j\}.$$

Since A_q is Borel, it follows that

$$(S_1/x_n) U_j \subseteq U_{j+1}.$$

We use induction on the number of variables, and assume that Theorem 41.7(1) holds for $n-1$ variables. Since $|L_j| = |U_j|$, by Theorem 41.7(1) it follows that

$$|(S_1/x_n) L_j| \leq |(S_1/x_n) U_j| \leq |U_{j+1}| = |L_{j+1}|.$$

As both $(S_1/x_n)L_j$ and L_{j+1} are lex monomial spaces, we conclude that $(S_1/x_n)L_j \subseteq L_{j+1}$. Let $x_n^{q-j}m$ be a monomial in C_q and $m \in L_j$. Then for each $1 \leq i < n$ we have that $x_i m \in (S_1/x_n)L_j \subseteq L_{j+1}$, so $x_n^{q-j-1}x_i m \in C_q$. If x_p divides m, then for each $1 \leq c \leq p$ we have that $\frac{x_c m}{x_p} \in L_j$ since L_j is lex. We proved that C_q is Borel. □

The main work for proving the Generalized Green's Theorem 44.5 is in the following lemma.

Lemma 44.3. *Let C_q be an n-compressed Borel S_q-monomial space, and let L_q be a lex monomial space in S_q with $|L_q| \leq |C_q|$. For each $1 \leq i \leq n$ and each $1 \leq j$ we have the inequality*

$$r_{i,j}(L_q) \leq r_{i,j}(C_q)\,.$$

Proof. Note that both L_q and C_q are Borel and n-compressed.

First, we consider the case $i = n$. Clearly, $r_{n,q+1}(L_q) = |L_q| \leq |C_q| = r_{n,q+1}(C_q)$. We induct on j decreasingly. Suppose that the inequality $r_{n,j+1}(L_q) \leq r_{n,j+1}(C_q)$ holds by induction.

If $\{C_q\}$ contains no monomial divisible by x_n^j then

$$r_{n,j}(L_q) \leq r_{n,j+1}(L_q) \leq r_{n,j+1}(C_q) = r_{n,j}(C_q)\,.$$

Suppose that $\{C_q\}$ contains a monomial divisible by x_n^j. Denote by $e = x_1^{e_1}\dots x_n^{e_n}$, with $e_n \geq j$, the lex-last monomial in C_q that is divisible by x_n^j.

Let $0 \leq p \leq j-1$. Let the monomial $v = x_1^{a_1}\dots x_{n-1}^{a_{n-1}}x_n^p \in S_q$ be lex-greater than e. Since C_q is Borel, it follows that $w = x_{n-1}^{e_n-p}\dfrac{e}{x_n^{e_n-p}} \in C_q$. This is the lex-last monomial that is lex-greater than e and x_n divides it at power p. Since C_q is n-compressed and v is lex-greater (or equal) than w, it follows that $v \in C_q$.

For a monomial u, we denote by $x_n^j \notin u$ the property that x_n^j does not divide u. By what we proved above, it follows that

(*)

$$\Big|\{u \in \{C_q\} \,|\, x_n^j \notin u,\ u \succ_{lex} e\}\Big| = \Big|\{u \in \{S_q\} \,|\, x_n^j \notin u,\ u \succ_{lex} e\}\Big|\,.$$

Therefore,

$$\begin{aligned}
& r_{n,j}(L_q) \\
&= |\{u \in \{L_q\} \,|\, x_n^j \notin u,\ u \succ_{lex} e\}| + |\{u \in \{L_q\} \,|\, x_n^j \notin u,\ u \prec_{lex} e\}| \\
&\le |\{u \in \{S_q\} \,|\, x_n^j \notin u,\ u \succ_{lex} e\}| + |\{u \in \{L_q\} \,|\, x_n^j \notin u,\ u \prec_{lex} e\}| \\
&\le |\{u \in \{S_q\} \,|\, x_n^j \notin u,\ u \succ_{lex} e\}| + |\{u \in \{L_q\} \,|\, u \prec_{lex} e\}| \\
&\le |\{u \in \{S_q\} \,|\, x_n^j \notin u,\ u \succ_{lex} e\}| + |\{u \in \{C_q\} \,|\, u \prec_{lex} e\}| \\
&= |\{u \in \{S_q\} \,|\, x_n^j \notin u,\ u \succ_{lex} e\}| + |\{u \in \{C_q\} \,|\, x_n^j \notin u,\ u \prec_{lex} e\}| \\
&= |\{u \in \{C_q\} \,|\, x_n^j \notin u,\ u \succ_{lex} e\}| + |\{u \in \{C_q\} \,|\, x_n^j \notin u,\ u \prec_{lex} e\}| \\
&= r_{n,j}(C_q)\,;
\end{aligned}$$

for the third inequality we used the fact that L_q is a lex monomial space in S_q with $|L_q| \le |C_q|$; for the equality after that we used the definition of e; for the next equality we used $(*)$. Thus, we have the desired inequality in the case $i = n$.

In particular, we proved that

$$(**) \qquad r_{n,1}(L_q) \le r_{n,1}(C_q)\,.$$

Finally, we prove the lemma for all $i < n$. Both $\{C_q/x_n\}$ and $\{L_q/x_n\}$ are lex monomial spaces in S_q/x_n since C_q is n-compressed. By (**) the inequality $r_{n,1}(L_q) \le r_{n,1}(C_q)$ holds, and it implies the inclusion $\{C_q/x_n\} \supseteq \{L_q/x_n\}$. The desired inequalities follow since

$$\begin{aligned}
r_{i,j}(C_q) &= r_{i,j}\big(C_q/(x_{i+1},\dots,x_n)\big) \\
r_{i,j}(L_q) &= r_{i,j}\big(L_q/(x_{i+1},\dots,x_n)\big)\,.
\end{aligned}$$

□

Comparison Theorem 44.4. *Let B_q be a Borel monomial space in S_q. Let L_q be a lex monomial space in S_q with $|L_q| \le |B_q|$. We have the inequalities*

$$\begin{aligned}
t_i(L_q) &\le t_i(B_q) \\
r_{i,j}(L_q) &\le r_{i,j}(B_q)\,.
\end{aligned}$$

for each $1 \leq i \leq n$ *and each* $1 \leq j$.

Proof. First, note that $t_i(A_q) = r_{i,q+1}(A_q)$ for any monomial space A_q. Thus, it suffices to prove the inequalities $r_{i,j}(L_q) \leq r_{i,j}(B_q)$.

We prove the inequalities by decreasing induction on the number of variables n. Let C_q be the n-compression of B_q. Since C_q is Borel and n-compressed by Lemma 44.2, we can apply Lemma 44.3 and we get

$$r_{i,j}(L_q) \leq r_{i,j}(C_q)$$

for each $1 \leq i \leq n$ and each $1 \leq j$. It remains to compare $r_{i,j}(C_q)$ and $r_{i,j}(B_q)$. For $i = n$, we have equalities $r_{n,j}(C_q) = r_{n,j}(B_q)$. Let $i < n$. Then $r_{i,j}(C_q) = r_{i,j}(C_q/x_n)$ and $r_{i,j}(B_q) = r_{i,j}(B_q/x_n)$, where $C_q/x_n = L_q$ is lex and $B_q/x_n = U_q$ is Borel in S/x_n. So, by induction the desired inequalities hold. □

Generalized Green's Theorem 44.5. *Let* B_q *be a Borel monomial space in* S_q. *Let* L_q *be a lex monomial space in* S_q *with* $|L_q| \leq |B_q|$. *The inequality*

$$\dim_k\Big(S_q/(L_q + x_n^j\, S_{q-j})\Big) \geq \dim_k\Big(S_q/(B_q + x_n^j\, S_{q-j})\Big)$$

holds for each $1 \leq j \leq q$.

Proof. Note that the desired inequality is equivalent to

$$r_{n,j}(L_q) \leq r_{n,j}(B_q)\,.$$

It holds by Theorem 44.4. □

Assume $\mathrm{char}(k) = 0$. Let I be a graded ideal in S and $R = S/I$. Fix an integer j. The affine space R_j is irreducible, so every nonempty Zariski-open subset is dense. We say that a property $\mathcal{P}$ ***holds for*** a ***generic*** j***-form*** if there exists a nonempty Zariski-open subset $\mathcal{U} \subseteq R_j$ such that the property $\mathcal{P}$ holds for every j-form in $\mathcal{U}$.

Lemma 44.6. *Assume* $\mathrm{char}(k) = 0$. *Suppose that* I *is a graded ideal in* S *and* $R = S/I$. *Fix integers* i *and* j. *Let*

$$t = \max\{\dim_k(g\,R_i)\,|\,g \in R_j\}$$

There exists a non-empty Zariski-open set $\mathcal{U} \subseteq R_j$ *such that* $\dim_k(h\, R_i) = t$ *for every generic* j*-form* $h \in \mathcal{U}$.

Proof. Let

$$\mathcal{U} = \{\, v \in R_j \mid \dim_k(vR_i) = t\} \subseteq R_j\,.$$

Choose a basis $f_1, \ldots, f_a$ of R_j and a basis $g_1, \ldots, g_c$ of R_i. The elements $f_p g_q$ span R_{i+j}, so we can choose a subset that is a basis. Write $v = \sum_{1 \leq p \leq a} \alpha_p f_p$, where the coefficients $\alpha_1, \ldots, \alpha_a$ are in k. The multiplication map $v:\ R_i \to R_{i+j}$ has a matrix M whose entries are linear forms in $\alpha_1, \ldots, \alpha_a$. A j-form v is in $\mathcal{U}$ if and only if the matrix M has a non-zero $(t \times t)$-minor. When we vary v, we can think of $\alpha_1, \ldots, \alpha_a$ as indeterminates which take values in k. Therefore, the complement of $V(I_t(M))$ is a Zariski-open set (here $I_t(M)$ is the ideal generated by all $(t \times t)$-minors of M, and $V(I_t(M))$ is the set on which all elements in $I_t(M)$ vanish). □

Exercise 44.7. *Assume* $\operatorname{char}(k) = 0$. *Let* I *be a graded ideal in* S *and* $R = S/I$. *Fix integers* i *and* j. *Let* $a = \min\{\dim_k((R/g)_i) \mid g \in R_j\}$. *Then* $\dim_k((R/h)_i) = a$ *for a generic* j*-form* h.

In Exercise 44.8 and Green's Theorem 44.9 by a generic j-form, we mean a j-form generic in the sense of Exercise 44.7.

Exercise 44.8. *Assume* $\operatorname{char}(k) = 0$. *Fix an integer* j. *Then* x_n^j *is a generic* j*-form for every Borel ideal in* S.

The following result is a straightforward corollary of Theorem 44.5 and Exercise 44.8.

Green's Theorem 44.9. (Herzog-Popescu), [Gasharov] *Assume that* $\operatorname{char}(k) = 0$. *Let* B_q *be a Borel monomial space in* S_q. *Let* L_q *be a lex monomial space in* S_q *with* $|L_q| \leq |B_q|$. *Let* g *be a generic form of degree* $j \geq 1$. *The inequality*

$$\dim_k \Big(S_q/(L_q + g\, S_{q-j}\Big) \geq \dim_k \Big(S_q/(B_q + g\, S_{q-j}\Big)$$

holds.

Green's Hyperplane Restriction Theorem 44.10. [Green] *Assume* char$(k) = 0$. *Let* J *be a graded ideal in* S*, and* L *be the lex ideal with the same Hilbert function as* J*. Let* h *be a generic linear form. For every* $q \geq 0$ *we have*

$$\dim_k \Big(S/(L,h)\Big)_q \geq \dim_k \Big(S/(J,h)\Big)_q .$$

Proof. Assume that we work in generic coordinates, so we can take $x_n = h$. Note that when we take the initial ideal with respect to revlex order we get $\mathrm{in}(J, x_n) = (\mathrm{in}(J), x_n)$. Therefore, we can replace J by $B = \mathrm{in}(J)$. By Theorem 28.4, the ideal B is Borel. Hence, Theorem 44.5 yields the desired result. □

Green's Theorem holds without the restriction char$(k) = 0$, see [Gasharov].

45 Proofs of Macaulay's Theorem

We are ready to prove Macaulay's Theorem 41.7; namely, we will prove that (1) in Proposition 41.6 holds. It is straightforward that (1) holds if $n \leq 2$. Consider the case $n \geq 3$. Applying Lemma 43.10, we conclude that there exist a (2)-multicompressed monomial space C_q such that $|C_q| = |A_q|$ and $|S_1 C_q| \leq |S_1 A_q|$. By Theorem 43.7 it follows that C_q is Borel. Let L_q be the lex monomial space for which $|C_q| = |L_q|$. We will prove that $|S_1 L_q| \leq |S_1 C_q|$.

We will present two different proofs. The former uses Green's Theorem. The latter uses the structure theorem for compressed ideals. A third proof by induction is given in [Mermin-Peeva].

Proof.
First Proof. This proof uses Green's Theorem. The monomial space C_q is Borel. For an S_q-monomial space D_q recall that $t_i(D_q) = \Big|\{ m \in \{D_q\} \mid \max(m) \leq i \}\Big|$. We apply Lemma 44.1 to conclude that

$$\Big|\{S_1 C_q\}\Big| = \sum_{i=1}^{n} t_i(C_q) \quad \text{and} \quad \Big|\{S_1 L_q\}\Big| = \sum_{i=1}^{n} t_i(L_q) .$$

Finally, we apply Theorem 44.4 and get $\big|\{S_1L_q\}\big| \le \big|\{S_1C_q\}\big|$.

Second Proof. (Mermin) This proof is by compression. Let $n > 3$. Applying Lemma 43.10, we conclude that there exist a (3)-multicompressed monomial space C_q such that $|C_q| = |A_q|$ and $|S_1C_q| \le |S_1A_q|$. By Theorem 43.7 it follows that C_q is lex, and we are done.

Suppose that $n = 3$. Let L_q be the lex S_q-monomial space such that $|L_q| = |C_q|$. As both L_q and C_q are Borel, we have

$$|S_1C_q| = |C_q| + |C_q \cap k[x_1, x_2]| + |C_q \cap k[x_1]|$$
$$|S_1L_q| = |L_q| + |L_q \cap k[x_1, x_2]| + |L_q \cap k[x_1]|$$

by Lemma 44.1. Note that $|L_q| = |C_q|$ by construction, and $|C_q \cap k[x_1]| = |L_q \cap k[x_1]| = 1$ as $\{C_q \cap k[x_1]\} = \{L_q \cap k[x_1]\} = x_1^q$. Therefore, we need to prove that $|L_q \cap k[x_1, x_2]| \le |C_q \cap k[x_1, x_2]|$. We will show that if a monomial $v \in L_q$ is not in C_q, then $v \notin k[x_1, x_2]$. Assume the opposite: let $v = x_1^a x_2^c \in L_q$ and $v \notin C_q$. As $L_q \neq C_q$ we conclude that there exists a monomial $x_1^{a'} x_2^{c'} x_3^{e'} \in C_q$ that is lex-smaller than v. Hence $a' \le a$. Since C_q is Borel, it follows that $v \in C_q$, which is a contradiction. □

46 Compression ideals

Proposition 41.6 makes it possible to work in our arguments by focusing on only two consecutive degrees at a time (instead of dealing with the whole ideal). In this section we show that the compressions can be assembled into an ideal.

Construction 46.1. Fix an $1 \le i \le n$. Let A be a monomial ideal in S. For each $q \ge 0$, let C_q be the i-compression of A_q. We call $C = \oplus_{0 \le q} C_q$ the ***i-compression*** of A.

As a corollary of Macaulay's Theorem, we will prove the following result.

Proposition 46.2. *Let A be a monomial ideal in S and fix an $1 \le i \le n$. Its i-compression C is an ideal.*

Proof. We use the following notation. For each $q \geq 0$ we have a disjoint union

$$\{A_q\} = \coprod_{0 \leq j \leq q} x_i^{q-j}\{U_j^q\}$$

where each U_j^q is a monomial space in $(S/x_i)_j$. Let

$$\{C_q\} = \coprod_{0 \leq j \leq q} x_i^{q-j}\{L_j^q\}$$

be the i-compression of A_q. Thus, L_j^q is the lexification of U_j^q in S/x_i. The i-compression of A is $C = \oplus_{0 \leq q} C_q$.

Fix a $q \geq 0$ and a $0 \leq j \leq q$. Let $m \in x_i^{q-j}L_j$ be a monomial. We will prove that $S_1 m \in C$.

We will show that $(S_1/x_i)L_j^q \subseteq L_{j+1}^{q+1}$. Both $(S_1/x_i)L_j^q$ and L_{j+1}^{q+1} are lex monomial spaces. So, in order to show that $(S_1/x_i)L_j^q \subseteq L_{j+1}^{q+1}$ it suffices to show that $|(S_1/x_i)L_j^q| \leq |L_{j+1}^{q+1}|$. This first inequality below follows from Macaulay's Theorem, and the second inequality holds since A is an ideal:

$$|(S_1/x_i)L_j^q| \leq |(S_1/x_i)U_j^q| \leq |U_{j+1}^{q+1}| = |L_{j+1}^{q+1}|\,.$$

Since $(S_1/x_i)L_j^q \subseteq L_{j+1}^{q+1}$, it follows that $(S_1/x_i)m \in C$.

It remains to prove that $x_i m \in C$. We will show that $L_j^q \subseteq L_j^{q+1}$. Both L_j^q and L_j^{q+1} are lex monomial spaces in $(S/x_i)_j$. So, in order to show that $L_j^q \subseteq L_j^{q+1}$ it suffices to show that $|L_j^q| \leq |L_j^{q+1}|$. Since A is an ideal, we have that $U_j^q \subseteq U_j^{q+1}$. Hence

$$|L_j^q| = |U_j^q| \leq |U_j^{q+1}| = |L_j^{q+1}|.$$

The inclusion $L_j^q \subseteq L_j^{q+1}$ implies that $x_i m \in C$. □

We will see that the situation is similar for multicompression.

Construction 46.3. Fix a set $\mathcal{A} \subset \{x_1, \ldots, x_n\}$. An S_q-monomial space A_q can be written uniquely in the form

$$A_q = \bigoplus_m m U_m,$$

where U_m is a monomial space in the ring $k[\mathcal{A}] = k[x_i \mid x_i \in \mathcal{A}]$. For each m, let L_m be the lexification of the space U_m in $k[\mathcal{A}]$. The monomial space C_q defined by

$$C_q = \bigoplus_m m L_m$$

is the ***$\mathcal{A}$-compression*** of A_q. Clearly, $|C_q| = |A_q|$.

Let A be a monomial ideal in S. For each $q \geq 0$, let C_q be the $\mathcal{A}$-compression of A_q. We call $C = \oplus_{0 \leq q} C_q$ the ***$\mathcal{A}$-compression*** of A.

The following result can be proved similarly to Proposition 46.2.

Proposition 46.4. *Let A be a monomial ideal in S and fix a set $\mathcal{A} \subset \{x_1, \dots, x_n\}$. The $\mathcal{A}$-compression C of A is an ideal.*

47 Ideals with a fixed Hilbert function

The problem *"What can be said about the properties of ideals with a fixed Hilbert function?"* has received a lot of attention. Evans raised the problem to study the properties of the Betti diagrams of all graded ideals in S with a fixed Hilbert function; since the problem is very complex in general, people focused on maximal and on minimal Betti numbers. We will show that a lex ideal attains the greatest Betti numbers among all ideals with a fixed Hilbert function.

For simplicity, we assume throughout this section that $\mathrm{char}(k) = 0$. If M is a monomial ideal, then $G(M)_j$ stands for the set of monomials of degree j in the minimal system of monomial generators of M, and furthermore we denote by $|G(M)_j|$ the number of monomials in $G(M)_j$.

Let J be a graded ideal in S. By Macaulay's Theorem 41.7, there exists a lex ideal L with the same Hilbert function as J. The next result follows by Proposition 41.6.

Proposition 47.1. *For every $j \geq 0$, the number of elements of degree j in a minimal system of homogeneous generators of J is $\leq |G(L)_j|$.*

This property extends to all graded Betti numbers as follows.

Theorem 47.2. (Bigatti, Hulett) *Assume* $\mathrm{char}(k) = 0$. *Let J be a graded ideal in S. If L is the lex ideal with the same Hilbert function as J, then*

$$b^S_{i,i+j}(J) \leq b^S_{i,i+j}(L) \quad \textit{for all } i, j.$$

Remark 47.3. It is proved in [Pardue] that Theorem 47.2 holds without the assumption $\mathrm{char}(k) = 0$.

Note that the minimal free resolution and the Betti numbers of a lex ideal are given by the Eliahou-Kervaire resolution, see Section 28.

Recall that for an S_q-monomial space A_q we set

$$t_i(A_q) = \Big|\{ m \in \{A_q\} \mid \max(m) \leq i \}\Big| .$$

Set

$$u_i(A_q) = t_i(A_q) - t_{i-1}(A_q) = \Big|\{ m \in \{A_q\} \mid \max(m) = i \}\Big| .$$

Lemma 47.4. *If M is a Borel ideal in S, then $b^S_{i,i+j}(M)$ is equal to*

$$|\, M_j \,| \binom{n-1}{i} - \sum_{p=1}^{n-1} t_p(M_j) \binom{p-1}{i-1} - \sum_{p=1}^{n} t_p(M_{j-1}) \binom{p-1}{i} .$$

Proof. By Corollary 28.12, we have that

$$b^S_{i,i+j}(M) = \sum_{m \in G(M)_j} \binom{\max(m)-1}{i} = \sum_{p=1}^{n} u_p(G(M)_j) \binom{p-1}{i} .$$

Since

$$G(M)_j = \{M_j\} \setminus \{S_1 M_{j-1}\}$$

we obtain

$$b^S_{i,i+j}(M) = \sum_{p=1}^{n} u_p(M_j)\binom{p-1}{i} - \sum_{p=1}^{n} u_p(S_1 M_{j-1})\binom{p-1}{i}.$$

Furthermore, since M is Borel, by Lemma 44.1 we have

$$\{S_1 M_{j-1}\} = \coprod_{p=1}^{n} x_p\{m \in \{M_{j-1}\} \mid \max(m) \leq p\},$$

and hence $u_p(S_1 M_{j-1}) = t_p(M_{j-1})$. Therefore,

$$\begin{aligned} &b^S_{i,i+j}(M) \\ &= \sum_{p=1}^{n} \Big(t_p(M_j) - t_{p-1}(M_j)\Big)\binom{p-1}{i} - \sum_{p=1}^{n} t_p(M_{j-1})\binom{p-1}{i} \\ &= |\,M_j\,|\binom{n-1}{i} - \sum_{p=1}^{n-1} t_p(M_j)\binom{p-1}{i-1} - \sum_{p=1}^{n} t_p(M_{j-1})\binom{p-1}{i}. \end{aligned}$$

□

Proof of Theorem 47.2. We will present the proof in [Chardin-Gasharov-Peeva]. Let M be the generic initial ideal of J with respect to a fixed term order (say, revlex). It is Borel, by Theorem 28.4. Thus, there exists a Borel ideal M with the same Hilbert function as J such that

$$b^S_{i,i+j}(J) \leq b^S_{i,i+j}(M) \quad \text{for all } i, j.$$

Both M and L are Borel ideals. Use the formula for the Betti numbers in Lemma 47.4 and apply 44.4 to obtain the inequalities

$$b^S_{i,i+j}(M) \leq b^S_{i,i+j}(L).$$

□

Theorem 47.5. *There is an upper bound on the regularities of all graded ideals with a fixed Hilbert function.*

Proof. By Remark 47.3, it follows that the regularity of the lex ideal with that Hilbert function is the smallest upper bound. □

Problem 47.6. [Geramita-Harima-Shin] *Does there exist an ideal that has greatest graded Betti numbers among all Gorenstein artinian graded ideals with a fixed Hilbert function?*

Next, we will discuss the following question: *Assume* char$(k) = 0$. *Let J be a graded ideal in S and let L be the lex ideal with the same Hilbert function. How do the graded Betti numbers of J and L differ?* We would like to obtain more precise information than Theorem 47.2.

The Hilbert function can be computed from the graded Betti numbers by Theorem 16.2 and we get

$$\begin{array}{ccc} \sum_{j=0}^{\infty} \dim_k(S/J)_j\, t^j & = & \dfrac{\sum_{j=0}^{\infty}\sum_{i=0}^{n}(-1)^i b_{i,j}^S(S/J)\, t^j}{(1-t)^n} \\ \| & & \\ \sum_{j=0}^{\infty} \dim_k(S/L)_j\, t^j & = & \dfrac{\sum_{j=0}^{\infty}\sum_{i=0}^{n}(-1)^i b_{i,j}^S(S/L)\, t^j}{(1-t)^n}\,. \end{array}$$

These equalities imply that the graded Betti numbers $b_{i,j}^S(S/J)$ and $b_{i,j}^S(S/L)$ are related as described below.

Given a sequence of numbers $\{c_{p,q}\}$, we obtain a new sequence by a ***cancellation*** as follows: fix a q, and choose $p < t$ so that one of the numbers is odd and the other is even; then replace $c_{p,q}$ by $c_{p,q} - 1$, and replace $c_{t,q}$ by $c_{t,q} - 1$. The equalities above imply that the graded Betti numbers $b_{i,j}^S(S/J)$ are related to the graded Betti numbers $b_{i,j}^S(S/L)$ by a sequence of cancellations. This has been observed and applied in order to study the Betti diagrams of ideals with a fixed Hilbert function. Recall the definition of a consecutive cancellation in Section 22.

Theorem 47.7. [Peeva 2] *Let J be a graded ideal and L be the lex ideal in S with the same Hilbert function. The graded Betti numbers*

$b^S_{i,j}(S/J)$ can be obtained from the graded Betti numbers $b^S_{i,j}(S/L)$ by a sequence of consecutive cancellations.

Extending Hartshorne's method [Hartshorne] Pardue proved the next result. There, by a sequence of deformations we mean a composition of deformations (applied one after another).

Theorem 47.8. [Pardue] *Every two graded ideals in a polynomial ring with the same Hilbert function are connected by a sequence of deformations over $\mathbf{A}^1_k$.*

More precisely, in the notation of 47.7 Pardue proved that J and L are connected by a sequence of deformations of the following three types:

(1) generic change of coordinates
(2) deformation between an ideal and an initial ideal; see Theorem 22.8
(3) polarization and then factoring out generic hyperplane sections; more precisely, applying $\sigma'_{\mathbf{L}}$ defined in [Pardue, Section 4].

We are ready to prove Theorem 47.7.

Proof. The graded Betti numbers are preserved under (1). For (2) we apply Theorem 22.9. By Theorem 21.10, we have that (3) preserves the graded Betti numbers as well. □

Theorem 47.7 can be used in order to prove that certain Hilbert functions are not attained within a given class of ideals.

It should be noted that the there are many examples where the existence of possible consecutive cancellations does not imply the existence of an ideal for which those cancellations are realized.

Corollary 47.9. *Let L be a lex ideal. Suppose that L does not have two minimal monomial generators in consecutive degrees. If J is a graded ideal with the same Hilbert function as L, then J has the same graded Betti numbers as L.*

The following can be explored.

Open-Ended Problem 47.10. (folklore) *Let J be a graded ideal in*

S and let L be the lex ideal with the same Hilbert function. Which consecutive cancellations occur as cancellations when we are comparing the graded Betti numbers of L and J, in the case when some additional properties of J (e.g. monomial, artinian, Gorenstein, compressed) are assumed?

Example 47.11. In contrast to Theorem 47.2, there exist examples where no ideal attains smallest Betti numbers among the ideals with a fixed Hilbert function. The following examples are proved in [Dodd-Marks-Meyerson-Richert] and were noted by Gelvin-LaVictore-Reed-Richert. Let

$$J = (x_1x_2, x_1x_3, x_2x_3, x_3x_4, x_3x_5, x_3x_6, x_4x_5)\,.$$

Then:

(1) Among the graded ideals with the same Hilbert function as J, there exists no ideal with smallest Betti numbers.
(2) Among the squarefree monomial ideals with the same Hilbert function as J, there exists no ideal with smallest Betti numbers.

48 Gotzmann's Persistence Theorem

Gotzmann's Persistence Theorem is a major result on Hilbert functions. It shows that once an ideal achieves minimal growth then it grows minimally forever after.

Gotzmann's Persistence Theorem 48.1. (Gotzmann) *Let J be a graded ideal in S, and L be the lex ideal with the same Hilbert function as J. Suppose that q is an integer such that the following two conditions are satisfied:*

(1) *J is generated in degrees $\leq q$.*
(2) $\dim_k(J_{q+1}) = \dim_k(S_1L_q)$.

We have that

$$\dim_k(J_{q+i}) = \dim_k(S_iL_q)$$

for all $i \geq 1$. Equivalently, L is generated in degrees $\leq q$.

Proof. The proof is from [Gahsarov-Murai-Peeva 2]. It uses consecutive cancellations. Assumption (2) means that L has no minimal generator in degree $q+1$. We will show that L has no minimal monomial generator in degree $q+2$. Assume the opposite, then we have that $b^S_{1,q+2}(S/L) \neq 0$. On the other hand, we know that J does not have a minimal monomial generator in degree $q+2$, so $b^S_{1,q+2}(S/J) = 0$. Since $b^S_{1,q+2}(S/J) = 0$ is obtained from $b^S_{1,q+2}(S/L) \neq 0$ by consecutive cancellations by Theorem 47.7, it follows that $b^S_{2,q+2}(S/L) \neq 0$.

The ideal L is Borel, so the minimal free resolution of S/L is the Eliahou-Kervaire resolution 28.6. Since L does not have a minimal monomial generator in degree $q+1$, it follows that $b^S_{2,q+2}(S/L) = 0$. This is a contradiction.

We proved that L does not have a minimal monomial generator in degree $q+2$. Therefore, $\dim_k(J_{q+2}) = \dim_k(S_1 L_{q+1})$. The theorem holds by induction on degree. □

Example 48.2. Consider the ideal $J = (y^2, z^2)$ in $A = k[x, y, z]$. We will compute the lex ideal L with the same Hilbert function as J. The k-vector space J_2 has basis y^2, z^2. Hence the k-vector space L_2 has basis x^2, xy. The k-vector space J_3 has basis y^3, y^2x, y^2z, z^2x, z^2y, z^3, so it is 6-dimensional. Therefore, L_3 has basis x^3, x^2y, x^2z, xy^2, xyz, xz^2. So far we have found that the lex ideal has minimal generators x^2, xy, xz^2.

The k-vector space $(A/J)_4$ has basis x^4, x^3y, x^3z, x^2yz. Hence the k-vector space $(A/L)_4$ has basis y^3z, y^2z^2, yz^3, z^4. Therefore, L_4 is spanned by A_1L_3 and y^4.

In degree 5, the k-vector space $(A/J)_5$ has basis x^5, x^4y, x^4z, x^3yz. Hence, the k-vector space $(A/L)_5$ has basis $y^3z^2, y^2z^3, yz^4, z^5$. Therefore, L_5 is spanned by A_1L_4. Thus, L has no minimal generators in degree 5.

By Gotzmann's Persistence Theorem 48.1 it follows that $L = (x^2, xy, xz^2, y^4)$.

Gotzmann's Regularity Theorem 48.3. *Let J be a graded ideal in S. Let q be an integer such that the following two conditions are satisfied:*

(1) *J is generated in degrees $\leq q$.*

(2) $\dim_k(J_{q+1}) = \dim_k(S_1 L_q)$.
Then $\operatorname{reg}_S(J) \leq q$.

Proof. By Remark 47.3, Corollary 28.13, and Theorem 48.1 we get $\operatorname{reg}_S(J) \leq \operatorname{reg}_S(L) \leq q$. □

Let J be a graded ideal in S, and L be the lex ideal with the same Hilbert function as J. We say that J is a ***Gotzmann ideal*** if the equality

$$\dim_k(S_1 J_q) = \dim_k(S_1 L_q)$$

holds for every $q \geq 0$.

Exercise 48.4. *Let J be a graded ideal in S, and L be the lex ideal with the same Hilbert function as J. The ideal J is Gotzmann if and only if J and L have the same number of minimal generators.*

Theorem 48.5. (Herzog-Hibi) *Let J be a graded Gotzmann ideal in S, and L be the lex ideal with the same Hilbert function as J. We have equalities of graded Betti numbers*

$$b^S_{i,j}(S/J) = b^S_{i,j}(S/L) \quad \textit{for all } i, j \geq 0\,.$$

Proof. Let p be the smallest degree in which L has a minimal monomial generator. For $q \geq p$, denote by $J(q)$ the ideal generated by all monomials in J of degree $\leq q$. Similarly, denote by $L(q)$ the ideal generated by all monomials in L of degree $\leq q$. By Gotzmann's Persistence Theorem 48.1, for each $q \geq p$ the ideals $J(q)$ and $L(q)$ have the same Hilbert function. Furthermore, by Remark 47.3 it follows that the graded Betti numbers of $S/L(q)$ are greater or equal to those of $S/J(q)$.

All Betti numbers in the proof are over S. By Theorem 16.2 the graded Betti numbers $b_{i,j}(S/T)$ for a homogeneous ideal T and its Hilbert function are related by

$$\sum_{j=0}^{\infty} \dim_k(S/T)_j\, t^j = \frac{\sum_{j=0}^{\infty} \sum_{i=0}^{n} (-1)^i b_{i,j}(S/T)\, t^j}{(1-t)^n}\,.$$

Therefore, for each $q \geq p$ we have that

$$(*) \qquad \sum_{j=0}^{\infty} \sum_{i=0}^{n} (-1)^i \Big(b_{i,j}(S/J(q)) - b_{i,j}(S/L(q)) \Big) t^j = 0 .$$

By induction on q we will show that the graded Betti numbers of $S/L(q)$ are equal to those of $S/J(q)$.

First, consider the case when $q = p$. By the Eliahou-Kervaire resolution, it follows that $L(p)$ has a p-linear minimal free resolution, that is, $b_{i,j}(S/L(p)) = 0$ for $j \neq i + p - 1$. Since the graded Betti numbers of $S/L(p)$ are greater or equal to those of $S/J(p)$, it follows that $b_{i,j}(S/J(p)) = 0$ for $j \neq i+p-1$. By $(*)$ we obtain the equalities of graded Betti numbers

$$b_{i,j}(S/J(p)) = b_{i,j}(S/L(p)) \quad \text{for all } i, j .$$

Suppose that the claim is proved for q. Now, we consider the ideals $L(q+1)$ and $J(q+1)$. For $j < i + q$, we have that

$$b_{i,j}(S/L(q+1)) = b_{i,j}(S/L(q)) = b_{i,j}(S/J(q)) ,$$

where the first equality follows from the Eliahou-Kervaire resolution and the second equality holds by induction hypothesis. As $J(q+1)_s = J(q)_s$ for $s \leq q$ by construction, and since $b_{i,j}(S/J(q)) = 0$ for $j \geq i + q$, we conclude that $b_{i,j}(S/J(q+1)) = b_{i,j}(S/J(q))$ for $j < i + q$. Hence,

$$b_{i,j}(S/L(q+1)) = b_{i,j}(S/J(q+1)) \text{ for } j < i + q$$
$$b_{i,j}(S/L(q+1)) = 0 \text{ for } j > i + q, \text{ by the Eliahou-Kervaire resolution.}$$

Since the graded Betti numbers of $S/L(q+1)$ are greater or equal to those of $S/J(q+1)$, we conclude that

$$b_{i,j}(S/J(q+1)) = b_{i,j}(S/L(q+1)) \text{ for } j < i + q$$
$$b_{i,j}(S/J(q+1)) = b_{i,j}(S/L(q+1)) = 0 \text{ for } j > i + q .$$

By $(*)$ it follows that

$$\sum_{i=0}^{n}(-1)^i\Big(b_{i,i+q}(S/J(q)) - b_{i,i+q}(S/L(q))\Big)\, t^{i+q} = 0\,.$$

Hence

$$b_{i,j}(S/J(q+1)) = b_{i,j}(S/L(q+1)) \quad \text{for all } i,j\,,$$

as desired. □

Let J be a graded ideal in S. We say that J is ***componentwise linear*** if for every $q \geq 0$ the ideal generated by J_q has a q-linear minimal free resolution. By Theorem 48.5, we have that a Gotzmann ideal is componentwise linear.

49 Numerical versions

Since lex ideals are highly structured, it is easy to derive the inequalities characterizing their Hilbert functions. As an application, we discuss numerical versions of some results proved earlier.

Note that by convention $\binom{a}{b} = 0$ if $a < b$.

Lemma 49.1. *Let q be a positive integer. For every $p \in \mathbf{N}$ there exist numbers $s_q > \ldots > s_1 \geq 0$ such that*

$$p = \binom{s_q}{q} + \binom{s_{q-1}}{q-1} + \ldots + \binom{s_1}{1}\,.$$

Proof. The proof is by induction. Set

$$s_q = \max\left\{ j \;\middle|\; \binom{j}{q} \leq p \right\}.$$

If $p = \binom{s_q}{q}$, then set $s_i = i-1$ for each $1 \leq i < q$. Suppose that $p - \binom{s_q}{q} > 0$. By induction, we can find $s_{q-1} > \ldots > s_1 \geq 0$ such that

$$p - \binom{s_q}{q} = \binom{s_{q-1}}{q-1} + \ldots + \binom{s_1}{1}\,.$$

It remains to show that $s_q > s_{q-1}$. Assume the opposite. Then we obtain

$$\binom{s_{q-1}}{q-1} \geq \binom{s_q}{q-1} = \binom{s_q+1}{q} - \binom{s_q}{q}$$
$$> p - \binom{s_q}{q} = \binom{s_{q-1}}{q-1} + \ldots + \binom{s_1}{1},$$

which is a contradiction. □

This is called the q'th ***Macaulay representation*** of p. The numbers $s_1, \ldots, s_q$ are called the q'th ***Macaulay coefficients*** of p.

Example 49.2. The 3'rd Macaulay representation of 14 is

$$14 = \binom{5}{3} + \binom{3}{2} + \binom{1}{1}.$$

Exercise 49.3. *The q'th Macaulay coefficients of p are unique.*

Set $0^{\langle q \rangle} = 0$ and

$$p^{\langle q \rangle} = \binom{s_q+1}{q+1} + \binom{s_{q-1}+1}{q-1+1} + \ldots + \binom{s_1+1}{1+1}.$$

Proposition 49.4. *Let L be an ideal generated by a lex segment in S_q. If $p = \dim_k (S/L)_q$, then $\dim_k (S/L)_{q+1} = p^{\langle q \rangle}$.*

Proof. Set $j = \min\{i |\, x_i^q \notin L\}$. We have that the monomials

$$\{\, u \in S_q \text{ is a monomial } |\, u \preceq_{lex} x_j^q \,\} = k[x_j, \ldots, x_n]_q$$

are non-zero monomials in $(S/L)_q$. The number of such monomials is

$$\dim_k k[x_j, \ldots, x_n]_q = \binom{n-j+q}{q} = \binom{s_q}{q},$$

where $s_q = n - j + q$. Furthermore,

$$(x_j, \ldots, x_n)\{\, u \,|\, u \preceq_{lex} x_{j-1}^q \,\} = k[x_j, \ldots, x_n]_{q+1}$$

are non-zero monomials in $(S/L)_{q+1}$. The number of such monomials is

$$\dim_k k[x_j,\dots,x_n]_{q+1} = \binom{n-j+q+1}{q+1} = \binom{s_q+1}{q+1}.$$

Let m be the lex-greatest monomial in S_q but not in L. Hence $x_{j-1}^q \in L$ and $x_{j-1}^q \succ_{lex} m \succ_{lex} x_j^q$. Set

$$\mathcal{D} = \{\, u \in \{S_q\} \mid m \succeq_{lex} u \succ_{lex} x_j^q \,\}$$
$$\mathcal{F} = \{\, u \in \{S_q\} \mid x_j^q \succeq_{lex} u \,\}.$$

All monomials in $\mathcal{D}$ are divisible by x_{j-1}, so we can write $\mathcal{D} = x_{j-1}\mathcal{D}'$. Now,

$$\dim_k (S/L)_q = |\mathcal{D}'| + |\mathcal{F}|$$
$$\dim_k (S/L)_{q+1} = |(x_j,\dots,x_n)_1\, \mathcal{D}'| + |(x_j,\dots,x_n)_1\, \mathcal{F}|\,.$$

We showed that $|\mathcal{F}| = \binom{s_q}{q}$ and $|(x_j,\dots,x_n)_1\, \mathcal{F}\,| = \binom{s_q+1}{q+1}$. By induction on the degree, we have that

$$|\mathcal{D}'| = \binom{s_{q-1}}{q-1} + \ldots + \binom{s_1}{1}$$

and

$$|(x_j,\dots,x_n)_1\, \mathcal{D}'| = \binom{s_{q-1}+1}{q} + \ldots + \binom{s_1+1}{1+1}.$$

In order to finish the proof, we need to verify that $s_q,\dots,s_1$ are the Macaulay coefficients, that is, we have to verify that $s_q > \ldots > s_1$. The inequalities $s_{q-1} > \ldots > s_1$ hold by induction. So we have to check that $s_q > s_{q-1}$.

We have that $s_q = n-j+q$, where $j = \min\{i \mid x_i^q \notin L\,\}$. Similarly, $s_{q-1} = n-c+(q-1)$, where

$$c = \min\{i \mid x_i^{q-1} \notin \mathcal{D}'\,\} = \min\{i \mid x_{j-1}x_i^{q-1} \notin \mathcal{D}\,\}$$
$$= \min\{i \mid x_{j-1}x_i^{q-1} \notin L,\, i \geq j\,\}.$$

Therefore, $c \geq j$. Hence $s_{q-1} = n - c + q - 1 < n - j + q = s_q$. □

The above proposition and Macaulay's Theorem imply the following result.

Numerical Version of Macaulay's Theorem 49.5. *Let J be a graded ideal in S. Then*

$$\dim_k (S/J)_{j+1} \leq (\dim_k (S/J)_j)^{\langle j \rangle} \quad \textit{for} \quad j \geq 0\,.$$

Similarly, the above proposition and Gotzmann's Persistence Theorem 48.1 imply the following result.

Numerical Version of Gotzmann's Theorem 49.6. *Let J be a graded ideal in S. Let q be an integer such that the following two conditions are satisfied:*

(1) *J is generated in degrees $\leq q$.*

(2) $\dim_k (S/J)_{q+1} = \left(\dim_k (S/J)_q\right)^{\langle q \rangle}$.

Then

$$\dim_k (S/J)_{j+1} = \left(\dim_k (S/J)_j\right)^{\langle j \rangle} \quad \textit{for } j \geq q\,.$$

Numerical Version of Gotzmann's Regularity Theorem 49.7. *Let J be a graded ideal in S. Let q be an integer such that the following two conditions are satisfied:*

(1) *J is generated in degrees $\leq q$.*

(2) $\dim_k (S/J)_{q+1} = \left(\dim_k (S/J)_q\right)^{\langle q \rangle}$.

Then $\operatorname{reg}_S(J) \leq q$.

Let $\alpha = \{\alpha_0, \alpha_1, \ldots\}$ be a sequence of non-negative integer numbers. We say that α is a ***Macaulay sequence*** if $\alpha_0 = 1$ and $\alpha_{q+1} \leq \alpha_q^{\langle q \rangle}$ for each $q \geq 1$.

Corollary 49.8. *Let $\alpha = \{\alpha_0 = 1,\ \alpha_1 \leq n,\ \alpha_2, \ldots\}$ be a sequence of non-negative integer numbers. There exists a graded ideal J in S with* $\dim_k (S/J)_i = \alpha_i$ *for all $i \geq 0$, if and only if, α is a Macaulay sequence.*

Proof. Note that $\alpha_0 = 1$ as $S_0 = k$. Furthermore, $\alpha_1 \leq n$ since $\dim_k S_1 = n$.

Suppose that there exists a graded ideal J in S with $\dim_k (S/J)_i = \alpha_i$ for all $i \geq 0$. By Macaulay's Theorem, there exists a lex ideal L with $\dim_k (S/L)_i = \alpha_i$ for all $i \geq 0$. Applying Proposition 49.4 we conclude that if L has no minimal monomial generators in degree $q+1$ then we have the equality $\alpha_{q+1} = \alpha_q^{\langle q \rangle}$, and otherwise we have the inequality $\alpha_{q+1} \leq \alpha_q^{\langle q \rangle}$.

Suppose that α is a Macaulay sequence. Let L_q be the lex segment in S_q such that $\dim_k (S/L)_q = \alpha_q$. By Proposition 49.4, it follows that $L_{q+1} \supseteq S_1 L_q$. Hence $L = \oplus_{q \geq 0} L_q$ is an ideal. It has the desired Hilbert function. □

Corollary 49.9. *Let J be a graded ideal in S. The Hilbert polynomial of S/J has the form*

$$h_{S/J}(t) = \binom{t + a_q}{a_q} + \binom{t + a_{q-1}}{a_{q-1}} + \ldots + \binom{t + a_1}{a_1}.$$

for some $a_q \geq \ldots \geq a_1 \geq 0$.

Proof. Let L be the lex ideal with the same Hilbert function as J. Let q be the maximal degree in which L has a minimal monomial generator. Denote by N the ideal generated by L_q. It follows that $\dim_k (J_i) = \dim_k (N_i)$ for $i \geq q$. Hence S/J and S/N have the same Hilbert polynomial.

Let $s_q > \ldots > s_1 \geq 0$ be the Macaulay's coefficients of the q'th Macaulay representation of the number $\dim_k (S/N)_q$.

By Proposition 49.4, it follows that the Hilbert polynomial of S/N is

$$h_{S/N}(q+t) = \binom{s_q + t}{q + t} + \binom{s_{q-1} + t}{q - 1 + t} + \ldots + \binom{s_1 + t}{1 + t}.$$

Set $a_i = s_i - i$ for each i. Hence

$$h_{S/N}(q+t) = \binom{t + q + a_q}{q + t} + \binom{t + q - 1 + a_q}{q - 1 + t} + \ldots + \binom{t + 1 + a_1}{1 + t}.$$

Therefore,

$$h_{S/N}(t) = \binom{t+a_q}{t} + \binom{t+a_{q-1}}{t} + \ldots + \binom{t+a_1}{t}$$
$$= \binom{t+a_q}{a_q} + \binom{t+a_{q-1}}{a_{q-1}} + \ldots + \binom{t+a_1}{a_1}.$$

□

Corollary 49.10. *Suppose that the field k is infinite. Let*

$$g(t) = a_r t^r + \ldots + a_1 t + a_0$$

$g(1) \neq 0$, and $a_i \in \mathbf{Z}$ for all i. There exists a $p \in \mathbf{N}$ such that $\dfrac{g(t)}{(1-t)^p}$ is equal to $\mathrm{Hilb}_{S/J}(t)$ *for some graded Cohen-Macaulay J if and only if a_0, $a_1 \leq n$, ..., a_r is a Macaulay sequence of positive numbers.*

Proof. Let $\mathrm{Hilb}_{S/J}(t) = \dfrac{g(t)}{(1-t)^{\dim(S/J)}}$. If S/J is Cohen-Macaulay, then by 20.1 there exists a regular sequence of linear forms of length $\dim(S/J)$. Hence, $g(t)$ is the Hilbert series of an artinian graded quotient of S. Therefore, $a_0, a_1 \leq n, \ldots, a_r$ is a Macaulay sequence of positive numbers.

On the other hand, suppose that $a_0, a_1 \leq n, \ldots, a_r$ is a Macaulay sequence of positive numbers. Therefore, there exists an artinian graded quotient S/J of S with Hilbert series $g(t)$. □

The following problems have been studied, cf. [Valla].

Problems 49.11.

(1) *Characterize the Hilbert functions of graded artinian Gorenstein quotients of S.*

(2) *Characterize the Hilbert functions of graded Cohen-Macaulay domains that are quotients of S.*

(3) *Characterize the Hilbert functions of sets of points in uniform position.*

A problem of this type is also the Eisenbud-Green-Harris Conjecture, discussed in Section 53. Another conjecture of this type is

Fröberg's conjecture.

Fröberg's Conjecture 49.12. (Fröberg) *Let $f_1, \dots, f_r$ be generic forms in S of degrees $a_1, \dots, a_r$, and let $T = (f_1, \dots, f_r)$. The Hilbert series of S/T is*

$$\mathrm{Hilb}_{S/T}(t) = \left| \frac{\prod_{1 \le i \le r} (1 - t^{a_i})}{(1-t)^n} \right| ,$$

where $|\ |$ means that a term $c_i t^i$ in the series is omitted if there exists a term $c_j t^j$ with $j \le i$ and negative coefficient c_j. (Here $r > n$ is the interesting case, since for $r \le n$ we have that $f_1, \dots, f_r$ is a regular sequence.)

Set $0_{\langle q \rangle} = 0$ and

$$p_{\langle q \rangle} = \binom{s_q - 1}{q} + \binom{s_{q-1} - 1}{q-1} + \dots + \binom{s_1 - 1}{1} .$$

Exercise 49.13. *Let L be an ideal generated by a lex segment in S_q. If $p = \dim_k (S/L)_q$, then*

$$\dim_k \big(S/(L, x_n)\big)_q = p_{\langle q \rangle} .$$

Green's Hyperplane Restriction Theorem 44.10 and 49.13 imply the next result.

Numerical Version of Green's Hyperplane Restriction Theorem 49.14. *Let J be a graded ideal in S, and h be a generic linear form. If $p = \dim_k (S/J)_q$, then*

$$\dim_k (S/(J, h))_q \le p_{\langle q \rangle} .$$

50 Hilbert functions over quotient rings

The main idea in Macaulay's Theorem is that every Hilbert function is attained by a lex ideal. One can wonder for what quotient rings this idea works out. If I is a monomial or toric ideal, then we can define the notion of a lex ideal in the quotient ring $R = S/I$. There might be other classes of rings for which one can introduce a meaningful notion of lex ideals, that is, *find a class of ideals which attain all Hilbert functions and which are defined in a nice way* (and call such ideals lex ideals).

It is easy to find quotient rings over which Macaulay's Theorem does not hold. For example, there exists no lex ideal with the same Hilbert function as the ideal (ab) in the quotient ring $k[a,b]/(a^2b, ab^2)$ by [Mermin-Peeva 2, Example 2.13]. Since the trouble is sometimes in the degrees of the minimal generators of I, it makes sense to relax the problem to Problem 50.1(1). Furthermore, in view of Hartshorne's Theorem that every graded ideal in S is connected by a sequence of deformations to a lex ideal, it is natural to raise Problem 50.1(2). Problem 50.1(3) is motivated by Theorem 47.2 which shows that a lex ideal attains the greatest graded Betti numbers among all graded ideals in S with the same Hilbert function.

Open-Ended Problems 50.1. [Mermin-Peeva, Mermin-Peeva 2]

(1) *Let p be the maximal degree of an element in a minimal homogeneous system of generators of I. Find classes of graded ideals I so that every Hilbert function over $R = S/I$ of a graded ideal generated in degrees $> p$ is attained by a lex ideal.*

(2) *Let J be a graded ideal in R, and L be a lex ideal with the same Hilbert function. When is J connected to L by a sequence of deformations? What can be said about the structure of the Hilbert scheme that parametrizes all graded ideals in R with the same Hilbert function as L?*

(3) *Let J be a graded ideal in R, and L be a lex ideal with the same Hilbert function. Find conditions on R and/or J so that the Betti numbers of J over R are less than or equal to those of L.*

Furthermore, one can also ask for generalizations or extensions of the Gotzmann's Persistence Theorem and the Lex-Plus-Powers Conjecture.

Open-Ended Problem 50.2. (Peeva) *Find classes of graded ideals*

I so that Gotzmann's Persistence Theorem holds over R.

Open-Ended Problem 50.3. [Mermin-Peeva] *Let J be a graded ideal in R, and L be a lex ideal with the same Hilbert function in R. Denote by $\widetilde{J}$ and $\widetilde{L}$ the preimages of J and L in S. Find conditions on R and/or J so that the Betti numbers of $\widetilde{J}$ over S are less than or equal to those of $\widetilde{L}$.* (We say that $\widetilde{L}$ is a ***lex-plus-I*** ideal.)

As we have seen in Section 51, Hilbert functions over an exterior algebra coincide with f-vectors of simplicial complexes. The situation over an exterior algebra is well-studied and we have the following results.

Theorem 50.4. *Let E be a standard graded exterior algebra on n variables of degree one.*

(1) (Kruskal-Katona) *For every graded ideal in E there exists a lex ideal with the same Hilbert function.*
(2) [Peeva-Stillman 3] *The Hilbert scheme, that parametrizes all graded ideals with a fixed Hilbert function, is connected. Each graded ideal in E is connected by a sequence of deformations to the lex ideal with the same Hilbert function.*
(3) [Aramova-Herzog-Hibi 2] *Each lex ideal in E attains maximal Betti numbers among all graded ideals with the same Hilbert function.*
(4) [Mermin-Peeva-Stillman] *Each lex-plus-$(x_1^2,\dots,x_n^2)$ ideal in S attains maximal Betti numbers among all graded ideals containing $(x_1^2,\dots,x_n^2)$ and with the same Hilbert function.*
(5) [Aramova-Herzog-Hibi 2] *Gotzmann's Persistence Theorem holds over E.*

51 Squarefree ideals plus squares

In this section, we study how the Hilbert function and the minimal free resolution change when we add the squares of the variables to a squarefree monomial ideal. This relates to (4) in Theorem 50.4.

Throughout the section Δ is a simplicial complex on the vertex set $\{x_1, \ldots, x_n\}$. Set $c = \dim(\Delta) + 1$.

The ***Stanley-Reisner ideal*** (in S) of Δ is

$$I_\Delta = (x_{i_1} \ldots x_{i_p} \,|\, \{x_{i_1}, \ldots, x_{i_p}\} \notin \Delta).$$

Each squarefree monomial ideal in S is the Stanley-Reisner ideal of some simplicial complex on vertex set $\{x_1, \ldots, x_n\}$.

The ***Stanley-Reisner ring*** of Δ is $R_\Delta = S/I_\Delta$. The ring

$$Q_\Delta = R_\Delta/\,(x_1^2, \ldots, x_n^2) = S/(I_\Delta + (x_1^2, \ldots, x_n^2))$$

is closely related to R_Δ. We say that $I_\Delta + (x_1^2, \ldots, x_n^2)$ is a ***squarefree-plus-squares*** ideal.

First, we study how the Hilbert function changes when we add the squares of the variables to a squarefree monomial ideal.

Construction 51.1. Consider the correspondence

$$\varphi : x_{i_1} \cdots x_{i_p} \to \text{ the face with vertices } \{x_{i_1}, \ldots, x_{i_p}\}.$$

from the set of squarefree monomials in n variables to the faces of the simplex on n vertices. Clearly, φ is a bijection.

The ***f-vector*** of Δ is $(f_{-1}, f_0, \ldots, f_{c-1})$, where f_i is the number of faces of dimension i in Δ. Note that $f_{-1} = 1$, since a simplicial complex has one empty face. The polynomial $f(t) = \sum_{0 \le i \le c} f_{i-1}\, t^i$ is called the ***f-polynomial***.

Proposition 51.2.

$$\mathrm{Hilb}_{Q_\Delta}(t) = \sum_{0 \le i \le c} f_{i-1}\, t^i = f(t).$$

Proof. The bijection in Construction 51.1 induces the bijection

$$\psi : \text{ the monomials in } Q_\Delta = R_\Delta/\,(x_1^2, \ldots, x_n^2) \longrightarrow \text{ the faces of } \Delta\,.$$

□

Proposition 51.3.

$$\mathrm{Hilb}_{R_\Delta}(t) = \mathrm{Hilb}_{Q_\Delta}\left(\frac{t}{1-t}\right).$$

Proof. Let m be a squarefree monomial in Q_Δ of degree i. Denote $\mathrm{supp}(m) = \{x_j | \, x_j \text{ divides } m\}$. All monomials in S with the same support as m are monomials in R_Δ and they are exactly the monomials in $m\,k[x_j | x_j \in \mathrm{supp}(m)]$; hence they contribute $\dfrac{t^i}{(1-t)^i}$ to $\mathrm{Hilb}_{R_\Delta}(t)$. Therefore,

$$\mathrm{Hilb}_{R_\Delta}(t) = \sum_{0 \le i \le c} f_{i-1} \frac{t^i}{(1-t)^i} = \mathrm{Hilb}_{Q_\Delta}\left(\frac{t}{1-t}\right).$$

□

Theorem 51.4. *If Δ and Δ' are simplicial complexes on n vertices, then*

$$\mathrm{Hilb}_{R_\Delta}(t) = \mathrm{Hilb}_{R_{\Delta'}}(t) \quad \Longleftrightarrow \quad \mathrm{Hilb}_{Q_\Delta}(t) = \mathrm{Hilb}_{Q_{\Delta'}}(t)\,.$$

Theorem 51.5.

$$\dim(R_\Delta) = \dim(\Delta) + 1\,.$$

Proof. We have that

$$\mathrm{Hilb}_{R_\Delta}(t) = \sum_{0 \le i \le c} f_{i-1} \frac{t^i}{(1-t)^i} = \sum_{0 \le i \le c} f_{i-1} \frac{t^i (1-t)^{c-i}}{(1-t)^c}\,.$$

Set $h(t) = \sum_{0 \le i \le c} f_{i-1} t^i (1-t)^{c-i}$. Then $\mathrm{Hilb}_{R_\Delta}(t) = \dfrac{h(t)}{(1-t)^c}$ and $h(1) = f_{c-1} \neq 0$. Hence $\dim(R_\Delta) = c = \dim(\Delta) + 1$. □

Recall that the polynomial $h(t)$ above is called the h-polynomial.

Corollary 51.6. $h(t) = (1-t)^c \cdot f\left(\dfrac{t}{1-t}\right)$.

Next, we study how the Betti numbers change when we add the squares of the variables to a squarefree monomial ideal.

Theorem 51.7. *Let N be a squarefree ideal. Set $P(i) = (x_1^2, \ldots, x_i^2)$ and $P(0) = 0$. For each $0 \le i < n$, the mapping cone of the short* *exact sequence*

$$0 \to S/\big((N+P(i)) : x_{i+1}\big) \xrightarrow{x_{i+1}^2} S/\big(N+P(i)\big) \to S/\big(N+P(i+1)\big) \to 0$$

yields a minimal free resolution of $S/\big(N+P(i+1)\big)$.

This theorem shows how to obtain the Betti numbers of $N + (x_1^2, \dots, x_n^2)$ starting from the minimal free resolution of N and adding the squares one after another. At each step, we use a mapping cone.

Proof. First, note that

$$\big((N+P(i)) : x_{i+1}^2\big) = \big((N+P(i)) : x_{i+1}\big)$$

because the ideal $N + P(i)$ is squarefree on the variable x_{i+1}. Thus, the sequence above is exact.

Since the ideal $N + P(i)$ is squarefree on the variable x_{i+1}, by Taylor's resolution, it follows that the Betti numbers of $S/\big(N+P(i)\big)$ are concentrated in multidegrees not divisible by x_{i+1}^2. On the other hand, the first map in the short exact sequence is multiplication by x_{i+1}^2. Therefore, there can be no cancellations in the mapping cone. Hence, the mapping cone yields a minimal free resolution. □

The disadvantage of the above theorem is that we may change the Hilbert function by adding some (but not all) of the squares. That is, if N and N' are two squarefree ideals with the same Hilbert function, then $N+P(i)$ and $N'+P(i)$ may not have the same Hilbert function. There are examples, when $N + (x_1^2)$ and $N' + (x_1^2)$ have different Hilbert functions. For example, consider the polynomial ring $k[a,b,c,e]$ and let T be the ideal generated by the squarefree cubic monomials; the ideal $N = (ab, ac, bc) + T$ is squarefree Borel and the ideal $N' = (ab, ac, ae) + T$ is squarefree lex. The ideals N and N' have the same Hilbert function, but $N + (a^2)$ and $N' + (a^2)$ have different Hilbert functions. The next theorem shows how to use mapping cones while preserving the Hilbert function.

For a $\sigma \subseteq \{x_1, \dots, x_n\}$, we set $\mathbf{x}_\sigma = \prod_{x_i \in \sigma} x_i$.

Theorem 51.8. [Mermin-Peeva-Stillman] *Let* $P = (x_1^2, \dots, x_n^2)$ *and* N *be a squarefree monomial ideal.*

(1) *We have the long exact sequence*

$$
(*)\quad
\begin{aligned}
0 &\to \oplus_{|\sigma|=n} S/(N:\mathbf{x}_\sigma) \xrightarrow{\varphi_n} \dots\\
&\to \oplus_{|\sigma|=i} S/(N:\mathbf{x}_\sigma) \xrightarrow{\varphi_i} \oplus_{|\sigma|=i-1} S/(N:\mathbf{x}_\sigma) \to \dots\\
&\to \oplus_{|\sigma|=1} S/(N:\mathbf{x}_\sigma) = \oplus_{1\le j\le n} S/(N:x_j) \xrightarrow{\varphi_1}\\
&\to \oplus_{|\sigma|=0} S/(N:\mathbf{x}_\sigma) = S/N \to S/(N+P) \to 0
\end{aligned}
$$

with maps φ_i the Koszul maps for the sequence $x_1^2,\dots,x_n^2$, and $\sigma \subseteq \{1,\dots,n\}$.

(2) *$S/(N+P)$ is minimally resolved by the iterated mapping cones from $(*)$.*

(3) *Each of the ideals $(N:\mathbf{x}_\sigma)$ in (1) is a squarefree monomial ideal.*

(4) *For the graded Betti numbers of $S/(N+P)$ we have*

$$b_{p,s}(S/(N+P)) = \sum_{0\le i\le p}\left(\sum_{|\sigma|=i} b_{p-i,s-2i}(S/(N:\mathbf{x}_\sigma))\right).$$

Proof. First, note that $(N:\mathbf{x}_\sigma^2) = (N:\mathbf{x}_\sigma)$ is squarefree since N is squarefree.

By Construction 14.1 and Theorem 14.7, the exact Koszul complex $\mathbf{K}$ for the sequence $x_1^2,\dots,x_n^2$ has the form

$$
\begin{aligned}
0 &\to \oplus_{|\sigma|=n} S \xrightarrow{\varphi_n} \dots\\
&\to \oplus_{|\sigma|=i} S \xrightarrow{\varphi_i} \oplus_{|\sigma|=i-1} S \to \dots\\
&\to \oplus_{|\sigma|=1} S = \oplus_{1\le j\le n} S \xrightarrow{\varphi_1}\\
&\to \oplus_{|\sigma|=0} S = S \to S/P \to 0\,.
\end{aligned}
$$

Write $\mathbf{K} = \mathbf{K}' \oplus \mathbf{K}''$, where $\mathbf{K}'$ consists of the components of $\mathbf{K}$ in all multidegrees $m \notin N$, and $\mathbf{K}''$ consists of the components of $\mathbf{K}$ in all multidegrees $m \in N$. Note that both $\mathbf{K}'$ and $\mathbf{K}''$ are exact by 3.7. We will show that $(*)$ coincides with $\mathbf{K}'$.

By 14.1, $\mathbf{K}$ is an exterior algebra on variables $e_1,\dots e_n$. Let $me_{j_1}\wedge\dots\wedge e_{j_i}$ be an element in $\mathbf{K}_i$ and m be a monomial. The multidegree of the variable e_j is x_j^2; hence, the multidegree of $me_{j_1}\wedge$

$\ldots \wedge e_{j_i}$ is $mx_{j_1}^2 \ldots x_{j_i}^2$. Now, $me_{j_1} \wedge \ldots \wedge e_{j_i} \in \mathbf{K}'$ if and only if $mx_{j_1}^2 \ldots x_{j_i}^2 \notin N$, if and only if $mx_{j_1} \ldots x_{j_i} \notin N$, if and only if $m \notin (N : x_{j_1} \ldots x_{j_i})$. Therefore,

$$\mathbf{K}'_i \to \oplus_{|\sigma|=i} S/(N : \mathbf{x}_\sigma)$$

$$me_{j_1} \wedge \ldots \wedge e_{j_i} \mapsto m \in S/(N : x_{j_1} \ldots x_{j_i})$$

is an isomorphism. We proved (1).

We will prove (2). Denote by V_i the kernel of $\varphi_i : \mathbf{K}'_i \to \mathbf{K}'_{i-1}$. We have the short exact sequence

$$0 \to V_i \to \oplus_{|\sigma|=i} S/(N : \mathbf{x}_\sigma) \to V_{i-1} \to 0\,.$$

Each of the ideals $(N : \mathbf{x}_\sigma)$ is squarefree. By Corollary 26.10, the Betti numbers of $\oplus_{|\sigma|=i} S/(N : \mathbf{x}_\sigma)$ are concentrated in squarefree multidegrees. On the other hand, the entries in the matrix of the map φ_i are squares of the variables. Therefore, there can be no cancellations in the mapping cone. Hence, the mapping cone yields a minimal free resolution.

(4) follows from (2). □

Furthermore, the following result is proved in [Mermin-Peeva-Stillman].

Proposition 51.9. *Let N and N' be two squarefree monomial ideals with the same Hilbert function. Fix an integer $1 \le p \le n$. The graded modules $\oplus_{|\sigma|=p} (N : \mathbf{x}_\sigma)$ and $\oplus_{|\sigma|=p} (N' : \mathbf{x}_\sigma)$ have the same Hilbert function.*

Proposition 51.10. *Let N be a squarefree monomial ideal, and Δ be its Stanley-Reisner simplicial complex. Let $\sigma \subseteq \{1, \ldots, n\}$. The Stanley-Reisner simplicial complex of $(N : \mathbf{x}_\sigma)$ is*

$$\mathrm{star}_\Delta(\sigma) = \{\tau \in \Delta \,|\, \tau \cup \sigma \in \Delta\}\,,$$

(recall 36.17).

Proof.

$$\begin{aligned}\mathrm{star}_\Delta(\sigma) &= \{\tau \in \Delta \,|\, \tau \cup \sigma \in \Delta\}\\ &= \{\tau \subseteq \{1,\ldots,n\} \,|\, \mathrm{lcm}(\mathbf{x}_\tau, \mathbf{x}_\sigma) \notin N\}\\ &= \{\tau \subseteq \{1,\ldots,n\} \,|\, \mathbf{x}_\tau \mathbf{x}_\sigma \notin N\}\\ &= \{\tau \subseteq \{1,\ldots,n\} \,|\, \mathbf{x}_\tau \notin (N : \mathbf{x}_\sigma)\}\\ &= \{\tau \subseteq \{1,\ldots,n\} \,|\, \mathbf{x}_\tau \notin (N : \mathbf{x}_\sigma)\}.\end{aligned}$$

□

52 Clements-Lindström rings

Counting faces of simplicial complexes (that is, counting in an exterior algebra) naturally generalizes to counting in multicomplexes. This leads to considering Clements-Lindström rings, which have the form

$$P = S/(x_1^{a_1}, \ldots, x_n^{a_n}),$$

where $2 \leq a_1 \leq \ldots \leq a_n$. We will prove the analogue of Theorem 50.4(1), that is, we will prove that Macaulay's Theorem holds over P. The remaining parts (2)-(5) of Theorem 50.4 hold over P as well and are proved in [Gasharov-Murai-Peeva 2] and [Mermin-Murai].

The notions of a P_q-monomial space, compression, Borel, and lex ideals easily extend over P. For example, we say that a P_q-monomial space B_q is ***Borel*** if whenever a non-zero monomial $x_j m \in B_q$ and $1 \leq i \leq j$ it follows that $x_i m \in B_q$ (note that $x_i m = 0$ is possible since P is a quotient ring).

We will need some lemmas. Minor modifications in the proofs of Structure Lemma 42.5, Lemma 44.1, the Comparison Theorem 44.4, Compression Lemma 42.9, and Proposition 41.6 lead to the following analogs (listed below) over P of these results. See [Mermin-Peeva] for detailed proofs.

Structure Lemma 52.1. *If a P_q-monomial space C_q is compressed and $n \geq 3$, then C_q is Borel.*

Lemma 52.2. *If a P_q-monomial space B_q is Borel, then*

$$\Big|\{P_1 B_q\}\Big| = \sum_{i=1}^{n} r_{i,a_i-1}(B_q)\,.$$

Comparison Theorem 52.3. *Let B_q be a Borel monomial space in P_q. Let L_q be a lex monomial space in P_q with $|L_q| \leq |B_q|$. Then*

$$r_{i,j}(L_q) \leq r_{i,j}(B_q)$$

for each $1 \leq i \leq n$ and each $1 \leq j \leq a_i$.

Compression Lemma 52.4. *Let A_q be a P_q-monomial space. There exists a compressed monomial space T_q in P_q such that $|T_q| = |A_q|$ and $|P_1 T_q| \leq |P_1 A_q|$.*

Proposition 52.5. *The following properties are equivalent.*

(1) *Let A_q be a P_q-monomial space and L_q be its lexification in P_q. Then $|\,P_1 L_q\,| \leq |\,P_1 A_q\,|$.*
(2) *For every graded ideal J in P there exists a lex ideal L with the same Hilbert function.*

Using the above results we will prove Macaulay's Theorem over the Clements-Lindström ring P.

Clements-Lindström's Theorem 52.6. *Let $P = S/(x_1^{a_1}, \ldots, x_n^{a_n})$, where $2 \leq a_1 \leq \ldots \leq a_n$. For every graded ideal in P there exists a lex ideal with the same Hilbert function.*

Proof. We will prove that (1) in Proposition 52.5 holds. We will use the argument in the first proof of Macaulay's Theorem from Section 45.

An easy calculation shows that the theorem holds provided $n = 2$ and we do not have $a_2 \leq q+1 < a_1$. But $a_1 \leq a_2$ by assumption, so the theorem holds for $n = 2$.

Consider the case $n \geq 3$. Applying 52.4, we conclude that there exist a compressed monomial space C_q such that $|C_q| = |A_q|$ and $|P_1 C_q| \leq |P_1 A_q|$. By Lemma 52.1 it follows that C_q is Borel. Let L_q be the lex monomial space for which $|C_q| = |L_q|$. We apply Lemma 52.2

to conclude that

$$\Big|\{P_1 C_q\}\Big| = \sum_{i=1}^{n} r_{i,a_i-1}(C_q)$$

$$\Big|\{P_1 L_q\}\Big| = \sum_{i=1}^{n} r_{i,a_i-1}(L_q)\,.$$

Finally, we apply Theorem 52.3 and get the inequality $\big|\{P_1 L_q\}\big| \leq \big|\{P_1 C_q\}\big|$. □

53 The Eisenbud-Green-Harris Conjecture

The most exciting currently open conjecture on Hilbert functions is the Eisenbud-Green-Harris Conjecture. It is wide open.

The Eisenbud-Green-Harris Conjecture 53.1. [Eisenbud-Green-Harris 1, Eisenbud-Green-Harris 2] *Let N be a graded ideal in S containing a maximal homogeneous regular sequence in degrees $2 \leq e_1 \leq \cdots \leq e_n$. There exists a monomial ideal T such that N and $T + (x_1^{e_1}, \cdots, x_r^{e_n})$ have the same Hilbert function.*

A monomial ideal $L + (x_1^{e_1}, \ldots, x_r^{e_n})$ is called ***lex-plus-powers*** if it is the preimage of a lex ideal in $S/(x_1^{e_1}, \ldots, x_r^{e_n})$. By Clements-Lindström's Theorem 52.6, it follows that the conjecture can be stated equivalently as follows.

Conjecture 53.2. *Let N be a graded ideal containing a maximal homogeneous regular sequence in degrees $2 \leq e_1 \leq \cdots \leq e_n$. There exists a lex-plus-powers ideal $L + (x_1^{e_1}, \ldots, x_r^{e_n})$ with the same Hilbert function.*

The original conjecture gives a numerical characterization of the Hilbert functions of graded ideals containing a maximal homogeneous regular sequence in degrees $2 \leq e_1 \leq \cdots \leq e_n$. It is well known that the numerical characterization is equivalent to the existence of a lex-plus-powers ideal $L + (x_1^{e_1}, \ldots, x_n^{e_n})$ with the same Hilbert function as the ideal N.

Another equivalent formulation of the conjecture is:

Conjecture 53.3. *Let $f_1, \ldots, f_n$ be a maximal homogeneous regular sequence in S in degrees $2 \leq e_1 \leq \cdots \leq e_n$. Let $\bar{N}$ be a graded ideal in the complete intersection ring $S/(f_1, \ldots, f_n)$. There exists a lex ideal $\bar{L}$ in the Clements-Lindström ring $S/(x_1^{e_1}, \ldots, x_r^{e_n})$ with the same Hilbert function as $\bar{N}$.*

The conjecture is especially interesting in the case $e_1 = \ldots = e_n = 2$ when the regular sequence consists of quadrics.

Next, we focus on problems based on the idea that the lex ideal has the greatest Betti numbers among all ideals with a fixed Hilbert function.

Conjecture 53.4. *Suppose that k is an infinite field (possibly, one should also assume* $\mathrm{char}(k) = 0$*). Let N be a graded ideal containing a homogeneous regular sequence $f_1, \ldots, f_n$ in S in degrees $2 \leq e_1 \leq \cdots \leq e_n$. Suppose that there exists a lex-plus-powers ideal $L + (x_1^{e_1}, \cdots, x_n^{e_n})$ with the same Hilbert function. Then:*

(1) *The Betti numbers of $\bar{N}$ over $S/(f_1, \ldots, f_n)$ are less than or equal to those of $\bar{L}$ over $S/(x_1^{e_1}, \ldots, x_n^{e_n})$, (where $\bar{N}$ and $\bar{L}$ are the images of N and L in the corresponding complete intersection rings).*

(2) **The LPP Conjecture (the lex-plus-powers conjecture).** (Evans) *The Betti numbers of N over S are less than or equal to those of $L + (x_1^{e_1}, \ldots, x_n^{e_n})$.*

The first part of the conjecture is about infinite resolutions, whereas the second part is about finite ones.

The LPP Conjecture was inspired by the Eisenbud-Green-Harris Conjecture. [Francisco-Richert] is an expository paper on the LPP Conjecture.

Chapter III

Monomial Resolutions

Abstract. In this chapter we discuss free resolutions of monomial ideals; we call them monomial resolutions. The problem to describe the minimal free resolution of a monomial ideal (over a polynomial ring) was posed by Kaplansky in the early 1960's. Despite the helpful combinatorial structure of monomial ideals, the problem turned out to be hard. The structure of a minimal free monomial resolution can be quite complex. There exists a minimal free monomial resolution which cannot be encoded in the structure of any CW-complex. In fact, even the minimal free resolutions of ideals generated by quadratic monomials are so complicated that it is beyond reach to obtain a description of them; we do not even know how to express the regularity of such ideals. In this situation, the guideline is to *introduce new ideas and constructions which either have strong applications or/and are beautiful.* Most proofs about monomial resolutions are easy. The key point is not to provide complicated proofs, but to introduce new beautiful ideas.

54 Examples and Notation

We will use the notation and terminology introduced in Section 26, and the tools from Section 36. Throughout the chapter M stands for a monomial ideal in S minimally generated by monomials $m_1, \ldots, m_r$. We denote by L_M the set of the least common multiples of subsets of $\{m_1, \ldots, m_r\}$. By convention, $1 \in L_M$ considered as the lcm of the empty set. Note that M is homogeneous with respect to the standard grading on S and with respect to the multigrading in 26.1.

I. Peeva, *Graded Syzygies*, Algebra and Applications 14,
DOI 10.1007/978-0-85729-177-6_3, © Springer-Verlag London Limited 2011

In the Running Example we will illustrate several definitions and constructions. The running example uses the ideal

$$Y = (x^2, xy, y^3)$$

in the ring $C = k[x, y]$.

Running Example 54.1. Consider the ideal $Y = (x^2, xy, y^3)$ in the ring $C = k[x, y]$. Computer computation shows that the minimal free resolution of C/Y is

$$\mathbf{F}_Y: \quad 0 \longrightarrow C^2 \xrightarrow{\begin{pmatrix} -y & 0 \\ x & -y^2 \\ 0 & x \end{pmatrix}} C^3 \xrightarrow{\begin{pmatrix} x^2 & xy & y^3 \end{pmatrix}} C \longrightarrow C/Y \longrightarrow 0 .$$

We consider the basis of the free modules in $\mathbf{F}_Y$ in which the above maps are given. Denote by h the basis element of C in homological degree 0, by f_1, f_2, f_3 the basis elements of C^3 in homological degree 1, and by g_1, g_2 the basis elements of C^2 in homological degree 2. Since h has multidegree 1 and the differential is supposed to be homogeneous, it follows that f_1, f_2, f_3 have multidegrees x^2, xy, y^3 respectively. Thus, in homological degree 1 we have the free module $C(x^2) \oplus C(xy) \oplus C(y^3)$. Furthermore, $d(g_1) = -yf_1 + xf_2$ has multidegree x^2y, hence g_1 has multidegree x^2y. Similarly, since $d(g_2) = -y^2 f_2 + xf_3$ has multidegree xy^3, we conclude that g_2 has multidegree xy^3. Thus, in homological degree 2 we have the free module $C(x^2y) \oplus C(xy^3)$. So the resolution can be written

$$0 \to C(x^2y) \oplus C(xy^3) \xrightarrow{\begin{pmatrix} -y & 0 \\ x & -y^2 \\ 0 & x \end{pmatrix}} C(x^2) \oplus C(xy) \oplus C(y^3) \xrightarrow{(x^2\ xy\ y^3)} C .$$

The non-zero multigraded Betti numbers are

$$b_{1,y^3}(C/Y) = b_{1,xy}(C/Y) = b_{1,x^2}(C/Y) = 1$$
$$b_{2,x^2y}(C/Y) = b_{2,xy^3}(C/Y) = 1 .$$

We will also write the component $(\mathbf{F}_Y)_{x^2y^2}$ of $\mathbf{F}_Y$ in multidegree x^2y^2. It is the exact sequence of k-vector spaces

$$0 \to C(x^2y)_{x^2y^2} \oplus C(xy^3)_{x^2y^2} \to C(x^2)_{x^2y^2} \oplus C(xy)_{x^2y^2} \oplus C(y^3)_{x^2y^2} \\ \to C_{x^2y^2} \to (C/Y)_{x^2y^2} \to 0\,.$$

Note that $C(x^2y)_{x^2y^2}$ is the 1-dimensional k-vector space with basis yg_1, so we can write it as $k\{yg_1\}$. Similarly, $C(xy^3)_{x^2y^3} = 0$, $C(x^2)_{x^2y^2} = k\{y^2f_1\}$, $C(xy)_{x^2y^2} = k\{xyf_2\}$, $C(y^3)_{x^2y^3} = 0$, $C_{x^2y^2} = k\{x^2y^2\}$. Furthermore, note that $(C/Y)_{x^2y^2} = 0$ because $x^2y^2 \in Y$. Therefore, $(\mathbf{F}_Y)_{x^2y^2}$ is the exact sequence of k-vector spaces

$$0 \to k\,\{yg_1\} \to k\,\{y^2f_1\} \oplus k\,\{xyf_2\} \to k\,\{x^2y^2\} \to 0 \to 0\,.$$

By Corollary 26.9 the entries in the matrices of the differentials in the minimal free resolution $\mathbf{F}_M$ of S/M are scalar multiples of monomials. After computing a few examples, one might get the feeling that the coefficients appearing in the differential matrices are only $0, \pm 1$. Unfortunately, this is not the case, as shown by the next example.

Example 54.2. [Reiner-Welker] Assume $\mathrm{char}(k) = 0$. Consider the monomial ideal

$$\begin{aligned} T = (\, & x_1x_4x_5x_6, x_2x_4x_5x_6, x_3x_4x_5x_6, x_2x_4x_5x_7, x_3x_4x_5x_7, \\ & x_1x_3x_5x_7, x_1x_2x_4x_7, x_1x_4x_6x_7, x_1x_5x_6x_7, \\ & x_3x_4x_6x_7, x_2x_5x_6x_7, x_2x_3x_6x_7, x_1x_2x_3x_7\,)\,. \end{aligned}$$

in $A = k[x_1, \ldots, x_7]$. Computer computation shows that the minimal free resolution of A/T is

$$0 \to A \to A^{10} \to A^{21} \to A^{13} \to A \to A/T \to 0$$

and the matrix of the last differential d_4 is $\begin{pmatrix} -2x_7 \\ 2x_1 \\ -x_4 \\ x_5 \\ x_1 \\ 2x_3 \\ -2x_2 \\ x_2 \\ x_3 \\ x_6 \end{pmatrix}$ (in some fixed basis). The last matrix contains coefficients ± 2. It is shown in [Reiner-Welker] that A/T does not have a minimal multigraded free resolution with coefficients only $0, \pm 1$ of the monomials in the differential matrices.

Another indication that a monomial resolution could be complicated is that it might not be independent of the characteristic of k; see Example 12.4.

55 Homogenization and dehomogenization

We will explore the idea to encode the structure of a monomial resolution in a complex of vector spaces. The encoding consists of the homogenization and dehomogenization described below. We will discuss the concept of a frame, which is a complex of vector spaces with a fixed basis. By Theorem 55.7 the minimal free resolution of any monomial ideal is encoded in any of its frames. The material in this section is from [Peeva-Velasco], which was motivated by several prior constructions on monomial resolutions.

Construction 55.1. A ***frame*** (or an r-***frame***) $\mathbf{U}$ is a complex of finite k-vector spaces with differential ∂ and a fixed basis that satisfies the following conditions:

(1) $U_i = 0$ for $i < 0$ and $i \gg 0$,

(2) $U_0 = k$,

(3) $U_1 = k^r$,

(4) $\partial(w_j) = 1$ for each basis vector w_j in $U_1 = k^r$.

An ***M*-complex** $\mathbf{G}$ is a multigraded complex of finitely generated free multigraded S-modules with differential d and a fixed multihomogeneous basis with multidegrees in L_M that satisfies the following conditions:

(1) $G_i = 0$ for $i < 0$ and $i \gg 0$,
(2) $G_0 = S$,
(3) $G_1 = S(m_1) \oplus \ldots \oplus S(m_r)$,
(4) $d(w_j) = m_j$ for each basis element w_j of G_1.

We need a correspondence between complexes of vector spaces and complexes of free S-modules. Such a correspondence is given by the homogenization and dehomegenization constructions described below.

Construction 55.2. Let $\mathbf{U}$ be an r-frame. We will construct an M-complex $\mathbf{G}$ of free S-modules with differential d and call it the ***M*-homogenization** of $\mathbf{U}$. The construction is by induction on homological degree. Recall that mdeg stands for multidegree.

Set

$$G_0 = S \quad \text{and} \quad G_1 = S(m_1) \oplus \ldots \oplus S(m_r) \, .$$

Let $\bar{v}_1, \ldots, \bar{v}_p$ and $\bar{u}_1, \ldots, \bar{u}_q$ be the given bases of U_i and U_{i-1} respectively. Let $u_1, \ldots, u_q$ be the basis of $G_{i-1} = S^q$ chosen on the previous step of the induction. Introduce $v_1, \ldots, v_p$ that will be a basis of $G_i = S^p$. If

$$\partial(\bar{v}_j) = \sum_{1 \le s \le q} \alpha_{sj} \bar{u}_s$$

with coefficients $\alpha_{sj} \in k$, then set

$$\mathrm{mdeg}(v_j) = \mathrm{lcm}\Big(\mathrm{mdeg}(u_s) \,\Big|\, \alpha_{sj} \neq 0\Big) , \text{ note that } \mathrm{lcm}(\emptyset) = 1$$

$$G_i = \oplus_{1 \le j \le p} \; S(\mathrm{mdeg}(v_j))$$

$$d(v_j) = \sum_{1 \le s \le q} \alpha_{sj} \frac{\mathrm{mdeg}(v_j)}{\mathrm{mdeg}(u_s)} u_s \, .$$

Clearly, $\operatorname{Coker}(d_1) = S/M$. Note that the differential d is multihomogeneous by construction. Lemma 55.4 shows that $\mathbf{G}$ is a complex. We say that the complex $\mathbf{G}$ is obtained from $\mathbf{U}$ by ***M-homogenization.***

Running Example 55.3. Consider the 3-frame

$$0 \longrightarrow k \xrightarrow{\begin{pmatrix}1\\1\\1\end{pmatrix}} k^3 \xrightarrow{\begin{pmatrix}-1 & 0 & 1\\ 1 & -1 & 0\\ 0 & 1 & -1\end{pmatrix}} k^3 \xrightarrow{\begin{pmatrix}1 & 1 & 1\end{pmatrix}} k\,.$$

The Y-homogenization of this frame is

$$\mathbf{G}: \qquad 0 \to C(x^2y^3) \xrightarrow{\begin{pmatrix}y^2\\x\\1\end{pmatrix}} C(x^2y) \oplus C(xy^3) \oplus C(x^2y^3)$$

$$\xrightarrow{\begin{pmatrix}-y & 0 & y^3\\ x & -y^2 & 0\\ 0 & x & -x^2\end{pmatrix}} C(x^2) \oplus C(xy) \oplus C(y^3) \xrightarrow{\begin{pmatrix}x^2 & xy & y^3\end{pmatrix}} C\,.$$

Lemma 55.4. $\mathbf{G}$ *in Construction* 55.2 *is a complex.*

Proof. Let $\bar{v}_1, \ldots, \bar{v}_p$, and $\bar{u}_1, \ldots, \bar{u}_q$, and $\bar{w}_1, \ldots, \bar{w}_t$ be the given bases of U_i, U_{i-1}, and U_{i-2} respectively. Let $v_1, \ldots, v_p$, and $u_1, \ldots, u_q$, and $w_1, \ldots, w_t$ be the corresponding bases of G_i, G_{i-1}, and G_{i-2} respectively. Fix a $1 \leq j \leq p$. Since $\mathbf{U}$ is a complex, we have that

$$0 = \partial^2(\bar{v}_j) = \partial\Big(\sum_{1\leq s\leq q} \alpha_{sj}\bar{u}_s\Big) = \sum_{1\leq s\leq q} \alpha_{sj}\Big(\sum_{1\leq l\leq t} \beta_{ls}\bar{w}_l\Big)$$

$$= \sum_{1\leq l\leq t}\Big(\sum_{1\leq s\leq q} \alpha_{sj}\beta_{ls}\Big)\bar{w}_l$$

with $\alpha_{sj}, \beta_{ls} \in k$. Hence $\sum_{1\leq s\leq q} \alpha_{sj}\beta_{ls} = 0$ for each $1 \leq l \leq t$.

Furthermore, in **G** we have

$$\begin{aligned}
d^2(v_j) &= d\Big(\sum_{1\le s\le q} \alpha_{sj}\frac{\mathrm{mdeg}(v_j)}{\mathrm{mdeg}(u_s)}u_s\Big)\\
&= \sum_{1\le s\le q} \alpha_{sj}\frac{\mathrm{mdeg}(v_j)}{\mathrm{mdeg}(u_s)}\Big(\sum_{1\le l\le t} \beta_{ls}\frac{\mathrm{mdeg}(u_s)}{\mathrm{mdeg}(w_l)}w_l\Big)\\
&= \sum_{1\le l\le t}\Big(\sum_{1\le s\le q} \alpha_{sj}\beta_{ls}\frac{\mathrm{mdeg}(v_j)}{\mathrm{mdeg}(u_s)}\frac{\mathrm{mdeg}(u_s)}{\mathrm{mdeg}(w_l)}\Big)w_l\\
&= \sum_{1\le l\le t}\Big(\sum_{1\le s\le q} \alpha_{sj}\beta_{ls}\Big)\frac{\mathrm{mdeg}(v_j)}{\mathrm{mdeg}(w_l)}w_l\\
&= 0\,.
\end{aligned}$$

□

Note that **G** in Construction 55.2 may not be exact even if the frame **U** is exact.

Construction 55.5. Let **G** be an M-complex. The complex

$$\mathbf{U} = \mathbf{G} \otimes S/(x_1 - 1, \dots, x_n - 1)$$

is called the ***frame*** of **G** or the ***dehomogenization*** of **G**. We also say that the complex **U** is obtained from **G** by ***dehomogenization***. Note that **U** is a finite complex of finite k-vector spaces with fixed basis and its differential matrices are obtained by setting $x_1 = 1, \dots, x_n = 1$ in the differential matrices of **G**.

Exercise 55.6. *If* **G** *is the* M*-homogenization of a frame* **U**, *then* **U** *is the frame of* **G**.

A fruitful approach for constructing minimal monomial resolutions is based on the fact that the minimal free resolution of any monomial ideal can be encoded in any of its frames; this was proved in [Peeva-Velasco, Theorem 4.14]:

Theorem 55.7. *The* M*-homogenization of any frame of the minimal multigraded free resolution* **F** *of* S/M *is* **F**.

This raises the following problem.

Open-Ended Problem 55.8. (folklore) *Find sources of frames that yield minimal free resolutions of monomial ideals.*

We will discuss some sources of frames later. In the rest of the section we provide a helpful criterion.

Construction 55.9. Let $\mathbf{G}$ be an M-complex, and let $m \in M$ be a monomial. Denote by $\mathbf{G}(\leq m)$ the subcomplex of $\mathbf{G}$ that is generated by the multihomogeneous basis elements of multidegrees dividing m.

Running Example 55.10. We continue the running example. Let $m = x^2y^2$. We have that

$$\mathbf{G}(\leq x^2y^2): \quad 0 \to C(x^2y) \xrightarrow{\begin{pmatrix} -y \\ x \end{pmatrix}} C(x^2) \oplus C(xy) \xrightarrow{(x^2 \quad xy)} C\,.$$

Proposition 55.11. *Let $m \in M$ be a monomial. Set*

$$m' = \operatorname{lcm}(m_i \,|\, m_i \textit{ divides } m)\,.$$

Then $\mathbf{G}(\leq m) = \mathbf{G}(\leq m')$.

Proof. By Construction 55.1, all the basis elements of $\mathbf{G}$ have multidegrees in L_M . □

Theorem 55.12. *Let $\mathbf{G}$ be an M-complex and $m \in M$ be a monomial. The component of $\mathbf{G}$ of multidegree m is isomorphic to the frame of the complex $\mathbf{G}(\leq m)$.*

Proof. Note that $\mathbf{G}_m$ has basis of the form

$$\left\{ \frac{m}{\operatorname{mdeg}(g)}\, g \,\middle|\, g \text{ is in the fixed basis of } \mathbf{G}, \text{ and } \operatorname{mdeg}(g) \text{ divides } m \right\}.$$

Therefore the component of $\mathbf{G}$ of multidegree m is isomorphic to the frame of the complex $\mathbf{G}(\leq m)$. □

The following criterion for exactness is very useful.

Theorem 55.13. *An M-complex $\mathbf{G}$ is a free multigraded resolution* *of S/M if and only if for all monomials $1 \neq m \in L_M$ the frame of the complex $\mathbf{G}(\leq m)$ is exact.*

Proof. Note that $G_0/d(G_1) = S/M$.

Since the complex $\mathbf{G}$ is multigraded, it suffices to check exactness in each multidegree, similarly to 3.7. As $(G_i)_m = 0$ for $i > 0$ and $m \notin M$, it suffices to check exactness in each multidegree $m \in M$. By Theorem 55.12, it suffices to check exactness of the frames $\mathbf{G}(\leq m)$ for all monomials $m \in M$.

Fix a monomial $m \in M$. Set $m' = \mathrm{lcm}(m_i \mid m_i \text{ divides } m)$ and apply Proposition 55.11. Hence, $\mathbf{G}(\leq m) = \mathbf{G}(\leq m')$. Therefore, it suffices to consider only the multidegrees in L_M. □

56 Subresolutions

We will present a first application of the approach in the previous section: we will show that the minimal free resolution of S/M contains as subcomplexes the minimal free resolutions of certain smaller monomial ideals.

Proposition 56.1. (Gasharov-Hibi-Peeva, Miller) *Let $u \in M$ be a monomial, and consider the monomial ideal $(M_{\leq u})$ generated by the monomials $\{m_i \mid m_i$ divides $u\}$. Fix a multihomogeneous basis of a multigraded free resolution $\mathbf{F}_M$ of S/M.*

(1) *The subcomplex $\mathbf{F}_M(\leq u)$ is a multigraded free resolution of $S/(M_{\leq u})$.*
(2) *If $\mathbf{F}_M$ is a minimal multigraded free resolution of S/M, then $\mathbf{F}_M(\leq u)$ is independent of the choice of basis.*
(3) *If $\mathbf{F}_M$ is a minimal multigraded free resolution of S/M, then the resolution $\mathbf{F}_M(\leq u)$ is minimal as well.*

Proof. Set $v = \mathrm{lcm}(m_i \mid m_i \text{ divides } u)$ and apply Proposition 55.11. Hence, $\mathbf{F}_M(\leq u) = \mathbf{F}_M(\leq v)$. Clearly, $(M_{\leq u}) = (M_{\leq v})$. Therefore, we can replace u by v.

By Theorem 55.13, we see that we have to show that for every monomial $1 \neq m \in L_{(M_{\leq v})}$ the frame of the complex $\big(\mathbf{F}_M(\leq v)\big)(\leq m)$ is exact. The frame of $\big(\mathbf{F}_M(\leq v)\big)(\leq m)$ is equal to the frame of

$\mathbf{F}_M(\leq w)$, where w is the maximal monomial that divides both v and m, and is in the set L_M. Since $\mathbf{F}_M$ is exact, by Theorem 55.13 it follows that the frame of $\mathbf{F}_M(\leq w)$ is exact. We proved (1).

(2) Note that the multidegrees of the basis elements in $\mathbf{F}_M$ are determined by the multigraded Betti numbers. Therefore, they are independent of the choice of basis.

(3) holds by construction. □

Theorem 56.2 is a useful particular case of the above result.

Theorem 56.2. *Let $\mathbf{F}_M$ be the minimal free multigraded resolution of S/M. Denote by N the ideal generated by the squarefree minimal monomial generators of M. The minimal free multigraded resolution of S/N is $\mathbf{F}_M(\leq x_1 \dots x_n) = \mathbf{F}_M(\leq u)$, where u is the product of the variables that appear in the minimal monomial generators of the ideal N.*

Example 56.3. We illustrate Theorem 56.2. Let $A = k[x, y, z]$, $T = (x^2, xy, xz, y^3)$, and $u = xyz$. Then $(T_{\leq xyz}) = (xy, xz)$. The minimal multigraded free resolution of A/T is

$$\mathbf{F}_T: \quad 0 \to A \xrightarrow{\begin{pmatrix} z \\ x \\ -y \\ 0 \end{pmatrix}} A^4 \xrightarrow{\begin{pmatrix} y & 0 & z & 0 \\ -x & z & 0 & y^2 \\ 0 & -y & -x & 0 \\ 0 & 0 & 0 & -x \end{pmatrix}} A^4$$
$$\xrightarrow{(x^2 \ \ xy \ \ xz \ \ y^3)} A\,.$$

The minimal multigraded free resolution of $A/(xy, xz)$ is the subcomplex

$$(\mathbf{F}_T)(\leq xyz): \quad 0 \to A \xrightarrow{\begin{pmatrix} z \\ -y \end{pmatrix}} A^2 \xrightarrow{(xy \ \ xz)} A\,.$$

As an application of Theorem 56.2, we will consider resolutions of squarefree Borel ideals. In the rest of this section, we will use the notation from Section 28.

Studying ideals in an exterior algebra has led to the study of squarefree Borel ideals. Their properties are similar to those of Borel ideals.

A squarefree monomial ideal N is ***squarefree Borel*** if it satisfies the ***squarefree Borel property***: whenever the conditions

- $i < j$
- g is a monomial such that $gx_j \in N$
- gx_i is squarefree,

are satisfied, we have $gx_i \in N$ as well.

Exercise 56.4. *A monomial ideal N is squarefree Borel if and only if whenever the conditions*

- $i < j$
- *g is such that gx_j is a minimal monomial generator of N*
- *gx_i is squarefree,*

are satisfied, we have $gx_i \in N$ as well.

Example 56.5. The ideal (wxy, wxz, wyz) is squarefree Borel in $k[w, x, y, z]$.

The interest in studying such special monomial ideals comes from the following result in [Aramova-Herzog-Hibi 2, Theorem 1.7].

Theorem 56.6. *The generic initial ideal of a graded ideal in an exterior algebra is squarefree Borel.*

Conjecture 56.7. (Aramova-Herzog-Hibi) *Let T' be a squarefree ideal in an exterior algebra on variables $x_1, \ldots, x_n$. Let B' be its generic initial ideal. Consider the monomial ideals T and B in S generated by the squarefree monomial generators of T' and B' respectively. For all $i \geq 0$, we have*

$$b_i^S(S/T) \leq b_i^S(S/B).$$

Note that by Theorem 51.4, the ideals B and T have the same Hilbert function.

Construction 56.8. Let N be squarefree Borel, and M be the smallest Borel ideal containing N. Consider the basis elements of the Eliahou-Kervaire resolution $\mathbf{E}_M$ of S/M that have squarefree multidegrees. In the notation of 28.6, these basis elements are

$$\{1\} \cup \left\{ (m_i; j_1, \dots, j_p) \,\middle|\, 1 \le j_1 < \ldots < j_p < \max(m_i)\ 1 \le i \le r, \right.$$
$$\left. m_i x_{j_1} \ldots x_{j_p} \text{ is squarefree} \right\}.$$

Let $\widetilde{\mathbf{E}}_N$ be the complex that is the essential subcomplex of $\mathbf{E}_M$ with basis the above elements (recall the definition of an essential subcomplex in Definition 3.5). We call $\widetilde{\mathbf{E}}_N$ the ***squarefree Eliahou-Kervaire resolution*** of S/N because of the next theorem, which follows immediately from Theorem 56.2 applied to the Eliahou-Kervaire resolution.

Theorem 56.9. (Aramova-Herzog) *Let N be squarefree Borel. Then $\widetilde{\mathbf{E}}_N$ is the minimal free resolution of S/N.*

Example 56.10. We will describe the squarefree Eliahou-Kervaire minimal free resolution of the squarefree Borel ideal

$$(x_1x_2,\ x_1x_3,\ x_2x_3)$$

in $A = k[x_1, x_2, x_3]$. The smallest Borel ideal, that contains it, is

$$(x_1^2,\ x_1x_2,\ x_1x_3,\ x_2^2,\ x_2x_3)\,.$$

The basis of the squarefree Eliahou-Kervaire resolution is

$$\begin{aligned} 1 &\text{ in homological degree } 0\\ (x_1x_2;\emptyset),\ (x_1x_3;\emptyset),\ (x_2x_3;\emptyset) &\text{ in homological degree } 1\\ (x_1x_3;2),\ (x_2x_3;1) &\text{ in homological degree } 2\,. \end{aligned}$$

The resolution is

$$0 \to A(x_1x_2x_3)^2 \xrightarrow{\begin{pmatrix} x_3 & x_3 \\ -x_2 & 0 \\ 0 & -x_1 \end{pmatrix}} A(x_1x_2) \oplus A(x_1x_3) \oplus A(x_2x_3)$$
$$\xrightarrow{(x_1x_2\ x_1x_3\ x_2x_3)} A\,.$$

Corollary 56.11. *Let N be a squarefree Borel ideal in S minimally generated by monomials $m_1, \ldots, m_r$. Use the notation in Section* 28. *Then*

$$\begin{aligned}
\operatorname{codim}(N) &= \max\{\min(m_i) \,|\, 1 \le i \le r\} \\
\operatorname{reg}(N) &= \text{highest degree of a minimal generator of } N \\
\operatorname{pd}(N) &= \max\{\max(m_i) - \deg(m_i) \,|\, 1 \le i \le r\} \\
b^S_{p,p+q}(N) &= \sum_{\deg(m_i)=q} \binom{\max(m_i) - \deg(m_i)}{p} \\
b^S_p(N) &= \sum_{i=1}^{r} \binom{\max(m_i) - \deg(m_i)}{p}.
\end{aligned}$$

Proof. First, we will prove the formula for the codimension by induction on the number of variables. The proof is from [Herzog-Srinivasan, Proposition 4.1]. Set

$$q = \max\{\min(m_i) \,|\, 1 \le i \le r\}.$$

It follows that $(x_1, \ldots, x_q) \supseteq N$. Hence, $\operatorname{codim}(N) \le q$.

We will show that $\operatorname{codim}(N) \ge q$. Write $N = x_1N'' \oplus N'$, where N' and N'' are squarefree Borel ideals in the ring $T = k[x_2, \ldots, x_n]$. If $N' = 0$, then $\operatorname{codim}(N) = 1$. In the rest of the proof we assume that $N' \neq 0$.

By induction hypothesis,

$$\operatorname{codim}(N') = \max\{\min(m_i') \,|\, 1 \le i \le r'\} - 1,$$

where $m'_1, \ldots, m'_{r'}$ are the minimal monomial generators of N'. The monomials $m'_1, \ldots, m'_{r'}$ are the minimal monomial generators of N not divisible by x_1. Hence,

$$q = \max\{\min(m_i) \,|\, 1 \le i \le r\} = \max\{\min(m'_i) \,|\, 1 \le i \le r'\}.$$

Thus, $\operatorname{codim}(N') = q - 1$.

If $v \in N'$ is a squarefree monomial, then $w = \dfrac{x_1 v}{x_{\min(v)}} \in x_1 N''$ and $\dfrac{v}{x_{\min(v)}} \in N''$. Clearly, $\min\Big(\dfrac{v}{x_{\min(v)}}\Big) > \min(v)$. By induction hypothesis, it follows that

$$\operatorname{codim}(N'') > \operatorname{codim}(N') = q - 1\,.$$

Hence, the squarefree Borel ideal $N' + N''$ has $\operatorname{codim}(N' + N'') \ge q$.

Let $P = (x_{i_1}, \ldots, x_{i_p})$ be a minimal prime containing N; we assume that $i_1 < \ldots < i_p$. We will show that $p \ge q$. If $i_1 = 1$, then $(x_{i_2}, \ldots, x_{i_p})$ contains N', so $p - 1 \ge \operatorname{codim}(N') = q - 1$. If $i_1 \ne 1$, then P is a prime ideal containing the ideal $N' + N''$ of codimension $\ge q$, so $p \ge q$.

Denote by $\operatorname{nsupp}(u) = \{j \,|\, x_j \text{ divides } u\}$ the numerical support of a monomial u. In order to prove the remaining formulas, note that the minimal free resolution of N has basis

$$\Big\{ (m_i; j_1, \ldots, j_p) \,\Big|\, 1 \le j_1 < \ldots < j_p < \max(m_i), \qquad 1 \le i \le r,\ m_i x_{j_1} \ldots x_{j_p} \text{ is squarefree} \Big\}$$

in homological degree p. For a fixed m_i, note that each j_s can take values in $\{1, \ldots, \max(m_i)\} \setminus \operatorname{nsupp}(m_i)$. Note that for the squarefree monomial m_i we have $|\operatorname{nsupp}(m_i)| = \deg(m_i)$. Therefore, for a fixed m_i, there are $\binom{\max(m_i) - \deg(m_i)}{p}$ choices for the sequence $j_1, \ldots, j_p$. □

57 Simplicial and cellular resolutions

We will explore the following idea: we will obtain frames from a standard construction in topology – homology of (simplicial) chain complexes.

Throughout this section Δ is a simplicial complex on vertices $\{m_1, \ldots, m_r\}$. Recall 36.1. Denote by $\widetilde{C}(\Delta, k)$ the augmented oriented simplicial chain complex of Δ over k; it is used in topology to compute the simplicial homology of Δ. This complex is

$$\widetilde{C}(\Delta; k) = \oplus_{\tau \in \Delta}\, k e_\tau\,,$$

where e_τ denotes the basis element corresponding to the face τ in homological degree $|\tau| - 1$, and the differential ∂ acts as

$$\partial(e_\tau) = \sum_{\tau' \text{ is a facet of } \tau} [\tau, \tau']\, e_{\tau'}\,,$$

where $[\tau, \tau']$ is the incidence (orientation) function: $[\tau, \tau'] = (-1)^i$ if $\tau \setminus \tau'$ is the $(i+1)$'st element in the sequence of the vertices of τ written in increasing order.

Definition 57.1. [Bayer-Peeva-Sturmfels] We use the notation above. After shifting $\widetilde{C}(\Delta; k)$ in homological degree, we get that $\widetilde{C}(\Delta; k)[-1]$ is a frame. Denote by $\mathbf{F}_\Delta$ the M-homogenization of $\widetilde{C}(\Delta; k)[-1]$ (see Construction 55.2). We say that $\mathbf{F}_\Delta$ is ***supported*** on Δ, or that Δ ***supports*** $\mathbf{F}_\Delta$. The complex $\mathbf{F}_\Delta$ is a ***simplicial resolution*** if it is exact. Simplicial resolutions are interesting because they are usually nicely combinatorially structured. They were introduced in [Bayer-Peeva-Sturmfels].

For each vertex m_i of Δ, we set that m_i has multidegree $\mathrm{mdeg}(m_i) = m_i$. We define that a face τ has multidegree

$$\mathrm{mdeg}(\tau) = \mathrm{lcm}(m_i \,|\, m_i \in \tau)\,.$$

By convention, $\mathrm{mdeg}(\emptyset) = 1$.

We think of Δ as a simplicial complex with labeled faces: each face is labeled by its multidegree.

Theorem 57.2. [Bayer-Peeva-Sturmfels] *For each face τ of dimension i the complex $\mathbf{F}_\Delta$ has the generator e_τ in homological degree $i+1$.*

(1) *We have* $\mathrm{mdeg}(e_\tau) = \mathrm{mdeg}(\tau)$.
(2) *The differential in* $\mathbf{F}_\Delta$ *is*

$$\partial(e_\tau) = \sum_{\tau' \text{ is a facet of } \tau} [\tau, \tau'] \, \frac{\mathrm{mdeg}(\tau)}{\mathrm{mdeg}(\tau')} \, e_{\tau'}$$

$$= \sum_{\tau' \text{ is a facet of } \tau} [\tau, \tau'] \, \frac{\mathrm{lcm}(m_i | m_i \in \tau)}{\mathrm{lcm}(m_i | m_i \in \tau')} \, e_{\tau'} \, .$$

Proof. (2) follows from (1) and the fact that the differential is multi-homogeneous. We will prove (1) by induction on homological degree. Clearly, $\mathrm{mdeg}(e_{m_i}) = m_i$ holds for each vertex m_i of Δ. Since

$$\partial(e_\tau) = \sum_{\tau' \text{ is a facet of } \tau} [\tau, \tau'] \, e'_\tau \, ,$$

it follows by Construction 55.2 that

$$\begin{aligned} \mathrm{mdeg}(e_\tau) &= \mathrm{lcm}(\mathrm{mdeg}(e_{\tau'}) \,|\, \tau' \text{ is a facet of } \tau\,) \\ &= \mathrm{lcm}(\mathrm{mdeg}(\tau') \,|\, \tau' \text{ is a facet of } \tau\,) \\ &= \mathrm{lcm}\big(\,\mathrm{lcm}(m_i | m_i \in \tau') \,\big|\, \tau' \text{ is a facet of } \tau\,\big) \\ &= \mathrm{lcm}(m_i | m_i \in \tau) = \mathrm{mdeg}(\tau) \, . \end{aligned}$$

□

Running Example 57.3. Consider the triangle Δ with vertices x^2, xy, y^3. We label each edge by the least common multiple of its vertices, so we get labels x^2y, xy^3, x^2y^3 on the edges. We label the triangle by the least common multiple x^2y^3 of its vertices. See Figure 8 below.

The following is an augmented oriented chain complex of the triangle:

$$0 \longrightarrow k \xrightarrow{\begin{pmatrix} 1 \\ 1 \\ 1 \end{pmatrix}} k^3 \xrightarrow{\begin{pmatrix} -1 & 0 & 1 \\ 1 & -1 & 0 \\ 0 & 1 & -1 \end{pmatrix}} k^3 \xrightarrow{\begin{pmatrix} 1 & 1 & 1 \end{pmatrix}} k \, .$$

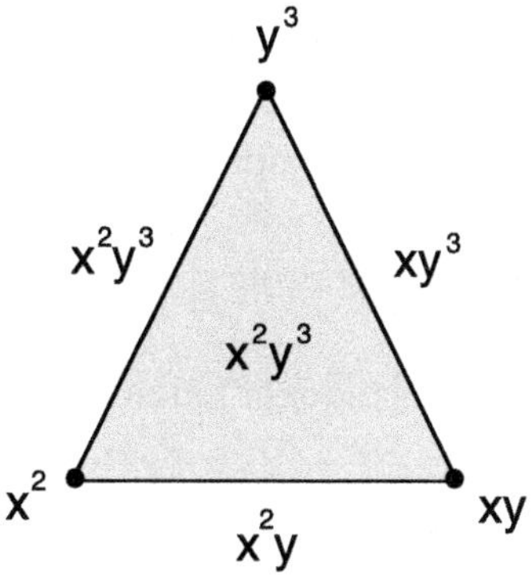

Figure 8.

The corresponding Y-homogenized complex $\mathbf{T}_Y$ is

$$\mathbf{T}_Y : 0 \longrightarrow A \xrightarrow{\begin{pmatrix} y^2 \\ x \\ 1 \end{pmatrix}} A^3 \xrightarrow{\begin{pmatrix} -y & 0 & y^3 \\ x & -y^2 & 0 \\ 0 & x & -x^2 \end{pmatrix}} A^3 \xrightarrow{\begin{pmatrix} x^2 & xy & y^3 \end{pmatrix}} A\,.$$

We see that $\mathbf{T}_Y = \mathbf{F}_Y \oplus \big(0 \to A \to A \to 0\big)$, so it is exact. Thus, $\mathbf{T}_Y$ is a simplicial resolution, which is non-minimal.

The minimal free resolution $\mathbf{F}_Y$ is also simplicial and corresponds to the simplicial complex with vertices labeled by x^2, xy, y^3 and with two edges $\{x^2, xy\}$ and $\{xy, y^3\}$.

We have

$$\mathbf{F}_Y : \quad 0 \longrightarrow A^2 \xrightarrow{\begin{pmatrix} -y & 0 \\ x & -y^2 \\ 0 & x \end{pmatrix}} A^3 \xrightarrow{\begin{pmatrix} x^2 & xy & y^3 \end{pmatrix}} A\,.$$

For each multidegree m, define the following two subcomplexes of Δ:

$$\Delta_{\leq m} = \{\tau \in \Delta \,|\, \mathrm{mdeg}(\tau) \text{ divides } m\}$$
$$\Delta_{<m} = \{\tau \in \Delta \,|\, \mathrm{mdeg}(\tau) \text{ strictly divides } m\}.$$

Running Example 57.4. See Figure 8 above. The subcomplex

$$\Delta_{\leq x^2y} = \{\tau \in \Delta \,|\, \mathrm{mdeg}(\tau) \text{ divides } x^2y\}$$

is the edge $\{x^2, xy\}$. The subcomplex

$$\Delta_{<x^2y^3} = \{\tau \in \Delta \,|\, \mathrm{mdeg}(\tau) \text{ strictly divides } x^2y^3\}$$

consists of the two edges $\{x^2, xy\}$ and $\{xy, y^3\}$.

Proposition 57.5. [Bayer-Peeva-Sturmfels] *The complex F_Δ is a free resolution of S/M if and only if for all multidegrees $1 \neq m \in L_M$ the complex $\Delta_{\leq m}$ is acyclic over k. Note that for $m \notin M$, the complex $\Delta_{\leq m}$ is empty.*

Proof. This follows from Theorem 55.13 because if $m \in L_M$ then the frame of $F_\Delta(\leq m)$ is $\widetilde{C}(\Delta_{\leq m}; k)[-1]$ by 57.1. □

Theorem 57.6. [Bayer-Sturmfels] *Let F_Δ be a resolution of S/M. For $i \geq 1$ and multidegree $m \neq 1$ we have*

$$b^S_{i,m}(S/M) = \begin{cases} \dim \widetilde{\mathrm{H}}_{i-2}(\Delta_{<m}; k) & \textit{if } \Delta_{\leq m} \neq \emptyset \\ 0 & \textit{if } \Delta_{\leq m} = \emptyset. \end{cases}$$

Proof. We will compute $\mathrm{Tor}_i(S/M, k)_m$ using F_Δ. Note that $(F_\Delta)_m$ has basis

$$\left\{ \frac{m}{\mathrm{mdeg}(e_\tau)}\, e_\tau \,\middle|\, \tau \in \Delta_{\leq m} \right\}.$$

Hence, $(F_\Delta)_m \otimes k$ has basis

$$\{\, e_\tau \,|\, \mathrm{mdeg}(\tau) = m \,\}.$$

The complex $(F_\Delta)_m \otimes k$ of k-vector spaces is isomorphic to the chain complex $\widetilde{C}(\Delta_{\leq m}, \Delta_{<m}; k)$, which computes the reduced relative simplicial homology with coefficients in k of the pair $(\Delta_{\leq m}, \Delta_{<m})$. We get

$$\mathrm{Tor}_i(S/M, k)_m = \widetilde{\mathrm{H}}_{i-1}(\Delta_{\leq m}, \Delta_{<m}; k).$$

If $\Delta_{\leq m} = \emptyset$, then $\Delta_{<m} = \emptyset$ and $\widetilde{\mathrm{H}}_{i-1}(\Delta_{\leq m}, \Delta_{<m}; k) = 0$.

If $\Delta_{\leq m} \neq \emptyset$, then $\Delta_{\leq m}$ is acyclic by Proposition 57.5. Therefore, the long exact sequence

$$\ldots \to \widetilde{\mathrm{H}}_{i-1}(\Delta_{\leq m}; k) \to \widetilde{\mathrm{H}}_{i-1}(\Delta_{\leq m}, \Delta_{<m}; k) \to \widetilde{\mathrm{H}}_{i-2}(\Delta_{<m}; k) \to \widetilde{\mathrm{H}}_{i-2}(\Delta_{\leq m}; k) \to \ldots$$

implies the isomorphism $\widetilde{\mathrm{H}}_{i-1}(\Delta_{\leq m}, \Delta_{<m}; k) \simeq \widetilde{\mathrm{H}}_{i-2}(\Delta_{<m}; k)$. □

Example 57.7. Consider $T = (ab, ac, ae, bc)$ in $A = k[a, b, c, e]$. The simplicial complex $\Theta_{<abce}$ is shown in Figure 9. We have $b^A_{3,abce}(A/T) = 1$.

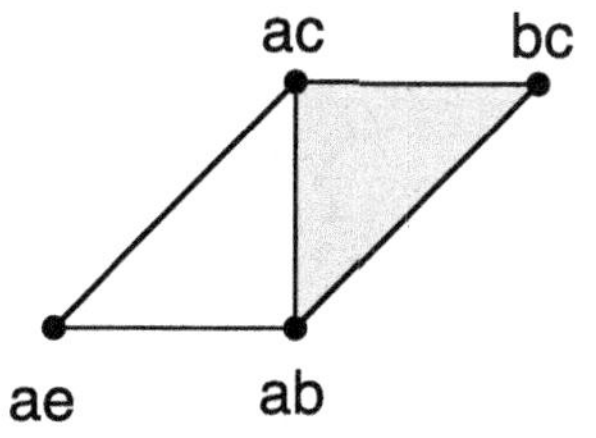

Figure 9.

Proposition 57.8. *Let F_Δ be a resolution of S/M, and let $u \in M$ be a monomial. Consider the monomial ideal $(M_{\leq u})$ generated by the monomials $\{m_i \,|\, 1 \leq i \leq r,\ m_i \text{ divides } u\}$. The complex $F_{\Delta_{\leq u}}$ is a resolution of $S/(M_{\leq u})$.*

Proof. This follows from Theorem 56.1 since $F_{\Delta_{\leq u}} = F_\Delta(\leq u)$. □

Theorem 57.9. [Bayer-Peeva-Sturmfels] *Denote by Θ the simplex with r vertices $m_1, \ldots, m_r$.*

(1) *Taylor's resolution* 26.5 *is supported on Θ.*
(2) *For $i \geq 1$, the Betti numbers of S/M are*

$$b^S_{i,m}(S/M) = \begin{cases} \dim \widetilde{\mathrm{H}}_{i-2}(\Theta_{<m}; k) & \text{if } m \text{ divides } \mathrm{lcm}(m_1, \ldots, m_r) \\ 0 & \text{otherwise.} \end{cases}$$

Proof. Apply Theorem 57.6 to Taylor's resolution. □

Example 57.10. Example 54.2 provides an example of a monomial ideal with a non-simplicial minimal free resolution since the coefficients of the monomials in the differential matrices cannot be chosen in $\{0, \pm 1\}$.

Cellular resolutions are a natural generalization of simplicial resolutions. They were introduced and studied in [Bayer-Sturmfels]. Let X be a finite regular cell complex on vertices $m_1, \ldots, m_r$. It is equipped with a (non-unique) incidence function, cf. [Bruns-Herzog, Lemma 6.2.1]. The augmented oriented chain complex $\widetilde{C}(X,;k)$ is used in topology to compute the reduced homology of X. This complex is $\widetilde{C}(X;k) = \oplus_{\tau \in X} k\, e_\tau$, where the basis element e_τ is placed in homological degree $\dim(\tau)$, and the differential ∂ acts as

$$\partial(e_\tau) = \sum_{\tau' \text{ is a facet of } \tau} [\tau, \tau']\, e_{\tau'}\,,$$

where $[\tau, \tau']$ is the incidence function. Clearly, $\widetilde{C}(X;k)[-1]$ is a frame. Denote by $\mathbf{F}_X$ the M-homogenization of $\widetilde{C}(X;k)[-1]$. We say that $\mathbf{F}_X$ is ***supported*** on X, or that X ***supports*** $\mathbf{F}_X$. The complex $\mathbf{F}_X$ is a ***cellular resolution*** if it is exact.

Example 54.2 provides an example of a monomial ideal with a non-cellular minimal free resolution.

A finite regular cell complex is a ***polyhedral cell complex*** if each closed cell is homeomorphic to a convex polytope on the vertices contained in the cell. If X is a polyhedral cell complex, then $\mathbf{F}_X$ is a ***polyhedral resolution***.

It is natural to consider CW-***cellular resolutions*** that are supported by CW-complexes. This is a significant generalization; the structure of CW-cellular resolutions is more complex than that of cellular resolutions. For example, it is no longer true that the coefficients appearing in the differential matrices are only 0, ± 1. CW-cellular resolutions are introduced and studied in [Batzies-Welker] and [Jöllenbeck-Welker]. But even this generalization is not sufficient to cover all minimal monomial resolutions: in [Velasco] it is shown that there exists a monomial ideal whose minimal free resolution does not admit any CW-cellular structure.

58 The lcm-lattice

We will discuss the lcm-lattice, which was introduced in [Gasharov-Peeva-Welker 2] and plays a key role in the study of monomial resolutions. The idea to use the lcm-lattice was inspired by the role of the intersection lattice in computing cohomology of subspace arrangements.

A ***lattice*** is a poset P for which every pair of elements has a join (least upper bound) and a meet (greatest lower bound). If P has a bottom element $\hat{0}$, then the elements in P covering $\hat{0}$ are called ***atoms.***

Lemma 58.1. *Let P be a finite poset with bottom element $\hat{0}$. If every pair of elements has a join, then P is a lattice.*

Proof. Let x and y be two elements in P. The set

$$T = \{\, z \in P \,|\, z \leq x,\ z \leq y \,\}$$

is finite and non-empty. The meet of x and y is the join of all elements in T. □

Construction 58.2. [Gasharov-Peeva-Welker 2] We denote by L_M the lattice with elements the least common multiples of subsets of $m_1, \ldots, m_r$ ordered by divisibility. The atoms in L_M are $m_1, \ldots, m_r$. The top element is $m_M = \mathrm{lcm}(m_1, \ldots, m_r)$. The bottom element is 1 regarded as the lcm of the empty set. The least common multiple of elements in L_M is their join. By Lemma 58.1, L_M is a lattice. We call L_M the ***lcm-lattice*** of M. For $m \in L_M$ we denote by $(1, m)_{L_M}$ the open interval in L_M below m; it consists of all non-unit monomials in L_M that strictly divide m.

Running Example 58.3. The lcm-lattice of (x^2, xy, y^3) is given in Figure 10.

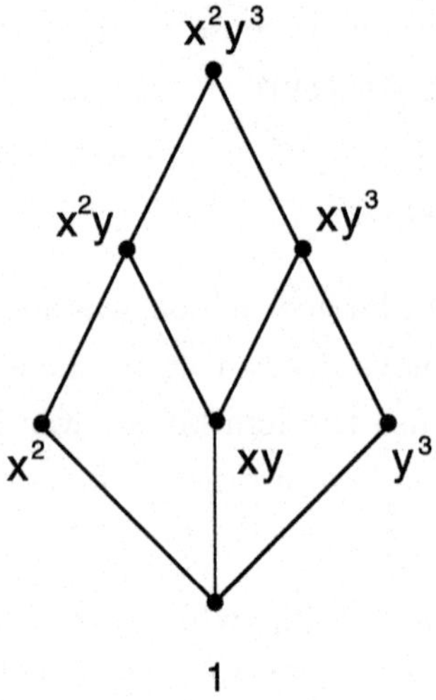

Figure 10.

Exercise 58.4. *If M_{pol} is the polarization of M, then $L_M \cong L_{M_{pol}}$.*

One might wonder what lattices appear as lcm-lattices. The answer to this question was given by Phan in his Ph.D. thesis.

Construction 58.5. [Phan] Let L be a finite atomic lattice (atomic means that each non-bottom element is a joint of atoms). An element in L is meet-irreducible if it is not the meet of two elements in L, and it is not the top or bottom element in L. Suppose that L has n meet-irreducible elements. Label them by $x_1, \ldots, x_n$. Now, label an element $c \in L$ by the monomial

$$\mathrm{mon}(c) = \frac{x_1 \ldots x_n}{\prod_{x_i \geq c} x_i} .$$

Let N_L be the monomial ideal generated by the labels of the atoms in L. This monomial ideal is called the ***L-ideal***.

Theorem 58.6. [Phan] *Let L be a finite atomic lattice. There exists a monomial ideal whose lcm-lattice is L.*

Proof. We use the notation in Construction 58.5. We will show that the L-ideal N_L has lcm-lattice L.

We will show that the monomial $\mathrm{mon}(c)$ is the least common multiple of the labels of the atoms below c. Let $p_1, \dots, p_q$ be the atoms below c. If x_i does not divide $\mathrm{mon}(c)$, then $x_i \geq c$; so $x_i \geq p_j$ for each $1 \leq j \leq q$; hence $\mathrm{mon}(p_j)$ divides $\mathrm{mon}(c)$ for each j.

On the other hand, if x_i divides $\mathrm{mon}(c)$, then $x_i \not\geq c$; so there exists an atom p_j such that $x_i \not\geq p_j$; hence, x_i divides some $\mathrm{mon}(p_j)$.

Denote by L' the lcm-lattice of N_L. Let $\psi : L \to L'$ be the map that maps an element $c \in L$ to $\mathrm{mon}(c) \in L'$. The map is order-preserving and surjective. We will show that it is injective. Let $c, c' \in L$ be such that $\mathrm{mon}(c) = \mathrm{mon}(c')$. Therefore, the set $\mathcal{M}_c$ of meet-irreducible elements over c coincides with the the set $\mathcal{M}_{c'}$ of meet-irreducible elements over c'. Note that c is the meet of the elements in $\mathcal{M}_c$, and c' is the meet of the elements in $\mathcal{M}_{c'}$. Hence, $c = c'$. □

Example 58.7. Consider the lattice in Figure 11. The meet-irreducible elements are labeled by variables.

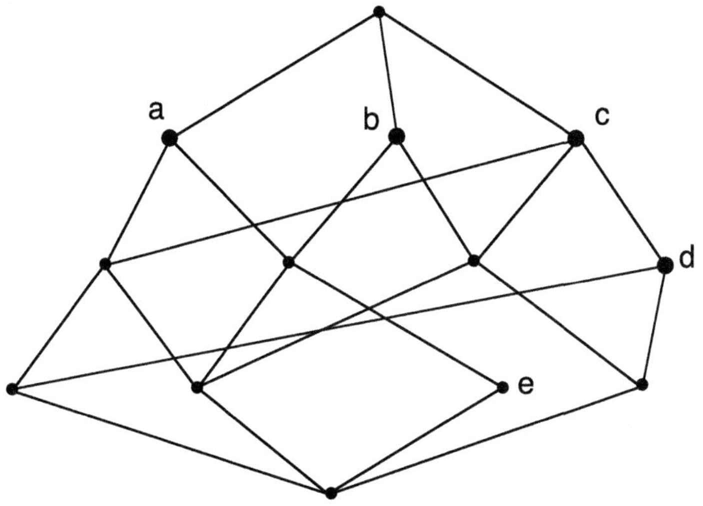

Figure 11.

In Figure 12 we show the same lattice but labeled as the lcm-lattice of the ideal constructed in the proof of the theorem above.

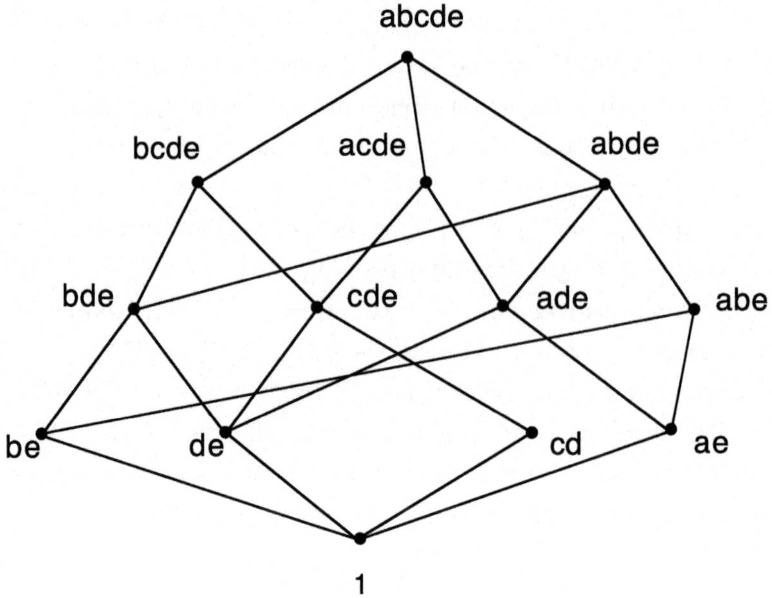

Figure 12.

Theorem 58.8 is the main result in this section.

Theorem 58.8. [Gasharov-Peeva-Welker 2] *For* $i \geq 1$ *we have*

$$b^S_{i,m}(S/M) = \begin{cases} \dim \widetilde{\mathrm{H}}_{i-2}\big((1,m)_{L_M}; k\big) & \text{if } 1 \neq m \in L_M \\ 0 & \text{if } m \notin L_M\ . \end{cases}$$

Note that $\widetilde{\mathrm{H}}_{i-2}\big((1,m)_{L_M}; k\big)$ means $\widetilde{\mathrm{H}}_{i-2}\big(O\big((1,m)_{L_M}\big); k\big)$; recall that $O\big((1,m)_{L_M}\big)$ is the order complex of $(1,m)_{L_M}$ (see Section 36).

Proof. By 57.9 we have that $b^S_{i,m}(S/M) = 0$ if $m \notin L_M$.

Let Θ be the simplex with r vertices labeled by $m_1, \dots, m_r$. Fix a monomial $m \in L_M$, $m \neq 1$. The formula for the Betti numbers in Theorem 57.9 is $b^S_{i,m}(S/M) = \dim \widetilde{\mathrm{H}}_{i-2}\big(\Theta_{<m}; k\big)$ for $i \geq 1$.

Recall 36.14. The set C of the minimal monomial generators of M that divide m forms a crosscut of the poset $(1,m)_{L_M}$. Its crosscut complex has faces the subsets of C whose lcm is in $(1,m)_{L_M}$, that is the lcm strictly divides m. So the crosscut complex coincides with the complex $\Theta_{<m}$. By Theorem 36.16 the crosscut complex $\Theta_{<m}$ is homotopic to the order complex of $(1, m)_{L_M}$. □

Running Example 58.9. See Figure 10. Consider the open interval $(1, x^2y^3)$ in the lcm-lattice of Y. Its order complex $O(1, x^2y^3)$ has 5 vertices x^2, xy, y^3, x^2y, xy^3 and 4 edges. It is contractible, so we get that $b^C_{i,x^2y^3}(A/Y)$ vanish for all i. Now, consider the open interval $(1, x^2y)$. Its order complex $O(1, x^2y)$ has 2 vertices and no edge. Hence $b^C_{2,x^2y}(C/Y) = 1$.

Forgetting about the multigrading in the Theorem 58.8 we obtain the following result.

Corollary 58.10. *For $i \geq 1$ we have*

$$b^S_i(S/M) \;=\; \sum_{\substack{m \in L_M \\ m \neq 1}} \dim \widetilde{\mathrm{H}}_{i-2}\big((1, m)_{L_M}; k\big)\,.$$

This formula is an analogue of the Goresky-MacPherson Formula, which expresses the dimensions of the cohomology groups of the complement of a subspace arrangement in terms of the dimensions of the homology groups of the lower intervals in the intersection lattice.

Next, we will show that in order to compute the Betti numbers one can use the lcm-lattice built on any set of monomial generators of the ideal M.

Proposition 58.11. [Gasharov-Peeva-Welker 2] *Let L' be the lattice of the least common multiples of subsets of a set of monomials generating M. For $i \geq 1$ we have*

$$b^S_{i,m'}(S/M) \;=\; \begin{cases} \dim \widetilde{\mathrm{H}}_{i-2}\big((1, m')_{L'}; k\big) & \text{if } m' \in L' \\ 0 & \text{if } m' \notin L'. \end{cases}$$

Proof. If $m' \in L' \setminus L_M$, then the order complex of $(1, m')_{L'}$ is a cone over $\mathrm{lcm}\{m_i | m_i \text{ divides } m'\}$. Hence,

$$\dim \widetilde{\mathrm{H}}_{i-2}\big((1, m')_{L'}; k\big) \;=\; 0 = b^S_{i,m'}(S/M)\,.$$

Consider the map

$$f: L' \to L_M \subseteq L'$$
$$m' \mapsto \operatorname{lcm}\{m_i | m_i \text{ divides } m'\}.$$

This map is order-preserving and it is a closure operator (see 36.6). By Theorem 36.6 it follows that $(1, m')_{L'}$ and $(1, f(m'))_{L_M}$ are homotopic. If $m' \in L_M$, then $f(m') = m'$ and

$$\dim \widetilde{\mathrm{H}}_{i-2}\big((1, m')_{L'}; k\big) = \dim \widetilde{\mathrm{H}}_{i-2}\big((1, m')_{L_M}; k\big) = b^S_{i,m'}(S/M).$$

□

Running Example 58.12. The lcm-lattice of Y is given in Figure 10. We can also use the lcm-lattice in Figure 13 consisting of the lcm's of the monomials $x^2, xy, y^3, x^2y^2, xy^5$.

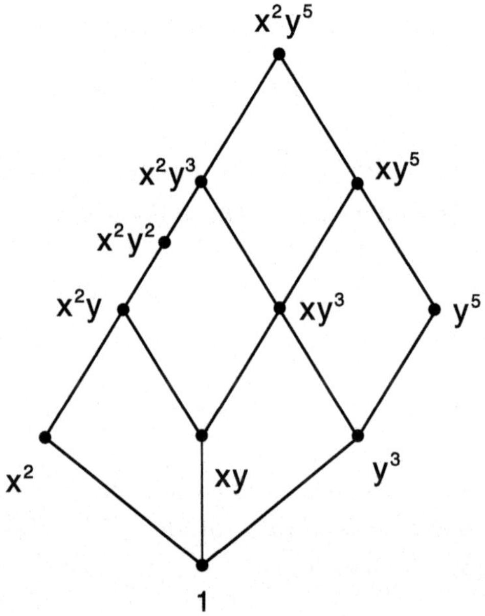

Figure 13.

59 The Scarf complex

We will show that generically the minimal free resolutions of monomial ideals are simplicial. For this purpose we will use the Scarf complex, introduced in [Bayer-Peeva-Sturmfels]. This complex is always contained in the minimal free resolution of a monomial ideal. We will discuss the ideals for which the Scarf complex provides the minimal free resolution; such ideals are called Scarf ideals. Unless otherwise stated, the material in this section is from [Bayer-Peeva-Sturmfels].

Construction 59.1. Recall that m_τ stands for $\mathrm{lcm}(m_i |\, m_i \in \tau)$. The ***Scarf complex*** of M is the simplicial complex

$$\Omega_M = \Big\{ \tau \subseteq \{m_1, \dots, m_r\} \mid \; m_\tau \neq m_\sigma \;\text{ for all }\; \sigma \subseteq \{m_1, \dots, m_r\} \\ \text{other than } \tau \Big\}.$$

In [Bayer-Peeva-Sturmfels] it is shown that Ω_M equals a simplicial complex introduced by Scarf in the context of mathematical economics. Denote by $\mathbf{F}_{\Omega_M}$ the M-homogenization of the augmented oriented simplicial chain complex of Ω_M (see 57.1).

The multidegree of a vertex m_i in Ω_M is the monomial m_i. The multidegree of a face $\tau \in \Omega_M$ is $\mathrm{mdeg}(\tau) = \mathrm{lcm}(\, m_i \mid m_i$ is a vertex of $\tau\,)$. By Theorem 57.2, the multidegree of the basis element e_τ in $\mathbf{F}_{\Omega_M}$ is $\mathrm{mdeg}(\tau)$. The multidegrees of the faces of Ω_M are called ***Scarf multidegrees***.

Theorem 59.2. *If* $\mathrm{mdeg}(\tau)$ *is a Scarf multidegree, then*

$$b^S_{i,\mathrm{mdeg}(\tau)}(S/M) = \begin{cases} 1 & \textit{if } i = \dim(\tau) + 1 \\ 0 & \textit{otherwise.} \end{cases}$$

Proof. Suppose $\mathrm{mdeg}(\tau) \neq 1$. The closed interval $[1, \mathrm{mdeg}(\tau)]$ in L_M is the face lattice of the simplex τ. Hence, the open interval $(1, \deg(\tau))$ is homotopic to the boundary of the simplex τ. Therefore,

$$b^S_{i,\mathrm{mdeg}(\tau)}(S/M) = \widetilde{\mathrm{H}}_{i-2}((1, \mathrm{mdeg}(\tau)); k) = \begin{cases} 1 & \text{if } i-2 = \dim(\tau) - 1 \\ 0 & \text{otherwise.} \end{cases}$$

□

Running Example 59.3. The Scarf complex Ω_Y of Y has the three vertices x^2, xy, y^3 and the two edges $\{x^2, xy\}$, $\{xy, y^3\}$.

Proposition 59.4. *The complex* $\mathbf{F}_{\Omega_M}$ *is an essential subcomplex of* *the minimal free resolution* $\mathbf{F}_M$ *of* S/M, *that is, there exists a basis of* $\mathbf{F}_{\Omega_M}$ *that is part of a basis of* $\mathbf{F}_M$.

Proof. Consider Taylor's resolution $\mathbf{T}_M$.

By Theorem 59.2, it follows that for each multidegree m, that divides a Scarf multidegree, we have that $\mathbf{F}_M$, $\mathbf{T}_M$, and $\mathbf{F}_{\Omega_M}$ have the same number (0 or 1) of basis elements in multidegree m.

Taylor's resolution is possibly non-minimal, so $\mathbf{T}_M \cong \mathbf{F}_M \oplus \mathbf{P}$, where $\mathbf{P}$ is a sum of trivial complexes of the form $0 \to S(u) \to S(u) \to 0$. It follows that such a multidegree u cannot divide a Scarf multidegree. Therefore, $\big(\mathbf{T}_M\big)_{\leq m} \cong \big(\mathbf{F}_M\big)_{\leq m}$ for each multidegree m, that divides a Scarf multidegree.

By the construction of Taylor's resolution, we have that $\mathbf{F}_{\Omega_M}$ is an essential subcomplex of $\mathbf{T}_M$. Hence, $\mathbf{F}_{\Omega_M}$ is an essential subcomplex of the minimal free resolution $\mathbf{F}_M$. □

We call M a ***Scarf ideal*** if $\mathbf{F}_{\Omega_M}$ is the minimal free resolution of S/M, and we say that $\mathbf{F}_{\Omega_M}$ is its ***Scarf resolution***. The definition of Ω_M immediately implies the following properties of the Scarf resolution.

Corollary 59.5. *Let* M *be a Scarf ideal.*

(1) *The number of* j*-faces of the Scarf complex* Ω_M *equals the Betti number* $b^S_{j+1}(S/M)$.
(2) $\mathbf{F}_{\Omega_M}$ *is multigraded and in each multidegree the Betti number is either* 0 *or* 1.

Theorem 59.6. *An ideal* M *is Scarf if and only if all non-zero Betti numbers of* S/M *are in Scarf multidegrees.*

Proof. Suppose that all non-zero Betti numbers of S/M are in Scarf multidegrees. If $\mathbf{F}_M$ is strictly larger than $\mathbf{F}_{\Omega_M}$, then there exists a face $\tau \in \Omega_M$ such that $\mathbf{F}_M$ has at least two basis elements in multidegree $\mathrm{mdeg}(\tau)$. This contradicts Theorem 59.2.

Theorem 59.7. [Peeva-Velasco]

(1) *A simplicial complex with* r *vertices is the Scarf complex of a*

monomial ideal if and only if it is not the boundary of the simplex with r vertices.

(2) *A finite simplicial complex Ω supports a Scarf resolution if and only if Ω is acyclic.*

In the proof we use the following construction.

Construction 59.8. (Mermin) Let $\Omega \neq \emptyset$ be a finite simplicial complex. For each face τ of Ω we introduce a variable x_τ, and then we set $B = k[x_\tau \,|\, \tau \in \Omega,\, \tau \neq \emptyset]$. We will construct a monomial ideal in this polynomial ring. Set z to be the product of all the variables.

Set the multidegree of a vertex v of Ω to be

$$\mathrm{mdeg}(v) = \prod_{v \notin \tau \in \Omega} x_\tau \,.$$

Denote by Θ the simplex on the vertices of Ω. It follows that a face $\sigma \in \Theta$ has multidegree

$$\mathrm{mdeg}(\sigma) = \mathrm{lcm}(\mathrm{mdeg}(v) \,|\, v \in \sigma) = \prod_{\sigma \not\subseteq \tau \in \Omega} x_\tau \,.$$

If $\sigma \notin \Omega$ then $\mathrm{mdeg}(\sigma) = z$. Every two faces in Ω have distinct multidegrees.

Denote by J_Ω the ideal generated by the multidegrees of the vertices. We say that J_Ω is the ***nearly Scarf ideal*** of Ω. It is easy to see that the lcm-lattice L_{J_Ω} consists of the top element z and the face poset of Ω.

Proof. Let Ω be a finite simplicial complex. Both (1) and (2) hold if Ω is either a point or $\emptyset$. We will assume that Ω has at least two vertices.

(1) The complex Ω is the Scarf complex of the monomial ideal J_Ω, constructed in Construction 59.8, if and only if Ω is a simplex or Ω has at least two non-faces. This happens if and only if Ω is not the boundary of the simplex Θ.

(2) If Ω supports a Scarf resolution, then it is acyclic by Theorem 55.13 applied to the multidegree m that is the lcm of all the minimal monomial generators of the ideal.

Suppose that the simplicial complex Ω is acyclic. We will show that the ideal J_Ω, constructed in Construction 59.8, is a Scarf ideal with Scarf complex Ω. The lcm of its minimal monomial generators is z.

We want to apply Theorem 59.6. Thus, we have to show that $b^S_{i,z}(S/J_\Omega) = 0$ for every i. Compute these Betti numbers using Theorem 58.8. The lcm-lattice of J_Ω consists of the Scarf multidegrees (including the bottom element 1) and the top element z. The interval $[1, z)$ is the face poset of Ω. The order complex of $(1, z)$ is the barycentric subdivision of Ω by 36.8, and is homotopic to Ω by 36.9, so it is acyclic. Therefore, the simplicial complex Ω supports the Scarf resolution of J_Ω. □

Theorem 59.9 provides a wide class of ideals which are Scarf ideals and which have the advantage of being defined by a simple combinatorial property; note that the ideals are defined by a generic condition on the exponents of the minimal monomial generators.

Theorem 59.9. *Suppose that no variable x_i appears at the same non zero exponent in two distinct minimal monomial generators of M. Then M is a Scarf ideal.*

Proof. For $\tau \subseteq \{m_1, \ldots, m_r\}$, set $m_\tau = \mathrm{mdeg}(\tau) = \mathrm{lcm}(m_i \,|\, i \in \tau)$. Consider a multidegree m_τ with $\tau \notin \Omega_M$. By Theorem 59.6 we have to show that all Betti numbers in multidegree m_τ vanish. We compute the Betti numbers of S/M using the Koszul complex $\mathbf{K}$ that is the minimal free resolution of k over S. We use the notation in Construction 26.4. The component of $\mathbf{K}$ in multidegree m_τ has basis

$$\left\{ \frac{m_\tau}{x_{j_1} \cdots x_{j_i}} e_{j_1} \wedge \cdots \wedge e_{j_i} \;\middle|\; x_{j_p} \text{ divides } m_\tau \text{ for } 1 \le p \le i, \right.$$
$$\left. 1 \le j_1 < \ldots < j_i \le n \right\}.$$

Fix an element $f = \dfrac{m_\tau}{x_{j_1} \cdots x_{j_i}} e_{j_1} \wedge \cdots \wedge e_{j_i}$ in this basis. Choosing τ minimal with respect to inclusion, we may assume $m_\tau = m_{\tau \cup m_s}$ for some $m_s \in \{m_1, \ldots, m_r\} \backslash \tau$. We have that m_s divides m_τ. On the other hand, by assumption the monomials m_s and m_τ have different non-zero exponents in each variable. Hence, the monomial m_s divides $\frac{m_\tau}{\prod_{\{i \,|\, x_i \text{ divides } m_\tau\}} x_i}$. Therefore, the image of f in $(S/M \otimes \mathbf{K}_i)_{m_\tau}$ vanishes. We conclude that $(S/M \otimes \mathbf{K}_i)_{m_\tau} = 0$. Thus, the Scarf complex is exact and $\mathbf{F}_{\Omega_M}$ is a fee resolution of S/M.

If $\sigma \in \Omega_M$ and $m_i \in \sigma$, then by the definition of the Scarf complex it follows that $\text{mdeg}(\sigma \setminus m_i)$ strictly divides $\text{mdeg}(\sigma)$. Therefore, $d(\mathbf{F}_{\Omega_M}) \subseteq (x_1, \ldots, x_n)\mathbf{F}_{\Omega_M}$. Thus, the resolution is minimal. □

60 Rootings and Lyubeznik's resolution

We will construct a simplicial resolution which is smaller than Taylor's resolution. The material in this section is from [Novik].

A ***rooting map*** on the lcm-lattice L_M is a map

$$h: \; L_M \setminus \{1\} \to \{m_1, \ldots, m_r\}$$

such that the following two conditions are satisfied:

(1) for every monomial m, we have that $h(m)$ divides m.

(2) if $m, m' \in L_M \backslash \{1\}$ are such that $h(m)$ divides m' and m' divides m, then $h(m) = h(m')$.

For every nonempty set $\mathcal{B} \subseteq \{m_1, \ldots, m_r\}$ we define

$$h(\mathcal{B}) = h(\text{lcm}(m_i | m_i \in \mathcal{B}))\,.$$

We say that $\mathcal{B}$ is ***unbroken*** if $h(\mathcal{B}) \in \mathcal{B}$. A set $\mathcal{B}$ is ***rooted*** if all nonempty subsets of $\mathcal{B}$ are unbroken. The collection of all rooted sets is the ***rooted complex*** of h, and is denoted RC_h. It is a subcomplex of the simplex Θ on vertices $m_1, \ldots, m_r$.

Exercise 60.1. *If h is a rooting map on L_M, then RC_h is a simplicial complex.*

Theorem 60.2. *If h is a rooting map on L_M, then the rooted complex RC_h supports a simplicial free resolution of S/M.*

Proof. We will apply Proposition 57.5.

Let $m \in L_M \setminus \{1\}$. We will prove that the simplicial complex $(\mathrm{RC}_h)_{\leq m}$ is a cone with apex $h(m)$.

Let $\tau \in (\mathrm{RC}_h)_{\leq m}$ be a face. We will show that either it contains the vertex $h(m)$ or $\tau \cup h(m) \in (\mathrm{RC}_h)_{\leq m}$. Suppose that $h(m) \notin \tau$. We will prove that every subset σ of $\tau \cup h(m)$ is unbroken. Note that σ either is a subset of τ and so is unbroken, or it contains $h(m)$. Suppose the latter case holds. Set $m' = \mathrm{lcm}(m_i | m_i \in \sigma)$. Then $h(m)$ divides m' since $h(m) \in \sigma$. As $\tau \in (\mathrm{RC}_h)_{\leq m}$ and $h(m)$ divides m, it follows that $m' = \mathrm{lcm}(m_i | m_i \in \sigma)$ divides m. By the definition of a rooting map it follows that $h(\sigma) = h(m') = h(m)$. Since $h(m) \in \sigma$, we conclude that σ is unbroken. Therefore, $\tau \cup h(m)$ is rooted. Thus, $\tau \cup h(m) \in (\mathrm{RC}_h)_{\leq m}$.

Since $(\mathrm{RC}_h)_{\leq m}$ is a cone, it is acyclic. The theorem follows by Proposition 57.5. □

For a set $\tau \subseteq \{m_1, \ldots, m_r\}$, set $\min(\tau) = \min\{i \,|\, m_i \in \tau)$ and recall that $\mathrm{mdeg}(\tau) = \mathrm{lcm}(m_i | m_i \in \tau)$.

Exercise 60.3. *Define a map*

$$h: \; L_M \setminus \{1\} \longrightarrow \{m_1, \ldots, m_r\}$$
$$m \mapsto \min\{i \,|\, m_i \textit{ divides } m) \,.$$

(1) *The map h is a rooting map.*
(2) *A set $\tau \subseteq \{m_1, \ldots, m_r\}$ is unbroken if and only if m_q does not divide* $\mathrm{mdeg}(\tau)$ *for every $q < \min(\tau)$.*

The following result is an immediate corollary of Theorem 60.2 and Exercise 60.3.

Theorem 60.4. *Let h be the rooting map constructed in Exercise* 60.3. *The rooted complex RC_h supports a simplicial free resolution of S/M denoted $\mathbf{L}_M$. It is the essential subcomplex of Taylor's resolution such that the free module in homological degree j in $\mathbf{L}_M$ has basis*

$$\left\{ e_\tau \,\middle|\, m_q \textit{ does not divide } \mathrm{mdeg}(\sigma) \textit{ for all } \sigma \subseteq \tau \textit{ and } q < \min(\sigma) \right\}.$$

This simplicial free resolution of S/M is called the ***Lyubeznik***

resolution. It was introduced in [Lyubeznik] and proved in a different way.

Running Example 60.5. Order the minimal monomial generators of Y by $m_1 = xy, m_2 = y^3, m_3 = x^2$. Consider the rooting map defined in Exercise 60.3. The edge $\{x^2, y^3\}$ is not unbroken, so this edge and the triangle $\{x^2, y^3, xy\}$ are not rooted. Therefore, the rooted complex has facets the edges $\{x^2, xy\}$ and $\{xy, y^3\}$. Thus, the rooted complex supports the minimal free resolution of C/Y. It is strictly smaller than Taylor's resolution.

61 Betti numbers via simplicial complexes

The Betti numbers of S/M can be computed using various simplicial complexes. It is helpful to have formulas based on different simplicial complexes since different complexes are useful in different situations.

Construction 61.1. Let $\Gamma(m)$ be the simplicial complex on vertices $x_1, \ldots, x_n$ and with faces

$$\left\{\tau \subseteq \{x_1, \ldots, x_n\} \;\middle|\; \frac{m}{\prod_{x_i \in \tau} x_i} \in M\right\}.$$

Sometimes, it is more convenient to denote the vertices of $\Gamma(m)$ by $\{1, \ldots, n\}$ and then the faces are

$$\left\{\tau \subseteq \{1, \ldots, n\} \;\middle|\; \frac{m}{\prod_{i \in \tau} x_i} \in M\right\}.$$

Let Θ be the simplex with r vertices $m_1, \ldots, m_r$. Also, let L_M be the lcm-lattice of M, and $O(1, m)$ be the order complex of the open interval $(1, m)$ in the lcm-lattice L_M.

Running Example 61.2. Let $m = x^2y^3$. The complex $\Theta_{<x^2y^3}$ has three vertices x^2, xy, y^3 and the two edges $\{x^2, xy\}$, $\{xy, y^3\}$. The complex $\Gamma(m)$ has two vertices x, y and the edge $\{x, y\}$. The complex $O(1, m)$ has 5 vertices x^2, xy, y^3, x^2y, xy^3 and 4 edges $\{x^2, x^2y\}$, $\{xy, x^2y\}$, $\{xy, xy^3\}$, $\{y^3, xy^3\}$. See Figure 14.

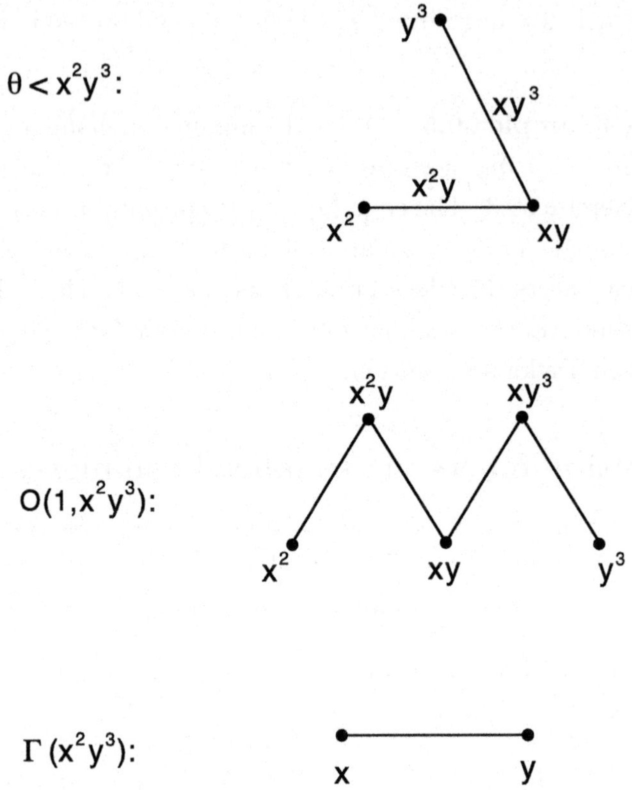

Figure 14.

Theorem 61.3. [Bayer-Sturmfels], [Bruns-Herzog 2], [Gasharov-Peeva-Welker 2] *The simplicial complexes* $\Theta_{<m}$, $\Gamma(m)$, *and* $O(1,m)$ *are homotopy equivalent. For* $i \geq 0$ *and any monomial* $m \neq 1$, *we have*

$$\begin{aligned} b^S_{i,m}(M) &= \dim \widetilde{\mathrm{H}}_{i-1}\big(\Theta_{<m}; k\big) \\ &= \dim \widetilde{\mathrm{H}}_{i-1}\big(\Gamma(m); k\big) \\ &= \dim \widetilde{\mathrm{H}}_{i-1}\big(O(1,m); k\big)\,. \end{aligned}$$

Proof. The proof of Theorem 58.8 shows that $O(1,m)$ and $\Theta_{<m}$ are homotopy equivalent. We will show that $\Theta_{<m}$ and $\Gamma(m)$ are homotopy equivalent. We are going to use the Nerve Theorem 36.11. For each

$1 \leq i \leq n$, consider the simplicial complex $\Lambda_{x_i} \subseteq \Theta$ with faces

$$\Big\{ \{m_{j_1}, \ldots, m_{j_q}\} \,\Big|\, m_{j_p} \text{ divides } \frac{m}{x_i} \text{ for } 1 \leq p \leq q \Big\}.$$

Each Λ_{x_i} is a simplex. The simplices $\Lambda_{x_1}, \ldots, \Lambda_{x_n}$ cover $\Theta_{<m}$. We have

$$\cap_{x_i \in \mathcal{A}} \Lambda_{x_i} = \Big\{ \{m_{j_1}, \ldots, m_{j_q}\} \,\Big|\, m_{j_p} \text{ divides } \frac{m}{\prod_{x_i \in \mathcal{A}} x_i} \text{ for } 1 \leq p \leq q \Big\}$$

so if an intersection is non-empty then it is a simplex, so contractible. The nerve of this cover is $\Gamma(m)$. Hence $\Gamma(m)$ and $\Theta_{<m}$ are homotopy equivalent by the Nerve Theorem 36.11.

The formula for the Betti numbers follows from Theorem 57.9.

Here is another proof following [Bruns-Herzog 2, 1.1]. Apply Construction 26.4 and use its notation. Denote by $\mathbf{T}$ the augmented oriented simplicial chain complex computing the reduced homology of the simplicial complex $\Gamma(m)$. Then

$$\begin{aligned} (M \otimes \mathbf{K}_i)_m &\longrightarrow \mathbf{T}_{i-1} \\ \frac{m}{x_{j_1} \ldots x_{j_i}} e_{j_1} \wedge \cdots \wedge e_{j_i} &\mapsto \text{the face with vertices } x_{j_1}, \ldots, x_{j_i} \end{aligned}$$

is an isomorphism of complexes. Hence, $b^S_{i,m}(M) = \dim \widetilde{\mathrm{H}}_{i-1}\big(\Gamma(m); k\big)$ as desired. □

62 The Stanley-Reisner correspondence

The first peak in the study of monomial resolutions was in the 1970's. The Stanley-Reisner theory was introduced by Hochster [Hochster] and Reisner [Reisner], and had applications in combinatorics, cf. [Stanley]. The main idea in the Stanley-Reisner theory is to use simplicial complexes in order to compute the Betti numbers of a squarefree monomial ideal. Polarization (see Section 21) can be used to reduce to the squarefree case, that is, to reduce the study of resolutions of monomial ideals to the study of resolutions of squarefree monomial ideals.

In this section we consider squarefree monomial ideals.

The ***support*** of a squarefree monomial m is the set

$$\mathrm{supp}(m) = \{x_i \,|\, x_i \text{ divides } m\}$$

For $\tau \subseteq \{x_1, \ldots, x_n\}$, set

$$\mathbf{x}_\tau = \prod_{i \in \tau} x_i \,.$$

Recall the definition of the Stanley-Reisner ideal (see Section 51). Let Δ be a simplicial complex with vertices $x_1, \ldots, x_n$. The Stanley-Reisner ideal in S of Δ is

$$I_\Delta = (x_{i_1} \ldots x_{i_p} \,|\, \{x_{i_1}, \ldots, x_{i_p}\} \notin \Delta)\,.$$

The Stanley-Reisner ring of Δ is S/I_Δ.

Note that any squarefree monomial ideal is the Stanley-Reisner ideal for some simplicial complex. In [Bruns-Gubeladze] it is proved that if two Stanley-Reisner rings are isomorphic as k-algebras, then their simplicial complexes are isomorphic.

The following basic equality is proved in Theorem 51.5.

Theorem 62.1. $\dim(S/I_\Delta) = \dim(\Delta) + 1$.

Construction 62.2. The ***Alexander dual complex*** of Δ is

$$\Delta^\vee = \Big\{ \{x_1, \ldots, x_n\} \setminus \tau \,\Big|\, \tau \notin \Delta \Big\}.$$

Exercise 62.3. $\Delta^{\vee\vee} = \Delta$.

The Stanley-Reisner ideal of the Alexander dual complex is

$$I_{\Delta^\vee} = \Big(\Big\{ \frac{x_1 \cdots x_n}{\mathbf{x}_\tau} \,\Big|\, \tau \in \Delta \Big\} \Big)$$

$$= \Big(\big\{ \text{monomial } m \,|\, \gcd(m, m') \neq 1 \text{ for each monomial } m' \in I_\Delta \big\} \Big).$$

Exercise 62.4. *The facets of* Δ *correspond bijectively to the minimal monomial generators of* $I_{\Delta^\vee}$.

The key idea in this section is that the algebraic properties of the minimal free resolution of I_Δ are closely related to the topological and combinatorial properties of $\Delta^\vee$.

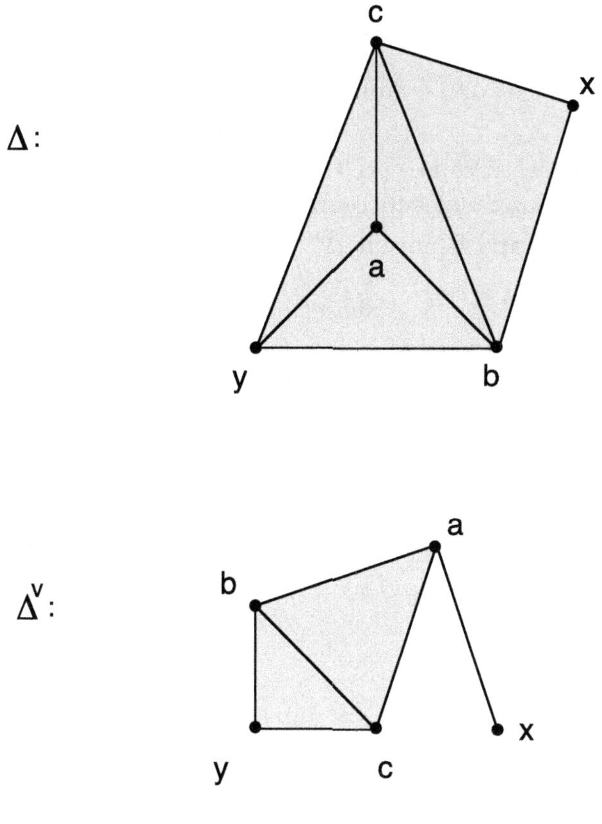

Figure 15.

Running Example 62.5. Polarizing Y we obtain the squarefree ideal (xa, xy, ybc). It is the Stanley-Reisner ideal of the simplicial complex Δ on vertices x, y, a, b, c and with facets the triangles $\{x, c, b\}$, $\{a, y, c\}$, $\{a, b, y\}$, $\{a, b, c\}$. The facets of the Alexander dual complex

$\Delta^\vee$ are the supports of the monomials $\dfrac{xyabc}{xa} = ybc$, $\dfrac{xyabc}{xy} = abc$, and $\dfrac{xyabc}{ybc} = xa$, so they are $\{y, b, c\}$, $\{a, b, c\}$, and $\{x, a\}$. See Figure 15 above.

Theorem 62.6 is presented without a proof, and is a useful tool from Algebraic Topology; cf. [Bayer-Charalambous-Popescu, 2.1].

Alexander Duality Theorem 62.6.

$$\dim \widetilde{\mathrm{H}}_{n-i-2}(\Delta; k) = \dim \widetilde{\mathrm{H}}^{i-1}(\Delta^\vee; k) = \dim \widetilde{\mathrm{H}}_{i-1}(\Delta^\vee; k)\,.$$

For a subset τ of $\{1, \ldots, n\}$, let Δ_τ be the ***restriction*** of Δ on τ, that is the maximal subcomplex of Δ on vertices τ. Note that $(\Delta_\tau)^\vee = (\Delta^\vee)_\tau$, and denote it $\Delta^\vee_\tau$.

Theorem 62.7. *Let τ be a subset of $\{1, \ldots, n\}$.*

(1) *Recall Construction* 61.1 *defining $\Gamma(\mathbf{x}_\tau)$ to be the simplicial complex with faces $\left\{\sigma \subseteq \tau \,\middle|\, \dfrac{\mathbf{x}_\tau}{\mathbf{x}_\sigma} \in I_\Delta\right\}$. We have*

$$\Gamma(\mathbf{x}_\tau) = \Delta^\vee_\tau$$

and

$$b^S_{i,\mathbf{x}_\tau}(I_\Delta) = \dim \widetilde{\mathrm{H}}_{i-1}\left(\Delta^\vee_\tau; k\right).$$

(2) *We have*

$$b^S_{i,\mathbf{x}_\tau}(I_\Delta) = \dim \widetilde{\mathrm{H}}_{|\tau|-i-2}(\Delta_\tau; k)\,.$$

Proof. First, we prove (1). The simplicial complex $\Gamma(\mathbf{x}_\tau)$ has faces

$$\left\{\sigma \subseteq \tau \,\middle|\, \frac{\mathbf{x}_\tau}{\mathbf{x}_\sigma} \in I_\Delta\right\} = \{\sigma \subseteq \tau \,|\, \sigma \in \Delta^\vee_\tau\}\,.$$

Therefore, $\Gamma(\mathbf{x}_\tau) = \Delta^\vee_\tau$. From Theorem 61.3 it follows that

$$b^S_{i,\mathbf{x}_\tau}(I_\Delta) = \dim \widetilde{\mathrm{H}}_{i-1}\left(\Gamma(\mathbf{x}_\tau); k\right) = \dim \widetilde{\mathrm{H}}_{i-1}\left(\Delta^\vee_\tau; k\right).$$

Now, we prove (2). By (1) and the Alexander Duality Theorem 62.6, we get

$$b^S_{i,\mathbf{x}_\tau}(I_\Delta) = \dim \widetilde{\mathrm{H}}_{i-1}(\Delta^\vee_\tau; k) = \dim \widetilde{\mathrm{H}}_{|\tau|-i-2}(\Delta_\tau; k)\,.$$

□

Theorem 62.8. [Terai], [Bayer-Charalambous-Popescu]

$$\mathrm{pd}(S/I_{\Delta^\vee}) = \mathrm{reg}(I_\Delta)\,.$$

Proof.

$$\begin{aligned}
\mathrm{reg}(I_\Delta) &= \max\{\, j \mid b^S_{i,i+j}(I_\Delta) \neq 0 \,\}\\
&= \max\{\, j \mid b^S_{i,\mathbf{x}_\tau}(I_\Delta) \neq 0 \text{ and } |\tau| = i+j \,\}\\
&= \max\{\, j \mid \widetilde{\mathrm{H}}_{|\tau|-i-2}(\Delta_\tau; k) \neq 0 \text{ and } |\tau| = i+j \,\} \quad \textit{by } 62.7(2)\\
&= \max\{\, j \mid \widetilde{\mathrm{H}}_{j-2}(\Delta_\tau; k) \neq 0 \,\}\\
&= \max\{\, j \mid b^S_{j-1,\mathbf{x}_\tau}(I_{\Delta^\vee}) \neq 0 \,\} \quad \textit{by Theorem } 62.7(1)\\
&= \mathrm{pd}(I_{\Delta^\vee}) + 1 = \mathrm{pd}(S/I_{\Delta^\vee})\,.
\end{aligned}$$

□

Corollary 62.9. (Eagon-Reiner) *The ideal I_Δ has a linear minimal free resolution if and only if $S/I_{\Delta^\vee}$ is Cohen-Macaulay.*

Proof. By 62.1, $\dim(S/I_{\Delta^\vee}) = \dim(\Delta^\vee) + 1$. Suppose that $S/I_{\Delta^\vee}$ is Cohen-Macaulay. Then,

$$\begin{aligned}
\mathrm{reg}(I_\Delta) &= \mathrm{pd}(S/I_{\Delta^\vee}) = n - \mathrm{depth}(S/I_{\Delta^\vee}) = n - \dim(S/I_{\Delta^\vee})\\
&= n - \dim(\Delta^\vee) - 1\\
&= \text{the minimal degree of a minimal monomial generator of } I_\Delta\,.
\end{aligned}$$

Hence, I_Δ has a linear minimal free resolution.

Suppose that I_Δ has a linear minimal free resolution. Then

$$\begin{aligned}
n - \mathrm{depth}(S/I_{\Delta^\vee}) &= \mathrm{pd}(S/I_{\Delta^\vee}) = \mathrm{reg}(I_\Delta)\\
&= \text{the minimal degree of a minimal monomial generator of } I_\Delta\\
&= n - \dim(\Delta^\vee) - 1\\
&= n - \dim(S/I_{\Delta^\vee})\,.
\end{aligned}$$

Hence, $\operatorname{depth}(S/I_{\Delta^\vee}) = \dim(S/I_{\Delta^\vee})$. Thus, the Stanley-Reisner ring $S/I_{\Delta^\vee}$ is Cohen-Macaulay. □

Corollary 62.10. [Stanley] *If $S/I_{\Delta^\vee}$ is Cohen-Macaulay, then the complex $\Delta^\vee$ is pure (that is, all maximal faces of $\Delta^\vee$ have the same dimension).*

Proof. This follows from the first part of the proof of Corollary 62.9. Since $\operatorname{reg}(I_\Delta)$ is equal to the minimal degree of a minimal monomial generator of I_Δ, it follows that all minimal monomial generators of I_Δ have the same degree. □

For $\sigma \subseteq \tau \in \Delta$ we define the ***closed interval***

$$[\sigma, \tau] = \{\mu \,|\, \sigma \subseteq \mu \subseteq \tau\}.$$

A ***partition*** of Δ, is a disjoint union

$$\Delta = \bigsqcup_{1 \le i \le s} [\sigma_i, \tau_i],$$

where $\tau_1, \ldots, \tau_s$ are the facets of Δ. Thus, the closed intervals $[\sigma_i, \tau_i]$ are disjoint and cover Δ. We say that Δ is ***partitionable*** if it has a partition.

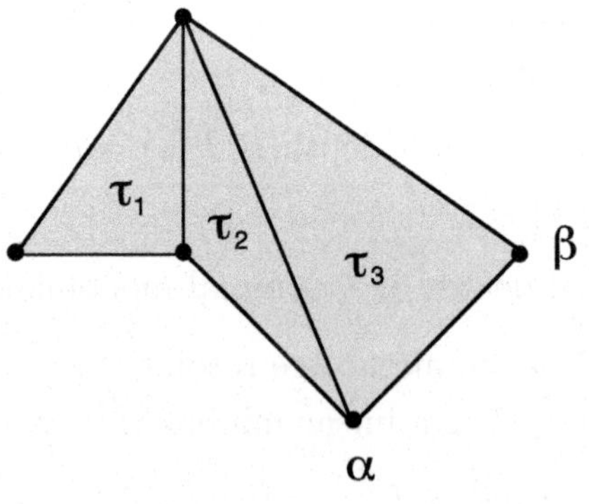

Figure 16.

Example 62.11. Consider the simplicial complex Δ in Figure 16 above. Let τ_1, τ_2, τ_3 be the three facets. Then Δ has the partition

$$\Delta = [\emptyset, \tau_1] \sqcup [\alpha, \tau_2] \sqcup [\beta, \tau_3],$$

where α and β are the vertices labeled in Figure 16.

A simplicial complex Δ is ***Cohen-Macaulay*** if its Stanley-Reisner ring is Cohen-Macaulay. An open conjecture, central in combinatorics, cf. [Stanley, Stanley 2], states that a Cohen-Macaulay simplicial complex is partitionable. By Corollary 62.9, we have the following equivalent conjecture.

Conjecture 62.12. [Stanley], [Stanley 2, Problem 6] *If $I_{\Delta^\vee}$ has a linear resolution, then Δ is partitionable.*

Exercise 62.13. *If $I_{\Delta^\vee}$ is squarefree Borel and has a linear resolution, then Δ is partitionable.*

In the next theorem we show how Alexander duality is related to the lcm-lattice. The ***proper part*** of a lattice P, with bottom element $\hat{0}$ and top element $\hat{1}$, is $P \setminus \{\hat{0}, \hat{1}\}$.

Theorem 62.14. [Gasharov-Peeva-Welker 2] *Let $L_{\Delta^\vee}$ be the lattice of all non-empty intersections of the facets of $\Delta^\vee$ ordered by reverse inclusion, and enlarged by an additional bottom element $\hat{0}$ and an additional top element $\hat{1}$. The lattices L_{I_Δ} and $L_{\Delta^\vee}$ are isomorphic. Furthermore, $\Delta^\vee$ is homotopy equivalent to the order complex of the proper part of L_{I_Δ}.*

Proof. Let $\tau_1, \ldots, \tau_p$ be the facets of $\Delta^\vee$. For any $\emptyset \neq A \subseteq \{1, \ldots, p\}$ we consider the bijective correspondence

$$\bigcap_{i \in A} \tau_i \quad \longleftrightarrow \quad \frac{x_1 \ldots x_n}{\bigcap_{i \in A} \operatorname{supp}(\mathbf{x}_{\tau_i})} = \operatorname{lcm}\left(\frac{x_1 \ldots x_n}{\mathbf{x}_{\tau_i}} \,\middle|\, i \in A \right).$$

The lattices $L_{\Delta^\vee}$ and L_{I_Δ} are isomorphic via the above correspondence.

In particular, for the minimal monomial generators of I_Δ we have

$$\tau_i \in \Delta^\vee \quad \longleftrightarrow \quad \frac{x_1 \ldots x_n}{\mathbf{x}_{\tau_i}} \in I_\Delta.$$

By Corollary 36.13 and Corollary 36.9 it follows that $\Delta^\vee$ is homotopy equivalent to the order complex of the proper part of L_{I_Δ}. □

Recall by 36.17 that for $\tau \in \Delta$, the link of τ is

$$\text{link}_\Delta(\tau) = \{\sigma \in \Delta | \sigma \cup \tau \in \Delta,\ \sigma \cap \tau = \emptyset\}.$$

Theorem 62.15. [Hochster] *Consider an open interval* $(1, m)$ *in the lcm-lattice* L_{I_Δ}. *Let* $\tau = \{x_1, \ldots, x_n\} \setminus \text{supp}(m)$ *be the complement of* $\text{supp}(m)$. *The order complex* $O(1, m)$ *is homotopic to the link* $\text{link}_{\Delta^\vee}(\tau)$. *In particular,*

$$\dim \widetilde{\text{H}}_j(O(1, m); k) = \dim \widetilde{\text{H}}_j(\text{link}_{\Delta^\vee}(\tau); k) \quad \textit{for all } j\,.$$

For $i \geq 0$, *we have that*

$$b^S_{i,m}(I_\Delta) = \dim \widetilde{\text{H}}_{i-1}(\text{link}_{\Delta^\vee}(\tau); k)\,.$$

The above formula for the Betti numbers is called the ***Hochster formula***.

Proof. As shown in the proof of Theorem 62.14, the open interval $(1, m)$ is isomorphic to the open interval $(\hat{0}, \tau)$ in the lattice $L_{\Delta^\vee}$ of all non-empty intersections of the facets of $\Delta^\vee$ ordered by reverse inclusion. By Corollary 36.13, the order complex of $(\hat{0}, \tau)$ is homotopic to the order complex of $(\tau, \hat{1})$ in the face lattice W of $\Delta^\vee$ (since W is ordered by inclusion, while $L_{\Delta^\vee}$ is ordered by reverse inclusion). Each element μ in the poset $(\tau, \hat{1})$ can be written as $\mu = \sigma_\mu \cup \tau \in \Delta$ so that $\sigma_\mu \cap \tau = \emptyset$. We get an isomorphism between $(\tau, \hat{1})$ and the face poset $F(\text{link}_\Delta(\tau))$ which maps μ to σ_μ. Therefore, the order complex of $(\tau, \hat{1})$ is homotopic to the simplicial complex $\text{link}_\Delta(\tau)$. The formula for the Betti numbers follows from Theorem 61.3:

$$b^S_{i,m}(I_\Delta) = \dim \widetilde{\text{H}}_{i-1}(O(1, m); k) = \dim \widetilde{\text{H}}_{i-1}(\text{link}_{\Delta^\vee}(\tau); k)\,.$$

□

Example 62.16. Consider $N = (ac, ah, be, ce)$ in $A = k[a, b, c, e, h]$. Its Stanley-Reisner simplicial complex Δ and the Alexander dual complex $\Delta^\vee$ are shown in Figure 17. Note that the non-faces are not shaded, so $\{c, e, h\}$ is not a face in $\Delta^\vee$.

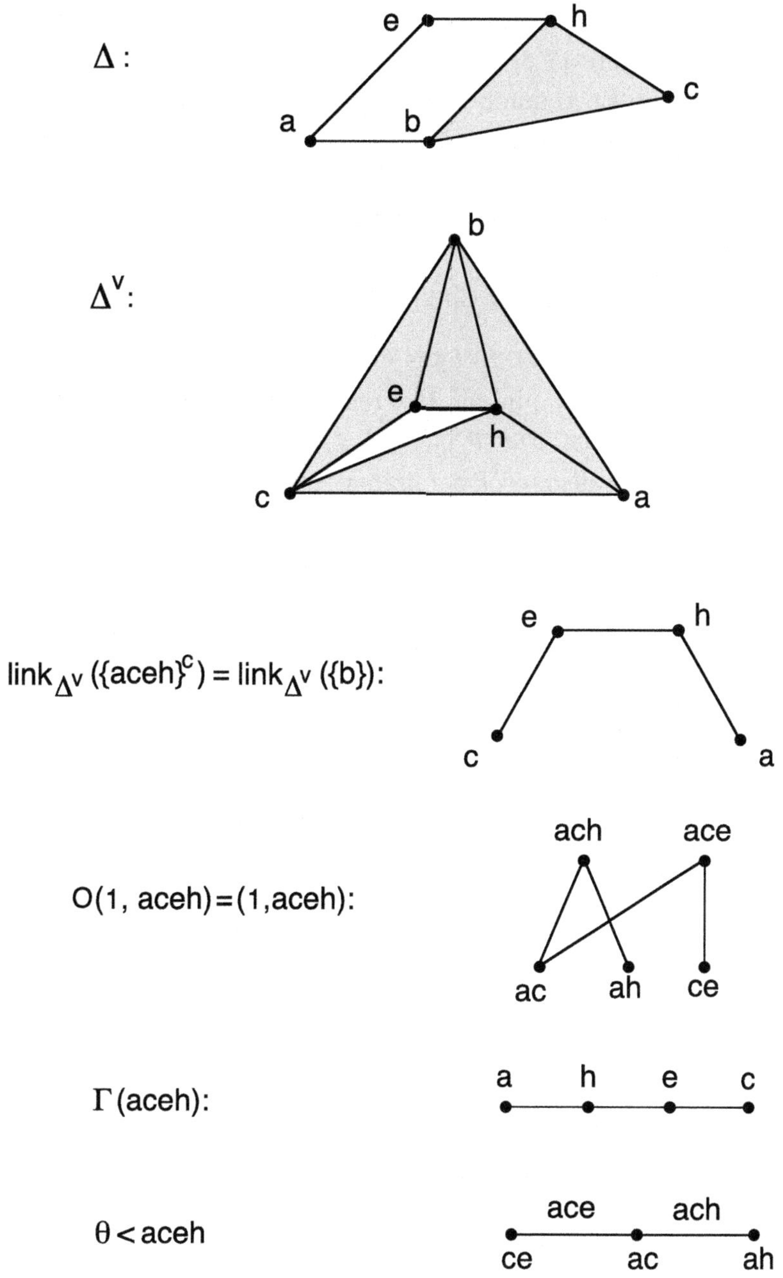
Δ :
e
h
c
a
b
Δ^v:
b
e
h
c
a
link_Δv ({aceh}^c) = link_Δv ({b}):
e
h
c
a
O(1, aceh)=(1,aceh):
ach
ace
ac
ah
ce
Γ(aceh):
a
h
e
c
θ<aceh
ace
ach
ce
ac
ah

Figure 17.

The complexes $\Theta_{<aceh}$, $\Gamma(aceh)$, $O(1, aceh)$, and $\mathrm{link}_{\Delta^\vee}(\{aceh\}^c)$ are shown in Figure 17 as well. By Theorem 61.3 and Theorem 62.15, the multigraded Betti numbers $b^A_{i,aceh}(N)$ can be computed using any of these simplicial complexes. We have that all the Betti numbers in multidegree $aceh$ vanish.

Corollary 62.17. [Reisner] *S/I_Δ is Cohen-Macaulay if and only if for each face $\tau \in \Delta$ we have $\widetilde{\mathrm{H}}_i(\mathrm{link}_\Delta(\tau); k) = 0$ for $i \neq \dim(\mathrm{link}_\Delta(\tau))$.*

Proof. By Corollary 62.9, S/I_Δ is Cohen-Macaulay if and only if the ideal $I_{\Delta^\vee}$ has a linear minimal free resolution, if and only if for every monomial m we have that $b^S_{i,m}(I_{\Delta^\vee}) = 0$ for $\deg(m) \neq p + i$, where p is the minimal degree of a minimal monomial generator of $I_{\Delta^\vee}$. By Theorem 62.15, this is equivalent to $\widetilde{\mathrm{H}}_i(\mathrm{link}_\Delta(\tau); k) = 0$ for $i \neq \dim(\mathrm{link}_\Delta(\tau))$, where $\tau = \{x_1, \ldots, x_n\} \setminus \mathrm{supp}(m)$. □

Define a Betti number $b_{i,m}$ to be ***i-extremal*** if $b_{i,m'} = 0$ for all monomials m' strictly divisible by m. The following result is proved in [Bayer-Charalambous-Popescu].

Theorem 62.18. *If $b^S_{i,m}(I_{\Delta^\vee})$ is i-extremal, then*

$$b^S_{i,m}(I_{\Delta^\vee}) \geq b^S_{\deg(m)-i-1,m}(I_\Delta).$$

Proposition 62.19. *Denote by* $\mathrm{mingens}(I_{\Delta^\vee})$ *the set of minimal monomial generators of $I_{\Delta^\vee}$. The irredundant primary decomposition of I_Δ is*

$$I_\Delta = \bigcap_{x_{j_1}\cdots x_{j_s} \in \mathrm{mingens}\,(I_{\Delta^\vee})} (x_{j_1}, \ldots, x_{j_s}).$$

Proof. The ideal I_Δ is radical, so it equals the intersection of its minimal prime divisors.The associated primes of S/I_Δ are its minimal prime divisors. An ideal P is an associated prime of S/I_Δ exactly when

$$P = \Big\{ (x_{i_1}, \ldots, x_{i_r}) \,|\, \{x_1, \ldots, x_n\} \setminus \{x_{i_1}, \ldots, x_{i_r}\} \text{ is a facet of } \Delta \Big\}.$$

□

63 Quadratic monomial ideals

One might expect that the simplest minimal free monomial resolutions are those of the ideals generated by quadratic monomials, and that it might be nearly an exercise to describe them. However, these resolutions are so complicated that it is beyond reach to obtain a description of them; we do not even know how to express the regularity. In this section we will apply the mapping cone construction to determine which quadratic monomial ideals have 2-linear free resolutions.

It is easy to encode a set of squarefree quadratic monomials in a graph. Throughout the section, we consider a simple (that is, with no loops and no multiple edges) graph G on vertices $x_1, \ldots, x_n$. The ***edge ideal*** I_G is

$$I_G = \big(x_i x_j \mid x_i x_j \text{ is an edge in } G\big)\,.$$

Exercise 63.1. *The polarization of any quadratic monomial ideal is an edge ideal.*

Thus, studying the minimal free resolutions of quadratic monomial ideals is equivalent to studying the minimal free resolutions of edge ideals. The following problems are open.

Problems 63.2. (folklore)

(1) *Express* $\operatorname{reg}(I_G)$ *in terms of properties of the graph* G.

(2) *Find upper (and lower) bounds on* $\operatorname{reg}(I_G)$ *in terms of properties of the graph* G.

(3) *Find upper (and lower) bounds on the Betti numbers of* I_G *in terms of properties of the graph* G.

For the next theorem we need a few definitions about graphs. The ***complement graph*** G^c of G is the graph on the same set of vertices, and with edges

$$\{x_i x_j \mid x_i x_j \text{ is not an edge in } G\}\,.$$

We say that a simple graph T contains a ***q-cycle*** $(x_{i_1} \ldots x_{i_q})$ if $x_{i_q} x_{i_1} \in T$ and $x_{i_j} x_{i_{j+1}} \in T$ for all $1 \le j \le q-1$. A ***chord*** in the cycle is an edge between two non-consecutive vertices. A cycle is called ***minimal*** (or ***induced***) if it has no chords. A cycle with three vertices is called a triangle.

Theorem 63.3. [Fröberg] *The following properties are equivalent.*
(1) *I_G has a 2-linear minimal free resolution.*
(2) $\operatorname{reg}(S/I_G) = 1$.
(3) *Every minimal cycle in G^c is a triangle.*

Proof. (1) and (2) are equivalent.

We will show that (3) implies (2). Dirac's Theorem, cf. [Herzog-Hibi-Zheng] and [Horwitz], states that if every minimal cycle in G^c is a triangle then there exists an order of the vertices so that the following property holds: if $x_i x_j \in G$ and x_p is a vertex with $i, j < p$, then either $x_i x_p$, or $x_j x_p$, or both are edges in G.

For $p \geq 1$, denote by G_p the induced subgraph of G on the vertices $x_1, \ldots, x_p$. Our proof is by induction on the number of vertices p.

Let $p \geq 2$. Set

$$J = (x_p x_q \,|\, 1 \leq q < p,\ x_p x_q \in G)\,.$$

Consider the short exact sequence

$$(*)\ 0 \to J/(I_{G_{p-1}} \cap J) \to S/I_{G_{p-1}} \to S/(I_{G_{p-1}} + J) = S/I_{G_p} \to 0\,.$$

We will show that $I_{G_{p-1}} \cap J = x_p I_{G_{p-1}}$. Consider a monomial $x_p x_q x_i x_j$ such that $x_p x_q \in J$ and $x_i x_j \in G_{p-1}$. Since $i, j < p$, by Dirac's order of the variables, we have that either $x_i x_p$, or $x_j x_p$, or both are edges in G_p. Therefore, $x_p x_i x_j \in I_{G_{p-1}} \cap J$. It follows that the ideal $I_{G_{p-1}} \cap J$ is generated by the monomials $\{x_p x_i x_j \,|\, x_i x_j \in G_{p-1}\}$. Hence,

$$I_{G_{p-1}} \cap J = x_p I_{G_{p-1}}$$
$$J = x_p (x_q \,|\, 1 \leq q < p,\ x_p x_q \in G)\,.$$

The minimal free resolution of the ideal $(x_q \,|\, 1 \leq q < p,\ x_p x_q \in G)$ is given by a Koszul complex. Therefore, we get

$$\operatorname{reg}(I_{G_{p-1}} \cap J) = 1 + \operatorname{reg}(I_{G_{p-1}}) = 3$$
$$\operatorname{reg}(J) = 1 + \operatorname{reg}(x_q \,|\, 1 \leq q < p,\ x_p x_q \in G) = 2\,.$$

Now, Corollary 18.6 applied to the short exact sequence

$$0 \to I_{G_{p-1}} \cap J \to J \to J/(I_{G_{p-1}} \cap J) \to 0$$

implies that $\operatorname{reg}\big(J/(I_{G_{p-1}} \cap J)\big) = 2$. By induction hypothesis, we have $\operatorname{reg}(S/I_{G_{p-1}}) = 1$. Therefore, Corollary 18.6 applied to the short exact sequence $(*)$ implies that $\operatorname{reg}(S/I_{G_p}) = 1$.

(1) implies (3) by the next exercise. □

Exercise 63.4. *If G^c contains a minimal cycle $(x_{i_1} \dots x_{i_q})$ with $q > 3$, then $b^S_{q-2, x_{i_1} \dots x_{i_q}}(S/I_G) \neq 0$.*

64 Infinite free monomial resolutions

Infinite free resolutions related to monomial ideals have been studied much less than finite ones. So far, the three main results in that area are Theorem 35.6 on the rate, Backelin's Theorem 64.2 on the rationality of the Poincaré series, and Berglund's Theorem 64.4 on computing the Betti numbers by simplicial complexes.

In this section, we study the multigraded minimal free resolution $\mathbf{G}$ of k over the quotient ring S/M. It is infinite (unless M is generated by variables) and starts with

$$\dots \to (S/M)^n \xrightarrow{\begin{pmatrix} x_1 & x_2 & \dots & x_n \end{pmatrix}} S/M \to k \to 0\,.$$

Theorem 64.1. *The entries in the matrices of the differentials in $\mathbf{G}$ are scalar multiples of monomials.*

Proof. Since $\mathbf{G}$ is multigraded we have that each entry f in the matrices of the differentials in $\mathbf{G}$ is homogeneous. Let m be the multidegree of f. Since R_m is one dimensional, it follows that f is a scalar multiple of the unique monomial in multidegree m. □

Problems of rationality of Poincaré and Hilbert series were stated by several mathematicians: by Serre and Kaplansky for local noetherian rings, by Kostrikin and Shafarevich for nilpotent algebras, by

Govorov for associative graded algebras, by Serre and Moore for simply-connected complexes. It is of interest to find explicit formulas in some cases and to establish rational relations between various Poincaré and Hilbert series. The Serre-Kaplansky problem, "*Is the total Poincaré series of a finitely generated commutative local Noetherian ring rational?*", was one of the central questions in Commutative Algebra for many years. An example of irrational Poincaré series was first constructed in [Anick]. In contrast, the following result is proved in [Backelin].

Theorem 64.2. *The multigraded Poincaré series of k over S/M can be written as*

$$\mathrm{P}_k^{S/M}(t, x_1, \ldots, x_n) = \frac{(1+tx_1)\cdots(1+tx_n)}{1+F(t, x_1, \ldots, x_n)},$$

where the denominator $1+F(t, x_1, \ldots, x_n)$ is a polynomial. The degree of the polynomial $F(t, x_1, \ldots, x_n)$ in t is bounded above by the degree of the monomial $\mathrm{lcm}(m_1, \ldots, m_r)$. The monomials in $x_1, \ldots, x_n$ appearing in $F(t, x_1, \ldots, x_n)$ (that is, the monomial coefficients of the powers of t) divide $\mathrm{lcm}(m_1, \ldots, m_r)$.

Open-Ended Problem 64.3. (folklore) *Understand the denominator $1 + F(t, \mathbf{x})$.*

Based on a substantial amount of computational evidence Charalambous and Reeves conjectured the form of the terms of the polynomial denominator $1 + F(t, \mathbf{x})$ of the Poincaré series $\mathrm{P}_k^{S/I}(t)$. Their conjecture is proved by Berglund, who provides a beautiful construction on how to compute the denominator using simplicial complexes as described in Theorem 64.4. For the formulation of that theorem, we need some terminology. Let G be the graph on vertices $m_1, \ldots, m_r$ and with edges $\{\, m_i m_j \,|\, \gcd(m_i, m_j) \neq 1 \,\}$; we call it the ***gcd-graph*** of the ideal M. If $\mathcal{M}$ is a nonempty subset of $\{m_1, \ldots, m_r\}$, then we denote by $G_{\mathcal{M}}$ the induced subgraph of G on the vertices in $\mathcal{M}$. Let $c_{\mathcal{M}}$ be the number of connected components of $G_{\mathcal{M}}$, and denote by $G_{\mathcal{M}}(1), \ldots, G_{\mathcal{M}}(c_{\mathcal{M}})$ the connected components; we say that $\mathcal{M}$ is

connected if $G_{\mathcal{M}}$ is. Set $m_{\mathcal{M}} = \operatorname{lcm}(\, m_i \mid m_i \in \mathcal{M}\,\}$. We say that $\mathcal{M}$ is ***saturated*** if for every m_i and every connected subset $\mathcal{N} \subseteq \mathcal{M}$ we have that m_i divides $m_{\mathcal{N}}$ implies $m_i \in \mathcal{M}$.

Theorem 64.4. [Berglund] *Suppose that M is generated by monomials of degree ≥ 2. For a subset $\mathcal{M}$ of $\{m_1, \ldots, m_r\}$ define the simplicial complex $\Delta_{\mathcal{M}}$ to have vertices the elements in $\mathcal{M}$ and faces*

$$\{\, \mathcal{K} \subseteq \mathcal{M} \mid m_{\mathcal{K}} \neq m_{\mathcal{M}} \text{ or } G_{\mathcal{K}} \cap G_{\mathcal{M}}(i) \text{ is disconnected for some } i\,\}.$$

(1) *The multigraded Poincarè series $P_k^{S/M}(t, u_1, \ldots, u_n)$ of k over S/M is*

$$\frac{\prod_{i=1}^{n}(1+tu_i)}{1+\sum_{\text{saturated } \mathcal{M}\subseteq\{m_1,\ldots,m_r\}}\left(m_{\mathcal{M}}(-t)^{c_{\mathcal{M}}+2}\sum_i \widetilde{\mathrm{H}}_i(\Delta_{\mathcal{M}};k)\,t^i\right)}.$$

The Poincarè series $P_k^{S/M}(t)$ of k over S/M is

$$\frac{(1+t)^n}{1+\sum_{\text{saturated } \mathcal{M}\subseteq\{m_1,\ldots,m_r\}}\sum_i(-1)^{c_{\mathcal{M}}}\widetilde{\mathrm{H}}_i(\Delta_{\mathcal{M}};k)\,t^{c_{\mathcal{M}}+2+i}}.$$

(2) *Let $\mathcal{P}$ be the poset of saturated subsets of $\{m_1, \ldots, m_r\}$ ordered by inclusion. If $\mathcal{M}$ is a saturated subset of $\{m_1, \ldots, m_r\}$, then*

$$\widetilde{\mathrm{H}}_*(\Delta_{\mathcal{M}};k) = \widetilde{\mathrm{H}}_*(\,(\emptyset, \mathcal{M})_{\mathcal{P}};k)\,,$$

where $(\emptyset, \mathcal{M})_{\mathcal{P}}$ is the open interval below $\mathcal{M}$ in $\mathcal{P}$.

Using simplicial complexes in order to compute the Betti numbers of finite monomial minimal free resolutions has a long and fruitful tradition. Very little is known about infinite resolutions. Theorem 64.4 shows that the Poincarè series of k over S/M can be computed using simplicial complexes.

Corollary 64.5. (Avramov) *Let M and M' be monomial ideals in the polynomial rings S and S' respectively. If there exists an isomorphism of the lcm-lattices of M and M' which induces an isomorphism of the gcd-graphs, then $\mathrm{P}_k^{S/M}(t)$ and $\mathrm{P}_k^{S'/M'}(t)$ have the same denominator (when written as in Theorem* 64.2).

Open-Ended Problem 64.6. (folklore) *Obtain information on the* *real roots of the polynomial denominator* $1 + F(t,(1,\dots,1))$ *of the Poincaré series* $\mathrm{P}_k^{S/M}(t)$.

The following construction provides the minimal free resolution of k over S/M explicitly in the case when M is a Borel ideal.

Construction 64.7. [Peeva] Let M be a Borel monomial ideal. The minimal free resolution of k over S/M can be described as follows.

Let $\mathbf{K}$ be the Koszul complex that resolves k over S. Consider $\mathbf{K}$ as the exterior algebra on basis $e_1, e_2, \dots, e_n$ with differential $d(e_i) = x_i$. Denote by E_{p+2} the k-space with basis

$$\Big\{ (m_i; j_1, \dots, j_p) \,|\, 1 \le j_1 < \dots < j_p < \max(m_i),\ 1 \le i \le r \Big\}.$$

Set $E = E_2 \oplus E_3 \oplus \dots \oplus E_{n+1}$. Define $\mathbf{G} = S/M \otimes \mathbf{K} \otimes T(E)$, where

$$T(E) = k \oplus E \oplus (E \otimes E) \oplus \dots$$

is the tensor algebra of E. A basis element in $\mathbf{G}$ has the form

$$t \otimes (z_1; i_1, \dots, i_{p_1}) \otimes (z_2; l_1, \dots, l_{p_2}) \otimes \dots \otimes (z_s; j_1, \dots, j_{p_s}),$$

where $t \in S/M \otimes \mathbf{K}$ and $z_1, \dots, z_s$ are among the minimal monomial generators of M. Define a differential ∂ on the basis elements in $\mathbf{G}$ as follows:

$$\begin{aligned}
&\partial\Big(t \otimes (z_1; i_1, \dots, i_{p_1}) \otimes (z_2; l_1, \dots, l_{p_2}) \otimes \dots \otimes (z_s; j_1, \dots, j_{p_s})\Big) \\
&\quad = d(t) \otimes (z_1; i_1, \dots, i_{p_1}) \otimes (z_2; l_1, \dots, l_{p_2}) \otimes \dots \otimes (z_s; j_1, \dots, j_{p_s}) \\
&\qquad + (-1)^{\deg(t)}\, t\, \frac{z_1}{x_{\max(z_1)}} e_{i_1} \wedge \dots \wedge e_{i_{p_1}} \wedge e_{\max(z_1)} \otimes (z_2; l_1, \dots, l_{p_2}) \otimes \\
&\qquad\qquad \dots \otimes (z_s; j_1, \dots, j_{p_s}),
\end{aligned}$$

where $d(t)$ is the differential in $S/M \otimes \mathbf{K}$ if $t \notin S/M \otimes K_0$, and we set $d(t) = 0$ in case $t \in S/M \otimes K_0$. Extend the differential by linearity. It is proved in [Peeva] that $\mathbf{G}$ is the minimal free resolution of k over S/M. The Poincaré series of the resolution is

$$\mathrm{P}_k^{S/M}(t) = \frac{(1+t)^n}{1 - t^2 \sum_{1 \le i \le r} (1+t)^{\max(m_i)-1}}.$$

Chapter IV

Syzygies of Toric Ideals

Abstract. In this chapter we study minimal free resolutions of toric ideals. Our goal is to use some ideas coming from monomial resolutions.

65 Basics and Notation

The notation, introduced in this section, will be used throughout the chapter.

Recall the concept of multigrading from Section 26. Let $\mathcal{A} = \{a_1, \dots, a_n\}$ be a subset of $\mathbf{N}^r \setminus \mathbf{0}$ and A be the matrix with columns a_i. Throughout the chapter, we suppose that $\text{rank}(A) = r$ for simplicity. Consider the polynomial ring $S = k[x_1, \dots, x_n]$ with variables $x_1, \dots, x_n$ in multidegrees $a_1, \dots, a_n$ respectively. For a monomial $x_1^{v_1} \dots x_n^{v_n}$, its ***multidegree*** is

$$\text{mdeg}(x_1^{v_1} \dots x_n^{v_n}) = \sum_{1 \le i \le n} v_i a_i \in \mathbf{N}\mathcal{A}\,.$$

On the other hand, for each i let $\mathbf{e}_i$ be the i'th standard vector in $\mathbf{N}^r$ and consider the polynomial ring $T = k[t_1, \dots, t_r]$ with variables $t_1, \dots, t_r$ in multidegrees $\mathbf{e}_1, \dots, \mathbf{e}_r$ respectively. The prime ideal $I_{\mathcal{A}}$, that is the kernel of the homomorphism

$$\begin{aligned} \varphi: \ S = k[x_1, \dots, x_n] &\to T = k[t_1, \dots, t_r] \\ x_i &\mapsto \mathbf{t}^{a_i} = t_1^{a_{i1}} \dots t_r^{a_{ir}} \end{aligned}$$

is called the ***toric ideal*** associated to $\mathcal{A}$. Sometimes, for simplicity we write I instead of $I_{\mathcal{A}}$. Note that

$$\text{mdeg}(x_i) = \text{mdeg}(\mathbf{t}^{a_i}) = a_i\,.$$

I. Peeva, *Graded Syzygies*, Algebra and Applications 14,
DOI 10.1007/978-0-85729-177-6_4,

Therefore, the homomorphism φ is multigraded. Hence, its kernel $I_{\mathcal{A}}$ is multigraded.

Set

$$\mathbf{x}^{\mathbf{v}} = x_1^{v_1} \dots x_n^{v_n} \text{ for a vector } \mathbf{v} = \begin{pmatrix} v_1 \\ \vdots \\ v_n \end{pmatrix} \in \mathbf{N}^n .$$

Exercise 65.1. $\varphi(\mathbf{x}^v) = \mathbf{t}^{Av}$.

The ***toric ring*** associated to $\mathcal{A}$ is

$$S/I_{\mathcal{A}} \cong k[\mathbf{t}^{a_1}, \dots, \mathbf{t}^{a_n}],$$

where the isomorphism is given by $\mathbf{x}^{\mathbf{v}} \mapsto \mathbf{t}^{A\mathbf{v}}$, for $\mathbf{v} \in \mathbf{N}^n$.

Note that the vector space $k[\mathbf{t}^{a_1}, \dots, \mathbf{t}^{a_n}]$ has basis $\mathbf{N}\mathcal{A}$ via the correspondence $\mathbf{t}^{\mathbf{a}} \mapsto \mathbf{a}$, where $\mathbf{a} \in \mathbf{N}^r$.

Therefore,

$$\dim_k (S/I_{\mathcal{A}})_\alpha = \begin{cases} 1 & \text{if } \alpha \in \mathbf{N}\mathcal{A} \\ 0 & \text{otherwise.} \end{cases}$$

If $\alpha \in \mathbf{N}\mathcal{A}$, then the set of all monomials in S of multidegree α is called the ***fiber*** of α.

In the running example we illustrate most of the definitions and constructions throughout the chapter.

Running Example 65.2. The matrix

$$A = \begin{pmatrix} 0 & 1 & 2 & 3 \\ 1 & 1 & 1 & 1 \end{pmatrix}$$

defines the ***twisted cubic curve***. The map φ is

$$\begin{aligned} k[a,b,c,h] &\to k[s,t] \\ a &\mapsto t \\ b &\mapsto st \\ c &\mapsto s^2 t \\ h &\mapsto s^3 t \,. \end{aligned}$$

Therefore, the toric ring is isomorphic to $k[t,\ st,\ s^2t,\ s^3t]$.

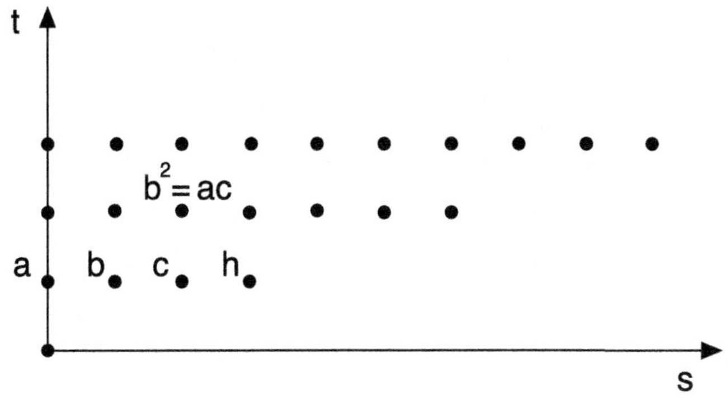

Figure 18.

In Figure 18 we show the monoid $\mathbf{N}\mathcal{A}$ inside $\mathbf{N}^2$ (we see the lower part of the monoid in the figure); the points in the monoid are depicted by big black dots. The monomials in the toric ring correspond bijectively to the elements in $\mathbf{N}\mathcal{A}$, so each big black dot corresponds bijectively to a monomial in the toric ring.

Each monomial m in S corresponds to paths from 0 to the black dot mdeg(m). We order $a < b < c < h$. Let $\alpha_1, \ldots, \alpha_i$ be a saturated path, that is $\alpha_{j+1} = \alpha_j + e_j$ where $e \in \{a, b, c, h\}$. We call this path ***increasing*** if $e_1 \leq e_2 \leq \ldots \leq e_i$. Then each monomial in S corresponds bijectively to a saturated increasing path from 0 to mdeg(m). Each difference $m - m'$ of two monomials in the toric ideal corresponds to two different saturated increasing paths leading to the same dot; this is illustrated in Figure 19 below. For example, b^2 and ac are the two monomials in $k[a, b, c, h]$ of multidegree $(2, 2)$. The element $b^2 - ac$ is in the toric ideal.

In order to list all monomials of multidegree $(3, 3)$, we look at the increasing saturated paths from $(0, 0)$ to $(3, 3)$. We see that the monomials are $b^3,\ a^2h,\ abc$. Thus, the fiber of $(3, 3)$ is $\{b^3, abc, a^2h\}$.

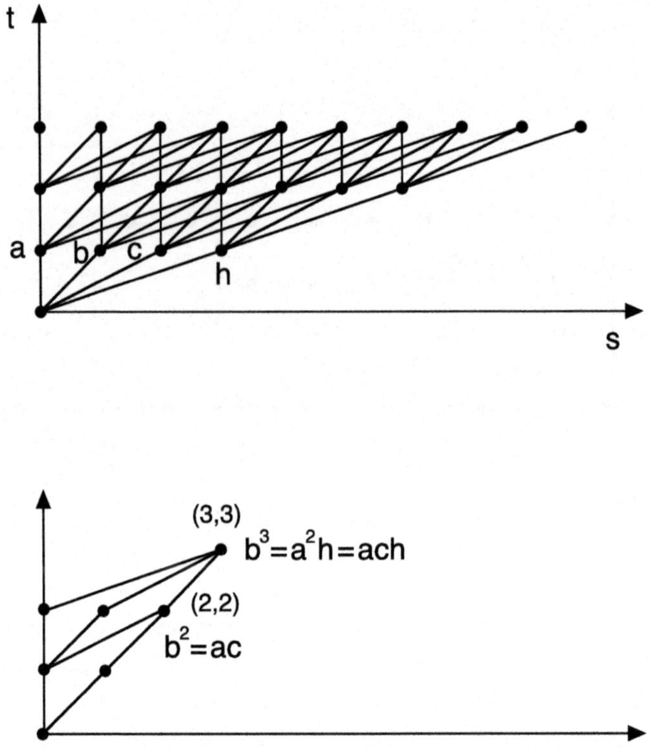

Figure 19.

Construction 65.3. Given a matrix A, one can compute the ideal $I_{\mathcal{A}}$ using elimination theory: eliminate the variables $t_1, \ldots, t_r$ from the ideal $(x_1 - \mathbf{t}^{a_1}, \ldots, x_n - \mathbf{t}^{a_n})$.

Running Example 65.4. We compute the defining ideal of the twisted cubic curve by eliminating the variables s and t from the ideal $(a-t,\, b-st,\, c-s^2t,\, h-s^3t)$ in $k[a,b,c,h,s,t]$. Straightforward computation shows that the defining ideal is

$$I = (ac - b^2,\ ah - bc,\ bh - c^2)$$

in $k[a,b,c,h]$.

A difference of two scalar multiples of monomials is called a

binomial. Binomials play major role in the theory of toric varieties. A ***pure binomial*** is a difference of two monomials. We will show that the ideal $I_{\mathcal{A}}$ has a minimal system of generators consisting of pure binomials.

Theorem 65.5. *The toric ideal $I_{\mathcal{A}}$ is generated by the pure binomials*

$$\{\mathbf{x}^{\mathbf{v}} - \mathbf{x}^{\mathbf{w}} \mid \mathbf{v}, \mathbf{w} \in \mathbf{N}^n,\ A(\mathbf{v}) = A(\mathbf{w})\}.$$

Proof. Set $I' = \left(\{\mathbf{x}^{\mathbf{v}} - \mathbf{x}^{\mathbf{w}} \mid \mathbf{v}, \mathbf{w} \in \mathbf{N}^n,\ A(\mathbf{v}) = A(\mathbf{w})\}\right)$. We have to show that $I_{\mathcal{A}} = I'$. The inclusion $I' \subseteq I_{\mathcal{A}}$ holds since

$$\varphi(\mathbf{x}^{\mathbf{v}}) = \mathbf{t}^{A\mathbf{v}} = \mathbf{t}^{A\mathbf{w}} = \varphi(\mathbf{x}^{\mathbf{w}}).$$

Suppose that $I_{\mathcal{A}} \neq I'$. Fix a monomial order in the polynomial ring S. Let f be a polynomial such that $f \in I_{\mathcal{A}}$, $f \notin I'$, and f has a minimal initial term among the polynomials satisfying these conditions. Denote by g the initial term of f. Since the monomials form a basis in $k[t_1, \ldots, t_r]$ and $\varphi(f) = 0$, it follows that there exists a term g' of f such that g and g' are mapped by φ to scalar multiples of the same monomial. Let $g = \alpha\mathbf{x}^{\mathbf{v}}$ and $g' = \beta\mathbf{x}^{\mathbf{w}}$, where $\alpha, \beta \in k^*$. We have that $(\mathbf{x}^{\mathbf{v}} - \mathbf{x}^{\mathbf{w}}) \in I'$ as $\mathbf{t}^{A\mathbf{v}} = \varphi(\mathbf{x}^{\mathbf{v}}) = \varphi(\mathbf{x}^{\mathbf{w}}) = \mathbf{t}^{A\mathbf{w}}$. Hence $\tilde{f} = f - \alpha(\mathbf{x}^{\mathbf{v}} - \mathbf{x}^{\mathbf{w}}) \notin I'$. This is a contradiction, since the initial term of $\tilde{f}$ is smaller than the initial term of f. □

We usually consider the matrix A as the matrix of a map from $\mathbf{Z}^n$ to $\mathbf{Z}^r$. When we write $\operatorname{Ker}_{\mathbf{Q}}(A)$ we mean $\{\mathbf{u} \in \mathbf{Q}^n \mid A\mathbf{u} = 0\}$.

Corollary 65.6. *If $A' \in GL_r(\mathbf{Z})A$ and has entries in $\mathbf{N}$, then A and A' determine the same toric ideal.*

Proof. Let $C \in GL_r(\mathbf{Z})$ be such that $A' = CA$. We have the equality

$$\{\mathbf{u} \mid \mathbf{u} \in \mathbf{Z}^n, \mathbf{u} \in \operatorname{Ker}_{\mathbf{Q}}(A)\} = \{\mathbf{u} \mid \mathbf{u} \in \mathbf{Z}^n, \mathbf{u} \in \operatorname{Ker}_{\mathbf{Q}}(CA)\}$$

since $\operatorname{Ker}_{\mathbf{Q}}(A) = \operatorname{Ker}_{\mathbf{Q}}(CA)$. □

Running Example 65.7. By 65.6, the matrix

$$A' = \begin{pmatrix} 3 & 2 & 1 & 0 \\ 0 & 1 & 2 & 3 \end{pmatrix}$$

defines the toric ideal of the twisted cubic curve. In Figure 20 we show the monoid $\mathbf{N}\mathcal{A}'$ inside $\mathbf{N}^2$ (we see the lower part of the monoid in the figure). The elements in the monoid are depicted by big black dots.

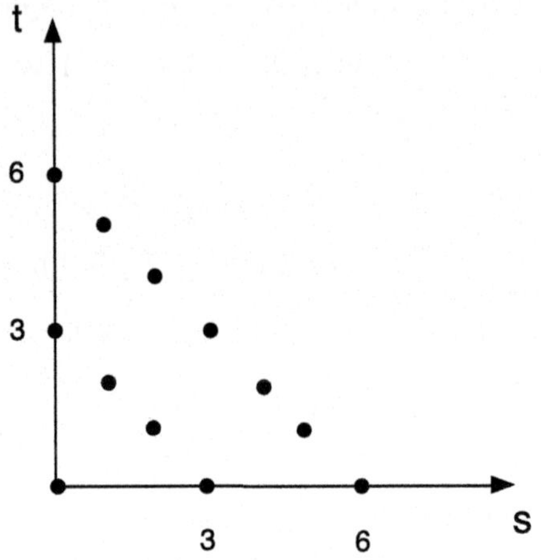

Figure 20.

The following result is an immediate consequence of Theorem 65.5.

Corollary 65.8. *With respect to any monomial order, the ideal $I_{\mathcal{A}}$ has a reduced Gröbner basis consisting of pure binomials.*

Proof. Choose a minimal system of pure binomial generators of $I_{\mathcal{A}}$. The new elements produced by Buchberger's Algorithm 39.9 are pure binomials. □

Running Example 65.9. With respect to the lex order with $b > c > a > h$, the ideal I has Gröbner basis

$$b^2 - ac,\ bc - ah,\ bh - c^2,\ c^3 - h^2 a\,.$$

Corollary 65.10. (Aramova-Herzog) *A binomial Gröbner basis of $I_{\mathcal{A}}$ does not depend on the base field.*

Proof. Fix a monomial order $<$. Take a Gröbner basis $\mathcal{T}$ of $I_{\mathcal{A}}$ over the field $\mathbf{Q}$ such that $\mathcal{T}$ consists of pure binomials. Let C be the ideal generated by the initial terms of the binomials in $\mathcal{T}$. The binomials in $\mathcal{T}$ are defined over any base field. Therefore, $C \subseteq \text{in}_< I_{\mathcal{A}}$ over any field.

We have that $C = \text{in}_< I_{\mathcal{A}}$ over the field $\mathbf{Q}$. Since the multigraded Hilbert function of the toric ring and the multigraded Hilbert function of S/C do not depend on the base field, it follows that $C = \text{in}_< I_{\mathcal{A}}$ over any field. Therefore, $\mathcal{T}$ is a Gröbner basis over any base field. □

We say that $I_{\mathcal{A}}$ ***defines a projective toric variety*** if $I_{\mathcal{A}}$ is homogeneous with respect to the standard grading with $\deg(x_i) = 1$ for all $1 \leq i \leq n$.

Exercise 65.11. *A pure binomial* $\mathbf{x}^{\mathbf{v}} - \mathbf{x}^{\mathbf{w}}$ *is homogeneous if and only if the scalar product* $(1, \ldots, 1) \cdot (\mathbf{v} - \mathbf{w})$ *vanishes.*

Proposition 65.12. *The following properties are equivalent.*

(1) $I_{\mathcal{A}}$ *defines a projective toric variety.*
(2) $(1, \ldots, 1)$ *is a* $\mathbf{Q}$*-linear combination of the rows of* A.
(3) *The points* $a_1, \ldots, a_n$ *lie on a hyperplane* $\sum_{1 \leq i \leq r} \beta_i w_i = 1$ *in* $\mathbf{Q}^r$ *with* $\beta_i \in \mathbf{Q}$*, where* $(w_1, \ldots, w_r)$ *are the coordinates in* $\mathbf{Q}^r$.
(4) *There exists an* $\mathcal{A}'$ *such that* $I_{\mathcal{A}'} = I_{\mathcal{A}}$ *and* ${a'}_1, \ldots, {a'}_n$ *lie on the hyperplane* $w_r = 1$*, where* $(w_1, \ldots, w_r)$ *are the coordinates in* $\mathbf{Q}^r$.

Proof. By Exercise 65.11 and Theorem 65.5, we see that $(1, \ldots, 1) \cdot \mathbf{u} = 0$ for every $\mathbf{u} \in \text{Ker}_{\mathbf{Q}}(A)$. Hence, (1) is equivalent to (2).

Denote by $\alpha_1, \ldots, \alpha_r$ the rows of A. If (2) holds, then there exist coefficients $\beta_i \in \mathbf{Q}$ such that $\sum_{1 \leq i \leq r} \beta_i \alpha_i = (1, \ldots, 1)$. Then the hyperplane $\sum_{1 \leq i \leq r} \beta_i w_i = 1$ contains the points $a_1, \ldots, a_n$. Hence, (2) implies (3). Thus, (2) and (3) are equivalent.

(2) implies (4) since we can replace one of the coordinates in each a_i by 1. If (4) holds, then $(1, \ldots, 1)$ is the last row of A', so (4) implies (2). □

Running Example 65.13. The ideal $I_{\mathcal{A}}$ is projective.

66 Examples

In this section we give a few simple, but important examples of toric ideals.

Monomial curves 66.1. A toric ideal associated to a set $\mathcal{A} \subseteq \mathbf{N} \setminus \mathbf{0}$ defines an ***affine monomial curve***. A toric ideal associated to a set $\mathcal{A} \subseteq \mathbf{N} \times 1$ defines a ***projective monomial curve***.

Running Example 66.2. The twisted cubic curve is a projective monomial curve.

Veronese varieties 66.3. Let $s \geq 0$ be an integer. Recall that $T = k[t_1, \dots, t_r]$. The *s'th **Veronese ring*** is

$$\oplus_{i=0}^{\infty} T_{is} = k[\,\text{all monomials of degree } s \text{ in } r \text{ variables}\,]\,.$$

It defines the s'th ***Veronese embedding*** of $\mathbf{P}^{r-1}$. The set of points defining the toric ideal can be taken to be

$$\mathcal{A}_{s,r} = \{(i_1, \dots, i_r) \in \mathbf{N}^r \setminus \mathbf{0} \mid \sum_{j=1}^{r} i_j = s\}\,.$$

The sum of all rows in $A_{s,r}$ is the vector with coordinates equal to s, so the toric ideal is projective.

In this case it is often convenient to denote the x-variables by $x_{\mathbf{c}}$ so that the variable $x_{\mathbf{c}}$ is mapped by the toric map to $\mathbf{t}^{\mathbf{c}}$ for $\mathbf{c} \in \mathcal{A}_{s,r}$.

Example 66.4. $k[s^3, s^2t, st^2, t^3]$ is the 3'rd Veronese ring in 2 variables.

Exercise 66.5. *Consider the standard grading on S with* $\deg(x_i) = 1$ *for each i. Let $I_{\mathcal{A}}$ be the defining ideal of the s'th Veronese ring in r variables. What is the Hilbert function of $S/I_{\mathcal{A}}$? (Note that* $\dim_k (S/I_{\mathcal{A}})_q$ *is the number of all monomials of degree sq in r variables.)*

Theorem 66.6. (Barcanescu and Manolache) *The toric ideal $I_{\mathcal{A}_{s,r}}$ has a quadratic binomial Gröbner basis. In particular, $I_{\mathcal{A}_{s,r}}$ is generated by quadratic binomials.*

Proof. Consider the lex order, denoted **lex**, on $\mathbf{N}^r$ such that $\mathbf{e_1} > \ldots > \mathbf{e_r}$, where $\mathbf{e_1}, \ldots, \mathbf{e_r}$ are the standard vectors. Denote the x-variables by $x_{\mathbf{c}}$ so that the variable $x_{\mathbf{c}}$ is mapped by the toric map to $\mathbf{t}^{\mathbf{c}}$ for $\mathbf{c} \in \mathcal{A}_{s,r}$.

Let $m = x_{\mathbf{c_1}} \ldots x_{\mathbf{c_p}}$ be a monomial written so that $\mathbf{c_1} \geq_{\mathbf{lex}} \ldots \geq_{\mathbf{lex}} \mathbf{c_p}$. Set $\mathbf{c} = \mathbf{c_1} + \ldots + \mathbf{c_p}$. We can write

$$\mathbf{c} = \mathbf{b_1} + \ldots + \mathbf{b_p} \,,$$

so that for every i we have that $\mathbf{b_i}$ is the lex-greatest vector in $\mathcal{A}_{s,r}$ such that $\mathbf{c} - \mathbf{b_1} - \ldots - \mathbf{b_i} \in \mathbf{N}\mathcal{A}_{s,r}$. Clearly, the monomial $\widetilde{m} = x_{\mathbf{b_1}} \ldots x_{\mathbf{b_p}}$ is in the same fiber as m.

Order the x-variables by

$$x_{\mathbf{c}} > x_{\mathbf{c}'} \text{ if } \mathbf{c} >_{\mathbf{lex}} \mathbf{c}' \,.$$

Consider the revlex order on the monomials in S. We will show that m can be reduced to $\widetilde{m}$ using the quadratic binomials in the ideal $I_{\mathcal{A}_{s,r}}$. In particular, it follows that $\widetilde{m}$ is the standard monomial in the fiber.

If $\mathbf{c_1} = \mathbf{b_1}$, then $\dfrac{m}{x_{\mathbf{c_1}}}$ can be reduced to $\dfrac{\widetilde{m}}{x_{\mathbf{b_1}}}$ by induction hypothesis on the degree of m, so we are done.

Suppose that $\mathbf{c_1} \neq \mathbf{b_1}$. Let i be the smallest number such that the i'th coordinates of $\mathbf{c_1}$ and $\mathbf{b_1}$ are different. It follows that $(\mathbf{c_1})_i < (\mathbf{b_1})_i$ since $\mathbf{b_1} >_{\mathbf{lex}} \mathbf{c_1}$. Hence, there exists a $q \neq 1$ such that $(\mathbf{c_q})_i > 0$. Also, there exists an $i < h \leq r$ such that $(\mathbf{b_1})_h < (\mathbf{c_1})_h$. Therefore $(\mathbf{c_1})_h > 0$. Now, we have

$$\mathbf{c_1} - \mathbf{e_h} + \mathbf{e_i} >_{\mathbf{lex}} \mathbf{c_1} \geq_{\mathbf{lex}} \mathbf{c_q} >_{\mathbf{lex}} \mathbf{c_q} - \mathbf{e_i} + \mathbf{e_h} \,.$$

Therefore, the inequalities

$$x_{\mathbf{c_1} - \mathbf{e_h} + \mathbf{e_i}} > x_{\mathbf{c_1}} \geq x_{\mathbf{c_q}} > x_{\mathbf{c_q} - \mathbf{e_i} + \mathbf{e_h}}$$

hold. So, in revlex order on the monomials in S we get

$$x_{\mathbf{c_1}} x_{\mathbf{c_q}} > x_{\mathbf{c_1} - \mathbf{e_h} + \mathbf{e_i}} x_{\mathbf{c_q} - \mathbf{e_i} + \mathbf{e_h}} \,.$$

Use the binomial $x_{\mathbf{c_1}}x_{\mathbf{c_q}} - x_{\mathbf{c_1-e_h+e_i}}x_{\mathbf{c_q-e_i+e_h}}$ to reduce m to a revlex-smaller monomial m'. Now, by induction hypothesis, we have that the monomial m' can be reduced to $\tilde{m}$. □

Currently, there is an open conjecture by Ottaviani-Paoletti for the property N_p (recall the definition of N_p in Section 17) of a Veronese toric ideal. The following problems are wide open. [Hering-Schenck-Smith] gives a bound on N_p in some cases.

Open-Ended Problems 66.7. (folklore) *What is the length of the 2-linear strand of a toric ideal generated by quadrics? Find tight upper/lower bounds on p for which N_p holds, for interesting classes of toric rings.*

Unimodular matrices 66.8. An $(r \times n)$-matrix A is ***unimodular*** if all non-zero $(r \times r)$-minors have the same absolute value. The Segre varieties, described below, are an interesting case of this type.

Segre varieties 66.9. Let Δ_s denote the simplex with $s+1$ vertices with $(s+1)$-coordinates $(1,0,\ldots,),\ldots,(0,\ldots,0,1,0,\ldots,0),\ldots,(0,\ldots,0,1)$. Let $\mathcal{A}$ consist of the vertices of the product of two simplices $\Delta_p \times \Delta_q$. This toric variety is called a ***Segre variety*** or the ***Segre embedding*** of $\mathbf{P}^p \times \mathbf{P}^q$ in $\mathbf{P}^{pq+p+q}$. It can be proved that the toric ideal is generated by the 2×2-minors of the $(p+1) \times (q+1)$-matrix whose entries are the indeterminates

$$\{x_{i,j} \,|\, 1 \le i \le p+1, 1 \le j \le q+1\}.$$

Example 66.10. Consider the matrix

$$\begin{pmatrix} x_{1,1} & x_{1,2} \\ x_{2,1} & x_{2,2} \\ x_{3,1} & x_{3,2} \end{pmatrix}.$$

The ideal generated by the 2×2-minors of this matrix is

$$(x_{1,1}x_{2,2} - x_{1,2}x_{2,1},\ x_{1,1}x_{3,2} - x_{1,2}x_{3,1},\ x_{2,1}x_{3,2} - x_{2,2}x_{3,1}).$$

This is the defining ideal in $k[x_{1,1}, x_{1,2}, x_{2,1}, x_{2,2}, x_{3,1}, x_{3,2}]$ of the Segre embedding of $\mathbf{P}^2 \times \mathbf{P}$ in $\mathbf{P}^5$. It is the toric ideal defined by the set

$$\mathcal{A} = \Delta_2 \times \Delta_1 = \Big\{(1,0,0,1,0), (0,1,0,1,0), (0,0,1,1,0), \\ (1,0,0,0,1), (0,1,0,0,1), (0,0,1,0,1)\Big\}.$$

Edge toric rings 66.11. Let W be a simple graph (i.e., with no double edges and no loops) on vertices $t_1, \ldots, t_r$. The ***edge toric ring*** of W is

$$k[t_i t_j \mid \{i,j\} \text{ is an edge in } W].$$

67 Betti numbers via simplicial complexes

In this section we discuss how to compute the Betti numbers of a toric ideal using simplicial complexes.

Construction 67.1. Since the ideal $I_{\mathcal{A}}$ is multigraded, there exists a minimal free resolution of $S/I_{\mathcal{A}}$ over S which is ***multigraded.*** We have ***multigraded Betti numbers***

$$b^S_{i,\alpha}(S/I_{\mathcal{A}}) = \dim_k \operatorname{Tor}^S_i(S/I_{\mathcal{A}}, k)_\alpha \quad \text{for } i \geq 0,\ \alpha \in \mathbf{N}^r .$$

Running Example 67.2. Set $Z = k[a,b,c,h]$. Computer computation shows that the minimal free resolution of Z/I over Z is

$$0 \to Z^2 \xrightarrow{\begin{pmatrix} h & c \\ c & b \\ b & a \end{pmatrix}} Z^3 \xrightarrow{\begin{pmatrix} ac-b^2 & bc-ah & bh-c^2 \end{pmatrix}} Z \to Z/I \to 0 .$$

Denote by d the differential in the resolution. We will illustrate how the multigrading 26.1 works. Use $\mathcal{A}$ to multigrade Z, so

$$\operatorname{mdeg}(a) = (0,1),\ \operatorname{mdeg}(b) = (1,1),\ \operatorname{mdeg}(c) = (2,1),\ \operatorname{mdeg}(h) = (3,1).$$

Note that the multigrading will be different if we use $\mathcal{A}'$.

First, we consider homological degree one. Denote by e_1, e_2, e_3 the basis of Z^3 such that

$$d(e_1) = ac - b^2, \quad d(e_2) = bc - ah, \quad d(e_3) = bh - c^2 .$$

Since the differential is multigraded, we have that

$$\begin{aligned} \mathrm{mdeg}(e_1) &= \mathrm{mdeg}(ac - b^2) = (2,2), \\ \mathrm{mdeg}(e_2) &= \mathrm{mdeg}(bc - ah) = (3,2), \\ \mathrm{mdeg}(e_3) &= \mathrm{mdeg}(bh - c^2) = (4,2). \end{aligned}$$

The Betti numbers in homological degree 1 are

$$b_{1,(2,2)} = 1, \quad b_{1,(3,2)} = 1, \quad b_{1,(4,2)} = 1.$$

Now consider homological degree two. Denote by f_1, f_2 the basis of Z^2 such that $d(f_1) = he_1 - ce_2 + be_3$ and $d(f_2) = ce_1 - be_2 + ae_3$. Since the differential is multigraded, we have

$$\begin{aligned} \mathrm{mdeg}(f_1) &= \mathrm{mdeg}(he_1 - ce_2 + be_3) = \mathrm{mdeg}(he_1) \\ &= \mathrm{mdeg}(h) + \mathrm{mdeg}(e_1) = (5,3) \\ \mathrm{mdeg}(f_2) &= \mathrm{mdeg}(ce_1 - be_2 + ae_3) \\ &= \mathrm{mdeg}(ce_1) = \mathrm{mdeg}(c) + \mathrm{mdeg}(e_1) = (4,3). \end{aligned}$$

The Betti numbers in homological degree 2 are $b_{2,(5,3)} = 1$ and $b_{2,(4,3)} = 1$. For completeness, we list that in homological degree 0 we have $b_{0,(0,0)} = 1$.

Recall that the set of all monomials in S of multidegree α is called the fiber of α.

Construction 67.3. [Stanley, Theorem 7.9] Recall that $\mathrm{rad}(m)$ is the maximal squarefree monomial dividing a given monomial m. For $\alpha \in \mathbf{N}^r$, denote by C_α the fiber of α. Let $\Gamma(\alpha)$ be the simplicial complex on vertices $x_1, \ldots, x_n$ and faces the radicals of the monomials in C_α and all their factors. We say that $\Gamma(\alpha)$ is ***generated*** by the radicals of the monomials in C_α.

Running Example 67.4. The fiber of $(3,3)$ is $C_{(3,3)} = \{b^3, abc, a^2h\}$. Therefore, $\Gamma((3,3))$ is the simplicial complex on vertices a, b, c, h that consists of the triangle abc and the edge ad. See Figure 21.

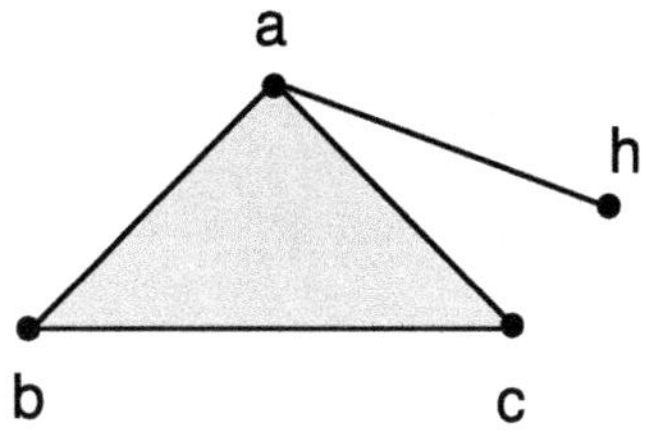

Figure 21.

Theorem 67.5. [Aramova-Herzog], (Campillo-Pison) *For* $\alpha \in \mathbf{N}^r$ *and* $i \geq 0$ *we have*

$$b^S_{i,\alpha}(I_\mathcal{A}) = \dim \widetilde{\mathrm{H}}_i\big(\Gamma(\alpha); k\big).$$

Proof. Note that $b^S_{i,\alpha}(I_\mathcal{A}) = b^S_{i+1,\alpha}(S/I_\mathcal{A})$. We compute the Betti numbers of $S/I_\mathcal{A}$ using the Koszul complex $\mathbf{K}$ that is the minimal free resolution of k over S. Let E be the exterior algebra over k on basis elements $e_1, \ldots, e_n$. The complex $\mathbf{K}$ equals $S \otimes E$ as an S-module and has differential

$$d(e_{j_1} \wedge \cdots \wedge e_{j_s}) = \sum\nolimits_{1 \leq i \leq s} (-1)^{i+1} \cdot x_{j_i} \cdot e_{j_1} \wedge \cdots \wedge \widehat{e}_{j_i} \wedge \cdots \wedge e_{j_s} ,$$

where $\widehat{e}_{j_i}$ means that e_{j_i} is omitted in the product. We have

$$b^S_{i+1,\alpha}(S/I_\mathcal{A}) = \dim_k \mathrm{Tor}_{i+1}(S/I_\mathcal{A}, k)_\alpha = \dim_k \mathrm{H}_{i+1}(S/I_\mathcal{A} \otimes \mathbf{K})_\alpha .$$

Recall that $\mathrm{mdeg}(x_j) = \mathrm{mdeg}(e_j) = a_j$. The component of $S/I_\mathcal{A} \otimes \mathbf{K}$ in homological degree $i+1$ and multidegree α has basis

$$\left\{ \frac{m}{x_{j_1} \cdots x_{j_{i+1}}} e_{j_1} \wedge \cdots \wedge e_{j_{i+1}} \;\middle|\; m \in C_\alpha,\ 1 \le j_1 < \ldots < j_{i+1} \le n \right.$$
$$\left. x_{j_1} \ldots x_{j_{i+1}} \text{ divides } m \right\}.$$

Note that the component of $S/I_{\mathcal{A}} \otimes \mathbf{K}$ in multidegree α is 0 if $\alpha \notin \mathbf{N}\mathcal{A}$. Denote by $\mathbf{T}$ the oriented chain complex computing the reduced homology of the simplicial complex $\Gamma(\alpha)$. We will show that the map

$$(S/I_{\mathcal{A}} \otimes \mathbf{K}_{i+1})_\alpha \longrightarrow \mathbf{T}_i$$
$$\frac{m_\alpha}{x_{j_1} \cdots x_{j_{i+1}}} e_{j_1} \wedge \cdots \wedge e_{j_{i+1}} \mapsto \text{the face with vertices } j_1, \ldots, j_{i+1}$$

is an isomorphism of complexes. The map is a surjection. In order to show that it is injective, we have to check that if m_α and m'_α are two monomials in the fiber of α and if both of them are divisible by $x_{j_1} \ldots x_{j_{i+1}}$, then the difference

$$\frac{m_\alpha}{x_{j_1} \cdots x_{j_{i+1}}} - \frac{m'_\alpha}{x_{j_1} \cdots x_{j_{i+1}}}$$

is in the ideal $I_{\mathcal{A}}$. The desired property holds because the monomials $\frac{m_\alpha}{x_{j_1} \cdots x_{j_{i+1}}}$ and $\frac{m'_\alpha}{x_{j_1} \cdots x_{j_{i+1}}}$ are in the same fiber.

Hence $b^S_{i+1,\alpha}(S/I_{\mathcal{A}}) = \dim \widetilde{\mathrm{H}}_i\big(\Gamma(\alpha); k\big)$. □

Running Example 67.6. The fiber of $(3,3)$ is $C_{(3,3)} = \{b^3, abc, a^2h\}$. The simplicial complex $\Gamma((3,3))$ consists of the triangle abc and the edge ad. Hence, all Betti numbers in multidegree $(3,3)$ vanish. Now consider the fiber $C_{(4,3)} = \{b^2c, abh, ac^2\}$ of $(4,3)$. The simplicial complex $\Gamma((4,3))$ consists of the triangle abh and the two edges ac and bc. Therefore, $b_{1,(4,3)}(I_{\mathcal{A}}) = 1$ and all other Betti numbers in multidegree $(4,3)$ vanish.

Corollary 67.7. *If the greatest common divisor of the monomials in a fiber C_α is not 1, then for all $i \ge 0$ we have*

$$b^S_{i,\alpha}(I_{\mathcal{A}}) = 0\,.$$

Proof. If the greatest common divisor of the monomials in a fiber C_α is not 1, then $\Gamma(\alpha)$ is a cone and so it has no reduced homology. Apply Theorem 67.5. □

The next result answers the question what simplicial complexes occur as $\Gamma(\alpha)$.

Theorem 67.8. [Bruns-Herzog 2] *Let Δ be a finite simplicial complex. There exist a set $\mathcal{A}$ and $\alpha \in \mathbf{N}\mathcal{A}$ such that the complex $\Gamma(\alpha)$ is homotopy equivalent to Δ.*

In particular, we see that the Betti numbers of $S/I_{\mathcal{A}}$ can depend on the characteristic of the field k.

Proof. Let Δ be a simplicial complex on a vertex set $\{x_1, \dots, x_p\}$. Denote by $F(\Delta)$ the set of facets of Δ. For each $\sigma \in F(\Delta)$ introduce a variable x_σ. Consider the homomorphism

$$\begin{aligned} k[x_1, \dots, x_p, \{x_\sigma \,|\, \sigma \in F(\Delta)\}] &\longrightarrow k[t, t_1, \dots, t_p] \\ x_i &\longmapsto t_i \\ x_\sigma &\longmapsto t \prod_{i \notin \sigma} t_i \,. \end{aligned}$$

The preimage of the monomial $tt_1 \dots t_p$ is the fiber

$$C_{(1,\dots,1)} = \left\{ x_\sigma \prod_{i \in \sigma} x_i \,\middle|\, \sigma \in F(\Delta) \right\}.$$

Therefore, $\Gamma((1, \dots, 1))$ is the simplicial complex on the vertex set $\{x_1, \dots, x_p, \{x_\sigma \,|\, \sigma \in F(\Delta)\}\}$ generated by the faces $\{\sigma \cup x_\sigma \,|\, \sigma \in F(\Delta)\}$. □

The multigraded Betti numbers can be computed using other simplicial complexes.

Construction 67.9. (Bayer-Peeva-Sturmfels) For $\alpha \in \mathbf{N}^r$, let $X(\alpha)$ be the simplicial complex on vertices the monomials in the fiber C_α and faces

$$\{\sigma \subseteq C_\alpha \,|\, \gcd(m | m \in \sigma) \neq 1\}.$$

Theorem 67.10. *For $\alpha \in \mathbf{N}^r$ and $i \geq 0$ we have*

$$b^S_{i,\alpha}(I_{\mathcal{A}}) = \dim \widetilde{\mathrm{H}}_i\big(X(\alpha); k\big).$$

The simplicial complexes $X(\alpha)$ and $\Gamma(\alpha)$ are homotopy equivalent.

Proof. We are going to apply the Nerve Theorem 36.11 to the simplicial complex $\Gamma(\alpha)$. For a monomial $m \in C_\alpha$, set Λ_m to be the simplex on vertices the variables that divide m. Then $\{\Lambda_m\}_{m \in C_\alpha}$ is a cover of $\Gamma(\alpha)$. If $\cap_{m\in\sigma} \Lambda_m \neq \emptyset$ for some $\sigma \subseteq C_\alpha$, then it is a simplex, so it is contractible. The nerve of that cover is $X(\alpha)$. By the Nerve Theorem 36.11, $X(\alpha)$ and $\Gamma(\alpha)$ are homotopy equivalent. Apply Theorem 67.5. □

68 Projective dimension

We will prove an upper bound on the projective dimension of $S/I_{\mathcal{A}}$ in terms of its codimension.

Exercise 68.1. $\mathrm{codim}(I_{\mathcal{A}}) = n - r$.

Theorem 68.2. [Peeva-Sturmfels 2]

$$\mathrm{pd}(S/I_{\mathcal{A}}) \leq 2^{\mathrm{codim}(I_{\mathcal{A}})} - 1\,.$$

Proof. Let $\alpha \in \mathbf{N}\mathcal{A}$. First, we will show that the simplicial complex $\Gamma(\alpha)$ has at most $2^{\mathrm{codim}(I_{\mathcal{A}})}$ facets. Denote by $F_1, \dots, F_s$ the facets of $\Gamma(\alpha)$. There exist elements $\mathbf{q}_1, \dots, \mathbf{q}_s$ in $\mathbf{N}^n$ such that $\mathrm{rad}(\mathbf{x}^{\mathbf{q}_i}) = F_i$ and $\mathbf{x}^{\mathbf{q}_i}$ is in the fiber of α for $i = 1, \dots, s$.

Suppose that $s > 2^{\mathrm{codim}(I_{\mathcal{A}})}$. Note that $\mathrm{codim}(I_{\mathcal{A}}) = n - r$. Let B be an integer $(n \times (n-r))$-matrix such that the sequence

$$0 \;\rightarrow\; \mathbf{Z}^{n-r} \xrightarrow{\;B\;} \mathbf{Z}^n \xrightarrow{\;A\;} \mathbf{Z}^r$$

is exact. For $\mathbf{v} \in \mathbf{Z}^{\mathrm{codim}(I_{\mathcal{A}})}$, we can write $B\mathbf{v}$ uniquely as $B\mathbf{v} = (B\mathbf{v})_+ - (B\mathbf{v})_-$, where $(B\mathbf{v})_+$ and $(B\mathbf{v})_-$ have non-negative coordinates and $\mathrm{supp}((B\mathbf{v})_+) \cap \mathrm{supp}((B\mathbf{v})_-) = \emptyset$. Set $\mathbf{q} = \mathbf{q}_s$. Let $\mathbf{r_i} = \gcd(\mathbf{x}^{\mathbf{q}}, \mathbf{x}^{\mathbf{q_i}})$ for $1 \leq i < s$. We have that $\mathbf{q} - \mathbf{q}_i \in \mathrm{Ker}(A) = \mathrm{Im}(B)$.

Let $\mathbf{v}_1, \ldots, \mathbf{v}_{s-1} \in \mathbf{Z}^{\operatorname{codim}(I_\mathcal{A})}$ be such that $\mathbf{q} = \mathbf{r_i} + (B\mathbf{v}_i)_+$ and $\mathbf{q}_i = \mathbf{r_i} + (B\mathbf{v}_i)_-$ for $1 \leq i < s$, that is, $\mathbf{x}^{\mathbf{q}} - \mathbf{x}^{\mathbf{q_i}} = \mathbf{x}^{\mathbf{r_i}}\left(\mathbf{x}^{(B\mathbf{v}_i)_+} - \mathbf{x}^{(B\mathbf{v}_i)_-}\right)$. Since $s > 2^{\operatorname{codim}(I_\mathcal{A})}$, it follows that there exist two vectors $\mathbf{v}_i$ and $\mathbf{v}_j$ such that the coordinates of $\mathbf{v}_i + \mathbf{v}_j$ are even numbers. Therefore, the coordinates of $\big(B(\mathbf{v}_i + \mathbf{v}_j)\big)_+$ and $\big(B(\mathbf{v}_i + \mathbf{v}_j)\big)_-$ are even numbers. Note that

$$\mathbf{x}^{2\mathbf{q}} - \mathbf{x}^{\mathbf{q}_i + \mathbf{q}_j} = \mathbf{x}^{\mathbf{r}}\left(\mathbf{x}^{(B(\mathbf{v}_i+\mathbf{v}_j))_+} - \mathbf{x}^{(B(\mathbf{v}_i+\mathbf{v}_j))_-}\right)$$

for some $\mathbf{r} \in \mathbf{N}^n$. Since $\mathbf{x}^{2\mathbf{q}} = \mathbf{x}^{\mathbf{r}}\mathbf{x}^{(B(\mathbf{v}_i+\mathbf{v}_j))_+}$, it follows that the coordinates of $\mathbf{r}$ are even numbers. Furthermore, $\mathbf{x}^{\mathbf{q}_i+\mathbf{q}_j} = \mathbf{x}^{\mathbf{r}}\mathbf{x}^{(B(\mathbf{v}_i+\mathbf{v}_j))_-}$ implies that the coordinates of $\mathbf{q}_i + \mathbf{q}_j$ are even numbers. Hence $\frac{1}{2}(\mathbf{q}_i + \mathbf{q}_j) \in \mathbf{N}^n$. The fact that $\mathbf{x}^{\mathbf{q}_i}$ and $\mathbf{x}^{\mathbf{q}_j}$ are in the fiber of α is equivalent to $A\mathbf{q}_i = A\mathbf{q}_j = \alpha$. We have that $A(\frac{1}{2}(\mathbf{q}_i + \mathbf{q}_j)) = \alpha$, so $\mathbf{x}^{\frac{1}{2}(\mathbf{q}_i+\mathbf{q}_j)}$ is in the fiber of α as well. Therefore, $\operatorname{rad}(\mathbf{x}^{\frac{1}{2}(\mathbf{q}_i+\mathbf{q}_j)}) = F_i \cup F_j$ is a face of $\Gamma(\alpha)$. This face properly contains both F_i and F_j. We obtained a contradiction to our choice that F_i and F_j are facets.

Therefore, $\Gamma(\alpha)$ has at most $2^{\operatorname{codim}(I_\mathcal{A})}$ facets. Hence the homology of $\Gamma(\alpha)$ vanishes in dimensions $\geq 2^{\operatorname{codim}(I_\mathcal{A})} - 1$. Note that $\operatorname{pd}(I_\mathcal{A}) + 1 = \operatorname{pd}(S/I_\mathcal{A})$ to obtain the desired upper bound. □

Exercise 68.3. *Find examples that show that the bound in the above theorem is sharp.*

69 The Scarf complex

We will discuss an algebraic complex, called the Scarf complex, which is always contained in the minimal free resolution of a toric ideal. The material in this section is from [Peeva-Sturmfels].

For simplicity, we denote a fiber by C instead of C_α if the multidegree α can be suppressed. Set $\operatorname{mdeg}(C) = \alpha$. In this case we write $b^S_{i,\operatorname{mdeg}(C)}(I_\mathcal{A})$ instead of $b^S_{i,\alpha}(I_\mathcal{A})$, and we write $\Gamma(C)$ instead of $\Gamma(\alpha)$.

We introduce a partial order $\prec$ on the set of all fibers as follows: if C and C' are fibers then $C' \preceq C$ if there exists a monomial m such that $m\,C' \subseteq C$.

We denote by $\gcd(C)$ the greatest common divisor of all monomials in a set C of monomials. We call a fiber C ***basic*** if $\gcd(C) = 1$ and $\gcd(C\backslash m) \neq 1$ for all $m \in C$.

Proposition 69.1. *If C is a basic fiber then*

$$b^S_{i,\mathrm{mdeg}(C)}(I_{\mathcal{A}}) = \begin{cases} 1 & \text{if } i = |C| - 2 \\ 0 & \text{otherwise.} \end{cases}$$

Proof. We will apply Theorem 67.10. Let α be the multidegree of C. For every monomial $m \in C$ we have that $\gcd(C\backslash m) \neq 1$. Therefore, $C\backslash m$ is a face in $X(\alpha)$. Since $\gcd(C) = 1$, we conclude that C is not a face in $X(\alpha)$. Hence, $X(\alpha)$ is the boundary of the simplex with vertices the monomials in C. □

Lemma 69.2. *Let C be a basic fiber and m a monomial in C. The monomials in $C \setminus m$ divided by their greatest common divisor form a basic fiber.*

Proof. Denote by g the greatest common divisor of the monomials in $C \setminus m$. Furthermore, denote by C' the set of the monomials in $C \setminus m$ divided by g. First, we will show that C' is a fiber. Suppose that there exists a monomial $\widetilde{m} \notin C'$ that is in the same fiber as C'. Then $\widetilde{m}g \in C$. It follows that $m = \widetilde{m}g$. Hence, $g \neq 1$ divides all monomials in the fiber C, which is a contradiction. Therefore, C' is a fiber.

Next, we will show that the fiber C' is basic. Clearly $\gcd(C') = 1$. Let $m' \in C'$ be a monomial. We will show that the monomials in $C'\backslash m'$ have a non-trivial greatest common divisor. Since C is basic we have that $g = \gcd(C\backslash m) \neq 1$, $h = \gcd(C\backslash m'g) \neq 1$ and $\gcd(g, h) = 1$. Hence $\gcd(C' \setminus m')$ is a multiple of h. □

Corollary 69.3. *Let C be a basic fiber. If C' is a fiber and $C' \preceq C$, then C' is a monomial multiple of a basic fiber.*

Proof. Let m be a monomial such that $mC' \subseteq C$. Then $mC' = C\backslash \{m_1, \ldots, m_s\}$ for some monomials $m_1, \ldots, m_s$. Apply Lemma 69.2 repeatedly. It follows that the monomials in $C\setminus \{m_1, \ldots, m_s\}$ divided by their greatest common divisor form a basic fiber. Therefore, C' is a monomial multiple of a basic fiber. □

Proposition 69.4. *If $C' \prec C$ are fibers and C is basic, then* $|C'| < |C|$.

Proof. Suppose that $|C'| = |C|$. Then $C = mC'$ for some monomial m. Since C is basic, it follows that $m = 1$, which is a contradiction.□

Construction 69.5. The basic fibers form a finite poset, which we call the ***Scarf poset*** and denote by $P_{\mathcal{A}}$. We will define the ***Scarf complex***. It is the complex of free S-modules

$$\mathbf{F}_{\mathcal{A}} = \oplus_{C \in P_{\mathcal{A}}} \ S \cdot e_C \,,$$

where e_C denotes a basis element in homological degree $|C| - 1$, and the sum runs over all basic fibers C. The differential is defined by

$$\partial(e_C) = \sum\nolimits_{m \in C} \operatorname{sign}(m, C) \cdot \gcd(C \setminus m) \cdot e_{C \setminus m} \,,$$

where $\operatorname{sign}(m, C)$ is $(-1)^{l+1}$ if m is in the l'th position (we order the monomials in C lexicographically and consider C as an ordered set here). Straightforward verification shows that $\partial^2 = 0$.

Exercise 69.6. *Show that* $\mathbf{F}_{\mathcal{A}}$*, in Construction* 69.5*, is a complex.*

Exercise 69.7. *In the notation of Construction* 69.5*, show that the multidegree of* e_C *is the multidegree of any of the monomials in the fiber* C*.*

Theorem 69.8. [Peeva-Sturmfels] *Let* $(\mathbf{G}, d)$ *be the minimal free resolution of* $S/I_{\mathcal{A}}$ *over* S*. Then* $\mathbf{F}_{\mathcal{A}}$ *is an essential subcomplex of* $\mathbf{G}$*.* (Recall Definition 3.5.)

Proof. The proof is by induction on homological degree.

The base of the induction is the first step in the complex $\mathbf{F}_{\mathcal{A}}$, which is $\oplus_C \, Se_C \rightarrow S$ and the sum runs over all fibers consisting of two relatively prime monomials. By Theorem 67.5 it follows that the difference of two such monomials is contained in every minimal system of homogeneous binomial generators of the toric ideal $I_{\mathcal{A}}$. Therefore, $\oplus_C Se_C \longrightarrow S$ is contained in $\mathbf{G}$ in the desired way.

Suppose that $\mathbf{F}_{\mathcal{A}}$ is an essential subcomplex of $\mathbf{G}$ in homological degrees less than i. Since $\mathbf{F}_{\mathcal{A}}$ is a complex we have that the elements

$$\{\, \partial(e_C) \,|\, |C| = i+1 \,\}$$

are syzygies. We have to prove that they are contained in a minimal system of generators of the syzygy module. Assume the opposite. Let C be a basic fiber with $|C| = i+1$ and $\partial(e_C)$ be an S-linear combination of syzygies of strictly smaller multidegrees. It follows that $\mathbf{G}$ has a basis element of homological degree i and multidegree α' whose fiber C' is strictly less than C. Note that $b^S_{i,\mathrm{mdeg}(C')}(S/I_{\mathcal{A}}) \neq 0$.

By Corollary 67.7 we get that $\gcd(C') = 1$. We can apply Corollary 69.3 since C is basic, and it implies that C' is a monomial multiple of a basic fiber. As $\gcd(C') = 1$ we conclude that C' is basic. Since $b^S_{i,\mathrm{mdeg}(C')}(S/I_{\mathcal{A}}) \neq 0$, Proposition 69.1 yields that $|C'| = i+1$.

By Proposition 69.4, $|C'| < |C|$. Therefore,

$$i+1 = |C'| < |C| = i+1$$

which is a contradiction. □

Using Proposition 69.1 we will get a criterion for the Scarf complex $\mathbf{F}_{\mathcal{A}}$ to be the minimal free resolution of $S/I_{\mathcal{A}}$.

Proposition 69.9. *Suppose that $b^S_{j,\mathrm{mdeg}(C)}(S/I_{\mathcal{A}}) = 0$ for all $j \geq 0$ and every non-basic fiber C. Then the Scarf complex $\mathbf{F}_{\mathcal{A}}$ is the minimal free resolution of $S/I_{\mathcal{A}}$.*

Proof. The complex $\mathbf{F}_{\mathcal{A}}$ is an essential subcomplex of the minimal free resolution by Theorem 69.8. Suppose that $\mathbf{F}_{\mathcal{A}}$ is strictly smaller than the minimal free resolution of $S/I_{\mathcal{A}}$. Since $\mathbf{F}_{\mathcal{A}}$ has one free generator in each multidegree that has a basic fiber, it follows by Proposition 69.1 that $b^S_{i,\mathrm{mdeg}(C)}(I_{\mathcal{A}}) \neq 0$ for some non-basic fiber C, which is a contradiction. □

70 Generic toric ideals

Definition 70.1. We say that $I_{\mathcal{A}}$ is ***generic*** if it has a minimal system of binomial generators such that each variable x_i appears in every generator. A result by Barany and Scarf, cf. [Peeva-Sturmfels, Theorem 4.1], explains the motivation for this definition.

Recall that $\mathrm{supp}(m) = \{x_j \,|\, x_j \text{ divides } m\}$ for a monomial m.

Theorem 70.2. [Peeva-Sturmfels] *If $I_{\mathcal{A}}$ is generic then $\mathbf{F}_{\mathcal{A}}$ is the minimal free resolution of $S/I_{\mathcal{A}}$.*

Proof. We will show that $b^S_{j,\mathrm{mdeg}(C)}(S/I_{\mathcal{A}}) = 0$ for all $j \geq 0$ and every non-basic fiber C, and then we will apply Proposition 69.9.

Suppose that C is a fiber which is not basic. We will show that the simplicial complex $\Gamma(C)$ is a cone, so we can apply Corollary 67.7.

If $\gcd(C) \neq 1$ then $\Gamma(C)$ is a cone over $\mathrm{rad}(\gcd(C))$. If C contains a monomial with support $\{x_1, \ldots, x_n\}$, then the simplicial complex $\Gamma(C)$ is a simplex.

Suppose that $\gcd(C) = 1$ and C contains no monomial with support $\{x_1, \ldots, x_n\}$. Since C is not basic, there exists a monomial $\mathbf{x}^{\mathbf{a}} \in C$ such that $\gcd(C \setminus \mathbf{x}^{\mathbf{a}}) = 1$. There exists an $x_i \notin \mathrm{supp}(\mathbf{x}^{\mathbf{a}})$. Since $\gcd(C \setminus \mathbf{x}^{\mathbf{a}}) = 1$, there exists a monomial $\mathbf{x}^{\mathbf{e}} \in C \setminus \mathbf{x}^{\mathbf{a}}$ such that $x_i \notin \mathrm{supp}(\mathbf{x}^{\mathbf{e}})$. Hence, there exist at least two monomials $\mathbf{x}^{\mathbf{a}}, \mathbf{x}^{\mathbf{e}} \in C$ that are not divisible by x_i. We will show that $\Gamma(C)$ is a cone over x_i. Let $F = \mathrm{supp}(\mathbf{x}^{\mathbf{f}})$ for some $\mathbf{x}^{\mathbf{f}} \in C$ be such that the face $F \in \Gamma(C)$ does not contain x_i. Let $\mathbf{x}^{\mathbf{b}}$ be any other monomial in C with $x_i \notin \mathrm{supp}(\mathbf{x}^{\mathbf{b}})$. Write $\mathbf{x}^{\mathbf{f}} = \mathbf{x}^{\mathbf{d}} \cdot \mathbf{x}^{\mathbf{f}'}$ and $\mathbf{x}^{\mathbf{b}} = \mathbf{x}^{\mathbf{d}} \cdot \mathbf{x}^{\mathbf{b}'}$ where $\gcd(\mathbf{x}^{\mathbf{f}'}, \mathbf{x}^{\mathbf{b}'}) = 1$. Since $I_{\mathcal{A}}$ is generated by binomials with full support, we can write $\mathbf{x}^{\mathbf{b}'} - \mathbf{x}^{\mathbf{f}'}$ as a combination of binomials with full support. Hence, there exists a monomial $\mathbf{x}^{\mathbf{c}}$ such that $\mathbf{x}^{\mathbf{b}'} - \mathbf{x}^{\mathbf{c}} \in I_{\mathcal{A}}$ and $\mathrm{supp}(\mathbf{x}^{\mathbf{b}'}) \cup \mathrm{supp}(\mathbf{x}^{\mathbf{c}}) = \{x_1, \ldots, x_n\}$. It follows that $\mathbf{x}^{\mathbf{c}}\mathbf{x}^{\mathbf{d}} \in C$ and $\{x_i\} \cup \mathrm{supp}(\mathbf{x}^{\mathbf{f}'}) \subseteq \mathrm{supp}(\mathbf{x}^{\mathbf{c}})$. The face $\mathrm{supp}(\mathbf{x}^{\mathbf{c}}\mathbf{x}^{\mathbf{d}}) \in \Gamma(C)$ contains both x_i and F. □

In the rest of this section we consider Gröbner basis.

Lemma 70.3. [Peeva-Sturmfels 2] *Let $\mathcal{G}$ be a minimal multigraded bi-*

nomial generating set of $I_{\mathcal{A}}$. *If a variable* x_i *appears in every binomial in* $\mathcal{G}$, *then* $\mathcal{G}$ *is a Gröbner basis with respect to any revlex monomial order which has* x_i *as the smallest variable.*

Proof. Consider the initial ideal of $I_{\mathcal{A}}$ with respect to the weight vector $-\mathbf{e}_i$ (where $\mathbf{e}_i$ is the i'th standard vector). The initial ideal $\text{in}_{-\mathbf{e}_i}(I_{\mathcal{A}})$ contains the set of monomials $\text{in}_{-\mathbf{e}_i}(\mathcal{G})$. We will prove that $\text{in}_{-\mathbf{e}_i}(I_{\mathcal{A}}) = (\text{in}_{-\mathbf{e}_i}(\mathcal{G}))$.

Let $f \in I_{\mathcal{A}}$ be not divisible by x_i. Hence, the variable x_i does not appear in $\text{in}_{-\mathbf{e}_i}(f)$. We will show that the initial term $\text{in}_{-\mathbf{e}_i}(f) \in (\text{in}_{-\mathbf{e}_i}(\mathcal{G}))$.

Since $I_{\mathcal{A}} = (\mathcal{G})$ we can write f as a combination of binomials in $\mathcal{G}$, and we see that every monomial in f is a multiple of one of the terms of some binomial in $\mathcal{G}$. Every non-initial term of a binomial in $\mathcal{G}$ contains x_i. Hence, if a term in $\text{in}_{-\mathbf{e}_i}(f)$ is not divisible by x_i then it is a multiple of the initial term of some of the binomials in $\mathcal{G}$. □

Choose positive integers $w_1, \ldots, w_n$ such that $I_{\mathcal{A}}$ is homogeneous with respect to the grading $\text{mdeg}(x_i) = w_i$ (the column vector with coordinates w_i is a combination of the columns in the matrix A). Fix the refined revlex order $<$ such that we order the monomials in S as follows: $x_1^{u_1} x_2^{u_2} \cdots x_n^{u_n} < x_1^{v_1} x_2^{v_2} \cdots x_n^{v_n}$ if $\sum_{i=1}^n u_i w_i < \sum_{i=1}^n v_i w_i$, or if $\sum_{i=1}^n u_i w_i = \sum_{i=1}^n v_i w_i$ and the last non-zero coordinate of $(u_1 - v_1, \ldots, u_n - v_n)$ is positive. Let $M = \text{in}_<(I_{\mathcal{A}})$ be the initial monomial ideal. Note that $<$ coincides with the revlex order on any binomial in the ideal $I_{\mathcal{A}}$.

Theorem 70.4. [Peeva-Sturmfels] *If* $I_{\mathcal{A}}$ *is generic, then the minimal generators of* $I_{\mathcal{A}}$ *form a Gröbner basis with respect to the order* $<$ *described above.*

Proof. This follows from the lemma above since every minimal generator contains x_n. □

Theorem 70.5. [Gasharov-Peeva-Welker] *Choose positive integers* $w_1, \ldots, w_n$ *such that* $I_{\mathcal{A}}$ *is homogeneous with respect to the grading* $\text{mdeg}(x_i) = w_i$. *Let* $<$ *be the refined revlex order in* S *described* *above. Let $I_{\mathcal{A}}$ be generic and set $M = \mathrm{in}_<(I_{\mathcal{A}})$. Then for any $i \geq 0$ and $\alpha \in \mathbf{N}^r$ we have the equality of multigraded Betti numbers $b^S_{i,\alpha}(S/M) = b^S_{i,\alpha}(S/I_{\mathcal{A}})$.*

Proof. Consider the monomial ideal $M' \subseteq k[x_1, \ldots, x_{n-1}]$ such that

$$S/M' = k[x_n] \otimes S/(I_{\mathcal{A}}, x_n).$$

We will prove that $M = M'$.

In the notation of Lemma 70.3, let $f \in \mathcal{G}$. Since $I_{\mathcal{A}}$ is generic, it follows that $f = m - m'x_n$. As $I_{\mathcal{A}}$ is prime, it follows that x_n does not divide m. Hence, the initial term of $m - m'x_n$ is m. By Theorem 70.4 $\mathcal{G}$ is a Gröbner basis with respect to the order $<$. Hence $M \subseteq M'$.

Recall by 26.1 that a multigraded finitely generated module T has a multigraded Hilbert series hilb_T. We have that

$$\mathrm{hilb}_{S/M'} = \frac{1}{(1 - x_n)}\mathrm{hilb}_{S/(I_{\mathcal{A}}, x_n)}$$

since $S/M' = k[x_n] \otimes S/(I_{\mathcal{A}}, x_n)$. As x_n is a non-zero divisor on $S/I_{\mathcal{A}}$ it follows that

$$\mathrm{hilb}_{S/(I_{\mathcal{A}}, x_n)} = (1 - x_n)\mathrm{hilb}_{S/I_{\mathcal{A}}} .$$

Therefore,

$$\mathrm{hilb}_{S/M'} = \mathrm{hilb}_{S/I_{\mathcal{A}}} .$$

The rings S/M and $S/I_{\mathcal{A}}$ have the same multigraded Hilbert series because M is an initial ideal, so we conclude that

$$\mathrm{hilb}_{S/M'} = \mathrm{hilb}_{S/M} .$$

Therefore, $M = M'$.

Since x_n is a non-zero divisor on $S/I_{\mathcal{A}}$, it follows that the minimal free resolution of S/M over S is obtained from the minimal free resolution of $S/I_{\mathcal{A}}$ by setting $x_n = 0$ in the matrices of the differential. In particular, the multigraded Betti numbers of S/M coincide with those of $S/I_{\mathcal{A}}$. $\square$

71 The Lawrence lifting

Construction 71.1. (cf. [Sturmfels, Chapter 7]) The ***Lawrence lifting*** of A is the matrix $L = \begin{pmatrix} A & \mathbf{O} \\ \mathbf{1} & \mathbf{1} \end{pmatrix}$, where $\mathbf{1}$ is the $(n \times n)$-identity matrix and $\mathbf{O}$ is the $(r \times n)$-zero matrix. The toric ideal I_L in $k[x_1, \ldots, x_n, y_1, \ldots, y_n]$ is called the ***Lawrence lifting*** of I_A. We have that

$$\mathrm{Ker}_{\mathbf{Q}}(L) = \left\{ \begin{pmatrix} \mathbf{u} \\ -\mathbf{u} \end{pmatrix} \,\middle|\, \mathbf{u} \in \mathrm{Ker}_{\mathbf{Q}}(A) \right\}.$$

Note that I_A and I_L have the same codimension $n - r$.

Exercise 71.2. *The ideal I_L is generated by*

$$\{ \mathbf{x}^{\mathbf{u}}\mathbf{y}^{\mathbf{v}} - \mathbf{x}^{\mathbf{v}}\mathbf{y}^{\mathbf{u}} \mid \mathbf{x}^{\mathbf{u}} - \mathbf{x}^{\mathbf{v}} \in I_A \}.$$

Running Example 71.3. The Lawrence lifting has matrix

$$L = \begin{pmatrix} 0 & 1 & 2 & 3 & 0 & 0 & 0 & 0 \\ 1 & 1 & 1 & 1 & 0 & 0 & 0 & 0 \\ 1 & 0 & 0 & 0 & 1 & 0 & 0 & 0 \\ 0 & 1 & 0 & 0 & 0 & 1 & 0 & 0 \\ 0 & 0 & 1 & 0 & 0 & 0 & 1 & 0 \\ 0 & 0 & 0 & 1 & 0 & 0 & 0 & 1 \end{pmatrix}$$

Computer computation shows that the Lawrence lifting ideal is

$$(acB^2 - b^2AC,\ ahBC - bcAH,\ bhC^2 - c^2BH,$$
$$b^3A^2H - a^2hB^3,\ c^3H^2A - h^2aC^3)$$

in $k[a, b, c, h, A, B, C, H]$.

Proposition 71.4.

(1) $k[x_1, \ldots, x_n, y_1, \ldots, y_n]/(I_L + (y_1 - 1, \ldots, y_n - 1)) = S/I_A$.

(2) *Every permutation of the elements $y_n - 1, \ldots, y_1 - 1$ forms a regular sequence on the toric ring $k[x_1, \ldots, x_n, y_1, \ldots, y_n]/I_L$.*

Proof. (1) follows from Theorem 65.5 and

$$\mathrm{Ker}_{\mathbf{Q}}(L) = \left\{ \begin{pmatrix} \mathbf{u} \\ -\mathbf{u} \end{pmatrix} \,\middle|\, \mathbf{u} \in \mathrm{Ker}_{\mathbf{Q}}(A) \right\}.$$

We will prove (2). Fix an $1 \leq i \leq n-1$. Denote by Y the ideal in $k[x_1, \ldots, x_n, y_1, \ldots, y_i]$ such that

$$k[x_1, \ldots, x_n, y_1, \ldots, y_i]/Y =$$
$$k[x_1, \ldots, x_n, y_1, \ldots, y_n]/\big(I_L + (y_{i+1} - 1, \ldots, y_n - 1)\big).$$

The ring $k[x_1, \ldots, x_n, y_1, \ldots, y_n]$ is $\mathbf{N}^{r+n}$-graded with the degrees of the variables given by the columns of the matrix L. Deleting the last $n-i$ coordinates in $\mathbf{N}^{r+n}$ we induce an $\mathbf{N}^{r+i}$-grading in which $\mathrm{mdeg}(y_{i+1}) = \ldots = \mathrm{mdeg}(y_n) = \mathbf{0}$. This induces an $\mathbf{N}^{r+i}$-grading on $k[x_1, \ldots, x_n, y_1, \ldots, y_i]$ and the ideal Y is $\mathbf{N}^{r+i}$-homogeneous. The elements 1 and y_i have different multidegrees, since the multidegree of 1 is the zero vector but the multidegree of y_i is the vector $(0, \ldots, 0, 1)$. This implies that $y_i - 1$ is a regular element on the quotient ring $k[x_1, \ldots, x_n, y_1, \ldots, y_i]/Y$; for completeness we include the proof.

Suppose that $(y_i - 1)f = 0$ for some non-zero element f in the quotient $k[x_1, \ldots, x_n, y_1, \ldots, y_i]/Y$. Let $f = f_1 + \ldots + f_q$, where $f_1, \ldots, f_q$ are the multihomogeneous non-zero components of f. Let f_1 be a component of minimal multidegree α. Then the component of multidegree α of $(y_i - 1)(f_1 + \ldots + f_q)$ is f_1 since $\mathrm{mdeg}(y_i f_j) > \mathrm{mdeg}(f_j)$ for every j. As $(y_i - 1)f = 0$, it follows that each of its multihomogeneous components vanishes. Hence, $f_1 = 0$. This contradicts to $f_1 \neq 0$. Thus, $(y_i - 1)f \neq 0$. □

Corollary 71.5. *If* $\mathbf{F}_L$ *is the minimal free resolution of* $k[x_1, \ldots, x_n, y_1, \ldots, y_n]/I_L$ *over* $k[x_1, \ldots, x_n, y_1, \ldots, y_n]$*, then*

$$\mathbf{F}_L \otimes k[x_1, \ldots, x_n, y_1, \ldots, y_n]/(y_n - 1, \ldots, y_1 - 1)$$

is a free resolution of $S/I_{\mathcal{A}}$ *over* S.

72 Hilbert functions

Throughout this section we assume that the toric ring $R = S/I_{\mathcal{A}}$ is projective. We will discuss Hilbert functions of graded ideals in R.

Construction 72.1. We define a partial monomial order $<_{toric}$ on S using the weight orders with respect to the rows in the matrix A. For $1 \leq i \leq r$, denote by $\mathbf{w}_i$ the weight order of the monomials in S with respect to the vector $((a_1)_i, \ldots, (a_n)_i)$. Let m and u be two monomials in S. We define that $m >_{toric} u$ if there exists a $1 \leq j \leq r$ such that

$$\mathbf{w}_j(m) > \mathbf{w}_j(u) \text{ and } \mathbf{w}_i(m) = \mathbf{w}_i(u) \text{ for } 1 \leq i < j\,.$$

This is a partial order on the monomials in S.

Two monomials m and m' are incomparable by $<_{toric}$ if and only if $\mathbf{w}_i(m) = \mathbf{w}_i(m')$ for all $1 \leq i \leq r$; this happens if and only if $m - m' \in I_{\mathcal{A}}$. Hence, the following two properties hold:

(1) $\text{in}_{<_{toric}}(I_{\mathcal{A}}) = I_{\mathcal{A}}$.

(2) if m and m' are incomparable monomials, then $m - m' \in I_{\mathcal{A}}$.

By (1) and (2) it follows that the order $<_{toric}$ induces a well-defined total monomial order $<$ in the quotient ring R.

Theorem 72.2. [Gasharov-Horwitz-Peeva] *Let P be a homogeneous ideal in a projective toric ring R. There exists a monomial ideal M in R such that the following properties hold.*

(1) *M has the same Hilbert function as P.*

(2) *The Betti numbers of M over R are greater than or equal to those of P.*

(3) *Let K and J be the preimages of M and P in S, respectively. The Betti numbers of K over S are greater than or equal to those of J.*

Proof. Let J be the preimage of P in S, and let M be the image in the ring $S/\text{in}_{<_{toric}}(I_{\mathcal{A}}) = S/I_{\mathcal{A}} = R$ of the initial ideal $\text{in}_{<_{toric}}(J)$ with respect to $<_{toric}$. Property (2) in Construction 72.1 implies that M is a monomial ideal.

(1) The ideals J and $\text{in}_{<_{toric}}(J)$ have the same Hilbert function. Therefore, the ideals P and M have the same Hilbert function. (3) follows from the fact that $\text{in}_{<_{toric}}(J)$ is an initial ideal of J. It remains to prove (2).

Let $N_0 = J$. For each $0 \leq i \leq r-1$ set $N_{i+1} = \mathrm{in}_{\mathbf{w_{i+1}}}(N_i)$. Note that $\mathrm{in}_{\mathbf{w_i}}(I_{\mathcal{A}}) = I_{\mathcal{A}}$ for each i. It follows that the the graded Betti numbers of S/N_i over $S/I_{\mathcal{A}}$ are smaller or equal to those of S/N_{i+1} over $S/I_{\mathcal{A}}$ by Theorem 22.9. Finally, note that $N_r = \mathrm{in}_{<_{toric}}(J)$. □

Proposition 72.3. *Let M be a monomial ideal in a projective toric ring R. The Hilbert function of M over R in degree $i \geq 0$ counts the number of fibers of degree i in M.*

Proof. Since all monomials in a fiber are equal in R, we conclude that

$$\dim_k (R/M)_\alpha \leq 1 \text{ for every } \alpha \in \mathbf{N}\mathcal{A}.$$

Hence, for a fixed $\alpha \in \mathbf{N}\mathcal{A}$, we have that either M contains the entire fiber of α or none of the monomials in it. Therefore, the Hilbert function of M (over R) in degree i counts the number of fibers of degree i in M. □

We close this section by pointing out that in the spirit of Section 50, one can consider the following problems on Betti numbers.

Problem 72.4. (Peeva-Stillman) *Fix the multigraded Hilbert function h of a toric ideal. Under what conditions does there exist an ideal with greatest Betti numbers among all ideals with Hilbert function h? Under what conditions does there exist an ideal with smallest Betti numbers among all ideals with Hilbert function h?*

We emphasize that the above problems are about Betti numbers, not about multigraded Betti numbers. It seems that often the toric ideal has the smallest Betti numbers. See [Gahsarov-Murai-Peeva] for results on ideals in Veronese rings.

73 Infinite free resolutions

The study of properties of infinite multigraded minimal free resolutions over toric rings is an area which emerged recently. Results are obtained in the following directions:

- computing the Betti numbers via simplicial complexes

- constructing explicit resolutions for special classes of toric rings
- obtaining bounds on the rate
- determining which toric rings are Koszul
- rationality of Poincaré series.

We study the multigraded minimal free resolution $\mathbf{G}$ of k over the toric ring $R = S/I_{\mathcal{A}}$. It is infinite (unless $I_{\mathcal{A}}$ is generated by linear forms) and starts with

$$\ldots \to R^n \xrightarrow{\begin{pmatrix} x_1 & x_2 & \ldots & x_n \end{pmatrix}} R \to k \to 0\,.$$

Theorem 73.1. *The entries in the matrices of the differentials in* $\mathbf{G}$ *are scalar multiples of monomials.*

Proof. Since $\mathbf{G}$ is multigraded we have that each entry f in the matrices of the differentials in $\mathbf{G}$ is homogeneous. Let α be the multidegree of f. Since R_α is at most one dimensional, it follows that f is a scalar multiple of the unique monomial in multidegree α. □

We have multigraded Betti numbers

$$\beta_{i,\alpha} = \dim_k \operatorname{Tor}_i^R(k,k)_\alpha \quad \text{for } \alpha \in \mathbf{N}^r, i \geq 0\,.$$

The generating function of the resolution

$$\mathrm{P}_k^R(t, \mathbf{z}) = \sum_{\substack{i \geq 0 \\ \alpha \in \mathbf{N}^d}} \beta_{i,\alpha} t^i \mathbf{z}^\alpha$$

is the ***multigraded Poincaré series***. Sometimes we consider the Poincaré series

$$\widetilde{\mathrm{P}}_k^R(t) = \sum_{i \geq 0} \Big(\sum_{\alpha \in \mathbf{N}^r} \beta_{i,\alpha} \Big) t^i = \mathrm{P}_k^R(t, (1, \ldots, 1))\,.$$

The Serre-Kaplansky problem, "Is the Poincaré series of a finitely generated commutative local Noetherian ring rational?", was one of the central questions in Commutative Algebra for many years. We have the following results on rationality of the Poincaré series in the toric case.

Theorem 73.2.

(1) [Roos-Sturmfels] *There exists a Cohen-Macaulay monomial curve in* $\mathbf{P}^8$ *whose toric ideal is generated by quadrics and whose Poincaré series is not rational.*

(2) [Gasharov-Peeva-Welker] *If the toric ideal* $I_{\mathcal{A}}$ *is generic, then* $\mathrm{P}_k^R(t, \mathbf{z})$ *is rational.*

We will show that the Betti numbers $\beta_{i,\alpha}$ can be computed by simplicial complexes. Consider the monoid $\mathbf{N}\mathcal{A}$ as a lattice with order defined by $\gamma < \mu$ if $\mu - \gamma \in \mathbf{N}\mathcal{A}$; following Herzog-Hibi, et al., this lattice is called the ***divisor poset*** of $I_{\mathcal{A}}$. For $\alpha \in \mathbf{N}^r$ denote by $(0, \alpha)$ the open interval $\{\gamma \in \mathbf{N}\mathcal{A} \,|\, 0 < \gamma < \alpha\}$. Furthermore, denote by Δ_α the order complex of $(0, \alpha)$, and set $\widetilde{\mathrm{H}}_{i-2}((0, \alpha); k) = \widetilde{\mathrm{H}}_{i-2}(\Delta_\alpha; k)$.

Theorem 73.3. [Peeva-Reiner-Sturmfels], [Herzog-Reiner-Welker], (Laudal-Sletsjøe) *For* $\alpha \in \mathbf{N}^r, i \geq 0$ *we have*

$$\beta_{i,\alpha} = \dim \widetilde{\mathrm{H}}_{i-2}((0, \alpha); k)\,.$$

Proof. Compute the Betti numbers using the Bar resolution $\mathbf{B}$ of k over R from Construction 32.2. Denote by Λ^i the set of ordered i-tuples of non-zero elements of $\mathbf{N}\mathcal{A}$. The basis elements have the form $[\lambda_1|\lambda_2|\cdots|\lambda_{i-1}|\lambda_i]$ with $\lambda_j \in \mathbf{N}\mathcal{A} \setminus \mathbf{0}$. The i-term B_i in $\mathbf{B}$ is the free R-module with basis Λ^i. The Bar resolution is multigraded and

$$\mathrm{mdeg}(\,[\lambda_1|\lambda_2|\cdots|\lambda_{i-1}|\lambda_i]\,) = \lambda_1 + \cdots + \lambda_i\,.$$

Tensoring the Bar resolution by k we obtain the complex

$$\mathbf{B} \otimes k : \ \ldots \to B_i \otimes k \xrightarrow{d_i \otimes k} B_{i-1} \otimes k \to \ldots \to B_0 \otimes k = k$$

where $B_i \otimes k$ is the k-vector space with basis Λ^i. The differential acts as follows

$$(d_i \otimes k)[\lambda_1|\cdots|\lambda_i] \ = \ \sum\nolimits_{1 \leq j \leq i-1} (-1)^j \cdot [\lambda_1|\cdots|\lambda_j + \lambda_{j+1}|\cdots|\lambda_i].$$

For $i = 1$ we have $(d_1 \otimes k)[\lambda] = 0$ for every $\lambda \in \mathbf{N}\mathcal{A} \setminus \mathbf{0}$.

Fix an $\alpha \in \mathbf{N}^r$. If $\lambda_1 + \cdots + \lambda_i = \alpha$, we identify $[\lambda_1|\lambda_2|\cdots|\lambda_i]$ with the chain $\lambda_1 \leq \lambda_1+\lambda_2 \leq \cdots \leq \lambda_1+\cdots+\lambda_{i-1}$ in the open interval $(0, \alpha)$ of $\mathbf{N}\mathcal{A}$. The differential $d \otimes k$ coincides with the boundary map in the simplicial complex Δ_α. Therefore, the reduced homology of Δ_α in dimension $i-2$ equals the i-th homology of $(\mathbf{B} \otimes k)_\alpha$. This implies the desired formula for the Betti numbers. □

Running Example 73.4. The lower part of the divisor poset of I is shown in Figure 22.

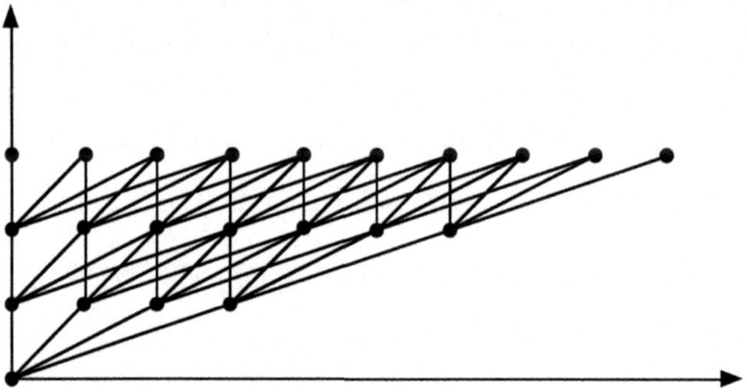

Figure 22.

The open interval $\big((0,0),(1,4)\big)$ is shown in Figure 23. Its order complex $\Delta_{(1,4)}$ has no reduced homology. By Theorem 73.3 it follows that all Betti numbers of k in multidegree $(1,4)$ vanish.

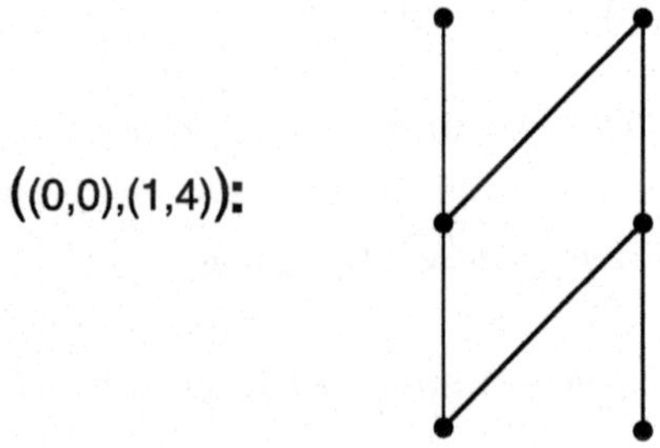

Figure 23.

The following problem is wide open.

Problem 73.5. (folklore) *Find the rate or sharp upper bounds on the rate of k over a projective toric ring.*

Theorem 73.6. [Gasharov-Peeva-Welker] *Assume $I_{\mathcal{A}}$ is projective. Denote by q the maximal degree of a binomial in a minimal system of binomial generators of $I_{\mathcal{A}}$. If $I_{\mathcal{A}}$ is generic, then* $\mathrm{rate}_{S/I_{\mathcal{A}}}(k) = q-1$.

74 Koszul toric rings

Note that R is standard graded if and only if $I_{\mathcal{A}}$ is projective. In the rest of this section, we assume that $I_{\mathcal{A}}$ is projective. We will discuss the following question.

Question 74.1. (folklore) *Which (classes of) projective toric rings are Koszul?*

If R is Koszul, then $I_{\mathcal{A}}$ is generated by quadratic binomials. However, this condition is not sufficient for Koszulness, as the next two examples show. In contrast, every ideal generated by quadratic monomials is Koszul by Theorem 34.11.

Example 74.2. [Hibi-Ohsugi] Consider the toric ring

$$k[t_1t_2, t_2t_3, t_3t_4, t_4t_5, t_5t_1, t_1t_6, t_2t_6, t_3t_6, t_4t_6, t_5t_6] \cong k[x_1, \dots, x_6]/I,$$

defined by the toric ideal

$$\big(x_4x_6 - x_5x_9,\ x_3x_{10} - x_4x_8,\ x_2x_9 - x_3x_7,\ x_1x_{10} - x_5x_7,\ x_1x_8 - x_2x_6\big).$$

The ideal I is generated by quadrics. On the other hand, computer computation shows that the toric ring is not Koszul. It should be noted that this toric ring is normal.

Example 74.3. [Roos-Sturmfels] Computer computation shows that the monomial curve in $\mathbf{P}^5$ defined by

$$\mathcal{A} = \{(0,1), (3,1), (5,1), (6,1), (7,1), (11,1)\}$$

with toric ideal

$$(x_2^2 - x_1x_4,\, x_3^2 - x_2x_5,\, x_3x_4 - x_1x_6,\, x_4^2 - x_3x_5,\, x_5^2 - x_2x_6)$$

is Cohen-Macaulay, but not Koszul.

The most interesting currently open conjecture on Koszul toric algebras seems to be the following conjecture.

Conjecture 74.4. (Bøgvad) *The toric ring of a smooth projectively normal toric variety is Koszul.* (Projectively normal means that $\mathrm{pos}(\mathcal{A}) \cap \mathbf{Z}\mathcal{A} = \mathbf{N}\mathcal{A}$.)

The existence of a quadratic Gröbner basis implies that the quotient ring is Koszul. By Proposition 66.6 every Veronese toric ideal has a quadratic Gröbner basis. There exist monomial orders with respect to which the quadrics generating the toric ideal of a Segre variety form a Gröbner basis, cf. [Peeva-Reiner-Sturmfels]. See [Bruns-Gubeladze-Trung] for results on Koszul polytopal toric rings. In [Caviglia], it is shown that the pinched Veronese, which does not have a quadratic Gröbner basis, is Koszul.

It is easy to see by 65.12 that $I_{\mathcal{A}}$ is projective if and only if the simplicial complexes Δ_α, introduced before Theorem 73.3, are pure for all α. The next proposition shows that the Koszul property for toric ideals amounts to Cohen-Macaulayness of certain open intervals.

A finite graded poset P is ***Cohen-Macaulay*** over k if for each open interval $(\mu_1, \mu_2) = \{\mu \in P \,|\, \mu_1 < \mu < \mu_2\}$ in P the reduced homology of its order complex $O((\mu_1, \mu_2))$ vanishes except in the top degree.

Proposition 74.5. (Herzog-Reiner-Welker), (Peeva-Reiner-Sturmfels) *Let $I_{\mathcal{A}}$ be projective. The toric ring R is Koszul if and only if for every $\alpha \in \mathbf{N}\mathcal{A}$ the open interval $(0, \alpha)$ is Cohen-Macaulay.*

Proof. Let the ***degree*** of α be the standard degree (with respect to the standard grading with $\deg(x_i) = 1$) of any monomial in the fiber of α. The toric ring R is Koszul if and only if $\mathrm{Tor}_i^R(k,k)_\alpha$ vanishes unless

$i = \deg(\alpha)$. By Theorem 73.3, R is Koszul if and only if $\widetilde{\mathrm{H}}_i(\Delta_\alpha; k) = 0$ for $i \neq \deg(\alpha) - 2$ for every α.

On the other hand consider the poset $(0, \alpha)$ in the monoid $\mathbf{N}\mathcal{A}$. Any open subinterval (μ_1, μ_2) can be written as $\mu_1 + (0, \mu_2 - \mu_1)$, which is topologically the same as the open interval $(0, \mu_2 - \mu_1)$. Therefore, $(0, \alpha)$ is Cohen-Macaulay for every α if and only if $\widetilde{\mathrm{H}}_i(\Delta_\alpha; k) = 0$ for $i \neq \deg(\alpha) - 2$ for every α. $\square$

The divisor poset of R is ***shellable*** if every open interval $(0, \alpha)$ is shellable (recall 36.22). The above proposition makes it possible to prove Koszulness by constructing shellings of the intervals $(0, \alpha)$; for examples, see [Peeva-Reiner-Sturmfels] and [Aramova-Herzog-Hibi].

References

[Anick] D. Anick: A counterexample to a conjecture of Serre, *Ann. Math. (2)* **115** (1982), 1–33.

[Aramova-Herzog] A. Aramova, J. Herzog: Koszul cycles and Eliahou-Kervaire type resolutions, *J. Algebra*, **181** (1996), 347–370.

[Aramova-Herzog 2] A. Aramova and J. Herzog: Almost regular sequences and Betti numbers, *Amer. J. Math.* **122** (2000), 689–719.

[Aramova-Herzog-Hibi] A. Aramova, J. Herzog, T. Hibi: Shellability of semigroup rings, *Nagoya Math. J.* **168** (2002), 65–84.

[Aramova-Herzog-Hibi 2] A. Aramova, J. Herzog, and T. Hibi: Gotzmann theorems for exterior algebras and combinatorics, *J. Algebra* **191** (1997), 174–211.

[Avramov] L. Avramov: Infinite resolutions, in *Six lectures on commutative algebra*, Birkhäuser, Progress in Mathematics **166** (1998), 119–185.

[Avramov 2] L. Avramov: Obstructions to the existence of multiplicative structures on minimal free resolutions, *Amer. J. Math.* **103** (1981), 1–31.

[Avramov-Eisenbud] L. Avramov and D. Eisenbud: Regularity of modules over a Koszul algebra, *J. Algebra* **153** (1992), 85–90.

[Avramov-Gasharov-Peeva] L. Avramov, V. Gasharov, and I. Peeva: Complete intersection dimension, *Publ. Math. IHES* **86** (1997), 67–114.

[Backelin] J. Backelin: Les Anneaux locaux à relations monomiales ont des séries de Poincaré-Betti rationnelles, *C. R. Acad. Sci.Paris*, **295** (1982), 605–610.

[Backelin 2] J. Backelin: Relations between rates of growth of homologies, preprint.

[Batzies-Welker] E. Batzies and V. Welker: Discrete Morse theory for cellular resolutions, *J. Reine Angew. Math.* **543** (2002), 147–168.

[Bayer] D. Bayer: *The division algorithm and the Hilbert scheme*, Ph.D. Thesis, Harvard University, 1982.

I. Peeva, *Graded Syzygies*, Algebra and Applications 14,
DOI 10.1007/978-0-85729-177-6,

[Bayer-Charalambous -Popescu] D. Bayer, H. Charalambous, and S. Popescu: Extremal Betti numbers and applications to monomial ideals, *J. Algebra* **221** (1999), 497–512.

[Bayer-Mumford] D. Bayer, D. Mumford: What can be computed in algebraic geometry? *Computational algebraic geometry and commutative algebra* (Cortona, 1991), Sympos. Math. XXXIV, (1993), Cambridge Univ. Press, 1–48.

[Bayer-Peeva-Sturmfels] D. Bayer, I. Peeva, and B. Sturmfels: Monomial resolutions, *Math. Res. Lett.* **5** (1998), 31–46.

[Bayer-Stillman] D. Bayer and M. Stillman: A criterion for detecting m-regularity, *Invent. Math.* **87** (1987), 1–11.

[Bayer-Sturmfels] D. Bayer, B. Sturmfels: Cellular resolutions of monomial modules, *J. Reine Angew. Math.* **502** (1998), 123–140.

[Berglund] A. Berglund: Poincaré series of monomial rings, *J. Algebra* **295** (2006), 211–230.

[Björner] A. Björner: Topological methods, in *Handbook of Combinatorics* (R. Graham, M. Grötschel, L. Lovász eds.), p. 1819-1872, North-Holland, 1994.

[Boij-Söderberg] M. Boij, J. Söderberg: Graded Betti numbers of Cohen-Macaulay modules and the multiplicity conjecture, *J. Lond. Math. Soc.* **78** (2008), 85–106.

[Boij-Söderberg 2] M. Boij, J. Söderberg: Betti numbers of graded modules and the Multiplicity Conjecture in the non-Cohen-Macaulay case, submitted.

[Bruns] W. Bruns: Jede endliche freie Auflsung ist freie Auflsung eines von drei Elementen erzeugten Ideals, *J. Algebra* **39** (1976), 429–439.

[Bruns-Herzog] W. Bruns and J. Herzog: *Cohen-Macaulay rings*, Cambridge University Press, 1996.

[Bruns-Herzog 2] W. Bruns and J. Herzog: Semigroup rings and simplicial complexes, *J. Pure Appl. Algebra* **122** (1997), 185–208.

[Bruns-Gubeladze] W. Bruns and J. Gubeladze: Combinatorial invariance of Stanley-Reisner rings, *Georgian Math. J.* **3** (1996), 315–318.

[Bruns-Gubeladze-Trung] W. Bruns, J. Gubeladze, and N. Trung: Normal polytopes, triangulations, and Koszul algebras, *J. Reine Angew. Math.* **485** (1997), 123–160.

[Buchsbaum-Eisenbud] D. Buchsbaum, D. Eisenbud: Algebra structures for finite free resolutions, and some structure theorems for ideals of codimension 3, *Amer. J. Math.* **99** (1977), 447–485.

[Caviglia] G. Caviglia: The pinched Veronese is Koszul, *J. Alg. Combin.*, **30** (2009), 539-548.

[Charalambous-Evans] H. Charalambous and G. Evans: Resolutions obtained by iterated mapping cones, *J. Algebra* **176** (1995), 750–754.

[Charalambous-Evans 2] H. Charalambous and G. Evans: Problems on Betti numbers of finite length modules, *Free resolutions in commutative algebra and algebraic geometry* (Sundance, UT, 1990), Res. Notes Math. **2** (1992), Jones and Bartlett, 25–33.

[Chardin] M. Chardin: Some results and questions on Castelnuovo-Mumford regularity, in *Syzygies and Hilbert functions*, I. Peeva editor, *Lect. Notes Pure Appl. Math.* **254** (2007), CRC Press.

[Chardin-Gasharov-Peeva] M. Chardin, V. Gasharov, and I. Peeva: Maximal Betti numbers, *Proc. Amer. Math. Soc.*, **130** (2002), 1877-1880.

[Clements-Lindström] G. Clements and B. Lindström: A generalization of a combinatorial theorem of Macaulay, *J. Combin. Theory* **7** (1969), 230–238.

[Conca-Herzog] A. Conca and J. Herzog: Castelnuovo-Mumford regularity of products of ideals, *Collect. Math.* **54** (2003), 137–152.

[Conca-Trung-Valla] A. Conca, N. Trung, G. Valla: Koszul property for points in projective spaces, *Math. Scand.* **89** (2001), 201–216.

[Cutkosky-Herzog-Trung] S. Cutkosky, J. Herzog, and N. Trung: Asymptotic behaviour of the Castelnuovo-Mumford regularity, *Compositio Math.* **118** (1999), 243–261.

[Derksen-Sidman] H. Derksen and J. Sidman: A sharp bound for the Castelnuovo-Mumford regularity of subspace arrangements, *Adv. Math.* **172** (2002), 151–157.

[Dodd-Marks-Meyerson -Richert] C. Dodd, A. Marks, V. Meyerson, B. Richert: Minimal Betti numbers, *Comm. Algebra* **35** (2007), 759–772.

[Eisenbud] D. Eisenbud: *Introduction to commutative algebra with a view towards algebraic geometry*, Springer Verlag, New York, 1995.

[Eisenbud 2] D. Eisenbud: Lectures on the geometry of syzygies, in Trends in commutative algebra, *Math. Sci. Res. Inst. Publ.* **51** (2004), Cambridge Univ. Press, 115–152.

[Eisenbud 3] D. Eisenbud: Homological algebra on a complete intersection, with an application to group representations, *Trans. Amer. Math. Soc.* **260** (1980), 35–64.

[Eisenbud-Goto] D. Eisenbud and S. Goto: Linear free resolutions and minimal multiplicity, *J. Algebra* **88** (1984), 89–133.

[Eisenbud-Green-Harris 1] D. Eisenbud, M. Green, and J. Harris: Cayley-Bacharach theorems and conjectures, *Bull. Amer. Math. Soc. (N.S.)* **33** (1996), 295–324.

[Eisenbud-Green-Harris 2] D. Eisenbud, M. Green, and J. Harris: Higher Castelnuovo theory, *Astérisque* **218** (1993), 187–202.

[Eisenbud-Koh] D. Eisenbud and J. Koh: Some linear syzygy conjectures, *Adv. Math.* **90**, (1991), 47–76.

[Eisenbud-Reeves-Totaro] D. Eisenbud, A. Reeves, and B. Totaro: Initial ideals, Veronese subrings, and rates of algebras, *Adv. Math.* **109** (1994), 168–187.

[Eisenbud-Schreyer] D. Eisenbud, F-O. Schreyer: Betti numbers of graded modules and cohomology of vector bundles, *J. Amer. Math. Soc.* **22** (2009), 859–888.

[Eliahou-Kervaire] S. Eliahou and M. Kervaire: Minimal resolutions of some monomial ideals, *J. Algebra*, **129** (1990), 1–25.

[Forman] R. Forman: Morse Theory for cell complexes, *Adv. Math.* **134** (1998), 90–145.

[Francisco-Richert] C. Francisco and B. Richert: Lex-plus-powers ideals, in *Syzygies and Hilbert functions*, I. Peeva editor, *Lect. Notes Pure Appl. Math.* **254** (2007), CRC Press.

[Fröberg] R. Fröberg: A note on a paper by M. Miyazaki, *Math. Scand.* **77** (1995), 17–18.

[Gasharov] V. Gasharov: Hilbert functions and homogeneous generic forms, *Compositio Math.* **116** (1999), 167–172.

[Gasharov-Horwitz-Peeva] V. Gasharov, N. Horwitz, I. Peeva: Hilbert functions over toric rings, special volume of the *Michigan Math. J.* dedicated to M. Hochster, **57** (2008), 339–359.

[Gasharov-Murai-Peeva] V. Gasharov, S. Murai, and I. Peeva: Hilbert schemes over Veronese rings, *Math. Zeit.*, to appear.

[Gasharov-Murai-Peeva 2] V. Gasharov, S. Murai, and I. Peeva: Applications of mapping cones over Clements-Lindström rings, submitted.

[Gasharov-Peeva-Welker] V. Gasharov, I. Peeva, and V. Welker: Rationality for generic toric rings, *Math. Zeit.* **233** (2000), 93–102.

[Gasharov-Peeva-Welker 2] V. Gasharov, I. Peeva, and V. Welker: The lcm lattice in monomial resolutions, *Math. Res. Lett.* **6** (1999), 521–532.

[Geramita-Harima-Shin] A. Geramita, T. Harima, and Y. Shin: Extremal point sets and Gorenstein ideals, *Adv. Math.* **152** (2000), 78–119.

[Golod] E. Golod: On the homologies of certain local rings, *Soviet Math. Dokl.* **3** (1962), 745–748.

[Gotzmann] G. Gotzmann: Eine Bedingung fr die Flachheit und das Hilbertpolynomeines graduierten Ringes, *Math. Zeit.* **158**, 61-70.

[Green] M. Green: Generic initial ideals, in *Six lectures on commutative algebra*, Birkhäuser, Progress in Mathematics **166**, (1998), 119–185.

[Hartshorne] R. Hartshorne: Algebraic vector bundles on projective spaces: a problem list, *Topology* **18** (1979), 117–128.

[Hartshorne 2] R. Hartshorne: Connectedness of the Hilbert scheme, *Publ. Math. IHES* **29** (1966), 5-48.

[Hering-Schenck-Smith] M. Hering, H. Schenck, G. Smith: Syzygies, multigraded regularity and toric varieties, *Compos. Math.* **142** (2006), 1499–1506.

[Herzog-Hibi-Zheng] J. Herzog, T. Hibi, X. Zheng: Dirac's theorem on chordal graphs and Alexander duality, *European J. Combin.* **25** (2004), 949–960.

[Herzog-Reiner-Welker] J. Herzog, V. Reiner, and W. Welker: The Koszul property in affine semigroup rings, *Pacific J. Math.* **186** (1998), 39–65.

[Herzog-Srinivasan] J. Herzog, H. Srinivasan: Bounds for multiplicities, *Trans. Amer. Math. Soc.* **350** (1998), 2879–2902.

[Hibi-Ohsugi] T. Hibi and H. Ohsugi: Toric ideals generated by quadratic monomials, *J. Algebra* **218** (1999), 509–527.

[Hilbert 1] D. Hilbert: Über the Theorie von algebraischen Formen, *Math. Ann.* **36** (1890), 473–534. An English translation is available in Hilbert's Invariant Theory Papers, Math. Sci. Press, 1978.

[Hilbert 2] D. Hilbert: Über die vollen Invariantensysteme, *Math. Ann.* **42** (1893), 313–373. An English translation is available in Hilbert's Invariant Theory Papers, Math. Sci. Press, 1978.

[Hochster] M. Hochster: Cohen-Macaulay rings, combinatorics, and simplicial complexes, in *Proc. of the 2nd Oklahoma Conf., Lecture Notes in Pure Appl. Math.* **26** (1977), Marcel Dekker, 171–224.

[Horrocks] G. Horrocks: Vector bundles on the punctured spectrum of a local ring, *Proc. London Math. Soc.* **14** (1964), 689–713.

[Horwitz] N. Horwitz: Linear resolutions of quadratic monomial ideals, *J. Algebra* **318** (2007), 981–1001.

[Huneke] C. Huneke: Hyman Bass and Ubiquity, Gorenstein rings, in Algebra, K-theory, groups, and education (New York, 1997), 55–78, *Contemp. Math.* **243**, Amer. Math. Soc., Providence, RI, 1999.

[Jöllenbeck-Welker] M. Jöllenbeck and V. Welker: Minimal resolutions via algebraic discrete Morse theory, *Memoirs Amer. Math. Soc.* **197** (2008).

[Kodiyalam] V. Kodiyalam: Asymptotic behaviour of Castelnuovo-Mumford regularity, *Proc. Amer. Math. Soc.* **128** (2000), 407–411.

[Lyubeznik] G. Lyubeznik: A new explicit finite free resolution of ideals generated by monomials in an R-sequence, *J. Pure Appl. Algebra* **51** (1988), 193–195.

[Mac Lane] S. Mac Lane: *Homology*, Springer-Verlag, 1995.

[Matsumura] H. Matsumura: *Commutative ring theory*, Cambridge Univ. Press, 1986.

[Mayr-Meyer] E. Mayr and A. Meyer: The complexity of the word problems for commutative semigroups and polynomial ideals, *Adv. Math.* **46** (1982), 305–329.

[Mermin] J. Mermin: Compressed ideals, *Bull. London Math. Soc.* **40** (2008), 77–87.

[Mermin-Murai] J. Mermin and S. Murai: The lex-plus-powers holds for pure powers, submitted.

[Mermin-Peeva] J. Mermin and I. Peeva: Hilbert functions and lex ideals, *J. Algebra* **313** (2007), 642–656.

[Mermin-Peeva 2] J. Mermin and I. Peeva: Lexifying ideals, *Math. Res. Lett.* **13** (2006), 409–422.

[Mermin-Peeva-Stillman] J. Mermin, I. Peeva, and M. Stillman: Ideals containing the squares of the variables, *Adv. Math.* **217** (2008), 2206–2230.

[Miller] M. Miller: Multiplicative structures on finite free resolutions, in *Free resolutions in commutative algebra and algebraic geometry* (Sundance, UT, 1990), 35–46, *Res. Notes Math.* **2**, Jones and Bartlett, 1992.

[Mora-Robbiano] T. Mora and L. Robbiano: The Gröbner fan of an ideal, Computational aspects of commutative algebra, *J. Symbolic Comput.* **6** (1988), 183–208.

[Murai-Peeva] S. Murai and I. Peeva: Hilbert schemes and Betti numbers over a Clements-Lindström ring, submitted.

[Northcott] D. Northcott: *An introduction to homological algebra*, Cambridge University Press, 1960.

[Novik] I. Novik: Lyubeznik's resolution and rooted complexes, *J. Alg. Combin.* **16** (2002), 97–101.

[Pardue] K. Pardue: Deformation classes of graded modules and maximal Betti numbers, *Ill. J. Math.* **40** (1996), 564–585.

[Peeva] I. Peeva: 0-Borel fixed ideals, *J. Algebra* **184** (1996), 945–984.

[Peeva 2] I. Peeva: Consecutive cancellations in Betti numbers, *Proc. Amer. Math. Soc.* **132** (2004), 3503–3507.

[Peeva-Reiner-Sturmfels] I. Peeva, V. Reiner and B. Sturmfels: How to shell a monoid, *Math. Ann.* **310** (1998), 379–393.

[Peeva-Stillman] I. Peeva and M. Stillman: Open problems on syzygies and Hilbert functions, *J. Comm. Alg.* **1** (2009), 159–195.

[Peeva-Stillman 2] I. Peeva, M. Stillman: The Minimal Free Resolution of a Borel Ideal, *Expositiones Math.* **26** (2008), 237–247.

[Peeva-Stillman 3] I. Peeva and M. Stillman: Connectedness of Hilbert schemes, *J. Algebraic Geom.* **14** (2005), 193–211.

[Peeva-Sturmfels] I. Peeva and B. Sturmfels: Generic lattice ideals, *J. Amer. Math. Soc.* **11** (1998), 363–373.

[Peeva-Sturmfels 2] I. Peeva and B. Sturmfels: Syzygies of codimension 2 lattice ideals, *Math. Zeit.* **229** (1998), 163–194.

[Peeva-Velasco] I. Peeva, M. Velasco: Frames and degenerations for monomial resolutions, *Trans. Amer. Math. Soc.*, to appear.

[Phan] J. Phan: *Monomial ideals and free resolutions*, Ph. D. Thesis, Columbia University, (2006).

[Polishchuk-Positselski] A. Polishchuk, L. Positselski: *Quadratic algebras*, University Lecture Series, **37**, Amer. Math. Soc., 2005.

[Priddy] S. Priddy: Koszul resolutions, *Trans. Amer. Math. Soc.* **152** (1970), 39–60.

[Reiner-Welker] V. Reiner and V. Welker: On the linear syzygies of a Stanley-Reisner ideal, *Math. Scand. 89* (2001), 117–132.

[Reisner] G. Reisner: Cohen Macaulay quotients of polynomial rings, *Adv. Math.* **21** (1976), 30–49.

[Roos] J.-E. Roos: Commutative non-Koszul algebras having a linear resolution of arbitrarily high order. Applications to torsion in loop space homology, *C. R. Acad. Sci. Paris Sr. I Math.* **316** (1993), 1123–1128.

[Roos-Sturmfels] J.-E. Roos and B. Sturmfels: A toric ring with irrational Poincar-Betti series, *C. R. Acad. Sci. Paris Sr. I Math.* **326** (1998), 141–146.

[Römer] T. Römer: Bounds for Betti numbers, *J. Algebra* **249** (2002), 20–37.

[Stanley] R. Stanley: *Combinatorics and Commutative Algebra*, Second edition, Birkhäuser, 1996.

[Stanley 2] R. Stanley: Positivity problems and conjectures in algebraic combinatorics, *Mathematics: frontiers and perspectives*, Amer. Math. Soc., 2000, 295–319.

[Sturmfels] B. Sturmfels: *Gröbner Bases and Convex Polytopes*, AMS University Lecture Series Vol. 8, 1995.

[Tate] J. Tate: Homology of Noetherian rings and local rings, *Illinois J. Math.* **1** (1957), 14–27.

[Taylor] D. Taylor: Ideals generated by monomials in an *R*-sequence, *Ph. D. Thesis*, University of Chicago, 1966.

[Terai] N. Terai: Alexander duality theorem and Stanley-Reisner rings, *Free resolutions of coordinate rings of projective varieties and related topics (Kyoto, 1998), Sūrikaisekikenkyūsho Kōkyūroku* **1078** (1999), 174–184.

[Valla] G. Valla: Problems and results on Hilbert functions of graded algebras, in *Six lectures on commutative algebra, Progr. Math.* **166** (1998), Birkhuser, Basel, 293–344.

[Velasco] M. Velasco: Minimal free resolutions that are not supported by a CW-complex, *J. Algebra* **319** (2008), 102–114.

Index

I. Peeva, *Graded Syzygies*, Algebra and Applications 14,
DOI 10.1007/978-0-85729-177-6, © Springer-Verlag London Limited 2011

E

F

G

H

I

K

L

Printed by Printforce, the Netherlands